History *of* Science Selections *from* ISIS

GENERAL EDITOR, Robert P. Multhauf

Science *in* America *since* 1820

Science *in* America *since* 1820

Nathan Reingold

Editor

Science History Publications
New York · 1976

First published in the United States by
Science History Publications
a division of
Neale Watson Academic Publications, Inc.
156 Fifth Avenue, New York 10010

First Edition 1976
Designed and manufactured in the U.S.A.

Library of Congress Cataloging in Publication Data
Main entry under title:

Science in America since 1820.

(Selections from Isis)
Bibliography: p.
1. Science--History--United States--Addresses, essays, lectures. I. Reingold, Nathan, 1927-
II. Isis.
Q127.U6S317 509'.73 76-3668
ISBN 0-88202-049-8

Publisher's Note: We should like to express our appreciation to ISIS for granting permission to reprint the articles in this volume. If the publishers have unwittingly infringed the copyright in any illustration reproduced they will gladly pay an appropriate fee on being satisfied as to the owner's title.

Contents

INTRODUCTION

Faced with the twenty-one articles that follow, it is hard to avoid the autobiographical mode. Most of the authors are known to me personally; some are close friends. The pieces appeared in a time-span roughly coincident with my professional career. As each contribution can stand or fall on its own merits, I will not bother to comment on particular topics. Rather, general remarks are called for, on *Isis*, on the group of historians interested in the sciences in the United States, and on the progress and prospects of the field.

Although published in the United States for most of its existence, *Isis* was never known as a journal with a special interest in American events. Neither George Sarton nor his three successors had any such intention. Yet, there is this volume (and the parallel volume for the earlier period edited by Brooke Hindle). The twenty-one articles represent a very substantial increment to what can be viewed as merely one of many sub-specialities in the history of science. I can, it is true, visualize a completely different anthology of articles of comparable size and quality taken from other sources. What is striking is that no other single journal yields so many and so varied pieces; only by dipping into conference proceedings could I construct a hypothetical rival volume. Clearly, *Isis* has had an important role in fostering the development of our knowledge of the history of the sciences in the United States.

A cynic might explain this away by observing that editors of journals in fledgling fields are sometimes grateful for contributions of publishable quality, even if outside the perceived core of a field. That explanation is unfair to the editors of *Isis*. While not Americanists, Sarton's three successors each had research interests involving American events. As for Sarton, his world view—a fervant positivism stressing the unity of the sciences and their pervasive cultural hegemony—resulted in a complacent hospitality to papers at least as bizarre as the occasional scribblings on the sciences in the American republic. Happily for all concerned, Sarton's *Isis* was (and is) a general, non-specialized history of science journal. Even more fortunate in the present context, *Isis* remained Sartonian in a pre-professional sense. By this I mean *Isis* never was exclusively the organ of the dominant clichés of-he years 1955 to 1968 when the history of science became an established professional specialty. Were this otherwise, many of these pieces might have gone elsewhere. Americanists (as I shall call students of the sciences in the United States) were a quite different breed with another set of animating clichés.

Of the seventeen authors, twelve were trained in American history. Quite commonly, they enlisted under Clio's banner waving little pennants bearing words like "cultural," "intellectual," and "social." Only two derive from history of science programs; one of those now commonly lists himself as an American intellectual historian. One is a journalist. The origins of the remaining two are unclear to me beyond the vague belief that they are scientists, rather than from the guild of historians. Pretty much the same pattern is disclosed by the authors of the pieces in my hypothetical rival collection.

This is a unique situation. In no other country in the world known to me is the writing of the nation's history in the sciences largely in the hands of general historians. Overwhelmingly, the historians of the sciences outside the United States derive from the communities of scientists and engineers. Overwhelmingly, it is their hands that fashion the writing of the national experience in science and technology. I know very few exceptions to these statements. In some instances the writing of the history of science devolved, at least in part, upon students of philosophy acting as intellectual historians. In contrast, in the United States until very recently the historians of science, similarly individuals largely with backgrounds in science and technology, were markedly disinclined to investigate events within the boundaries of the United States. The few exceptions only underline the strangeness of the situation. I can only suggest answers to the two apparent historiographic questions: why were there historians of U.S. history willing and able to study the national experience in science and technology, and why were Americans in the history of science reluctant, if not hostile, to the very idea?

The answer to the first is rather straight-forward. There is a long-standing historiographic tradition in the United States admitting to the scope of national history all sorts of topics outside the conventional areas of diplomatic, political, constitutional, and military history. Certainly by the early decades of this century, this tolerance extended to the history of the sciences. Although very few American historians actually worked in these heterodox fields, a recognition of legitimacy grew, particularly under the aegis of some rather eminent scholars, such as A.M. Schlesinger, Sr. This American attitude extended to the study of European history—for example, the work of Preserved Smith of Cornell, an example from the tradition of intellectual history. Perhaps the attitude is vestigial, a left-over from the previous century's positivistic belief in a historical progress necessarily fueled by science and technology. But in some locations, like Columbia University under the influence of men like James Harvey Robinson, there was a conscious view of a history far broader than the traditional scope.

Well before World War II, the written history of the American nation was understood to comprehend, in theory if not in practice, an extraordinary range of past experiences. Since the pure and applied sciences were obviously important in modern societies, a sprinkling of references appeared in general histories, monographs, and articles. Here and there, specialized studies were produced.

After World War II no one questioned that science and technology had a role in American history. A few young historians in the years 1945-1965 could turn to the history of the sciences in America as a field of specialization, usually with the approbation of their elders. When this occurred, motivations differed from those of their contemporaries going into conventional history of science. Americanists wanted to place scientists and engineers, as well as science and technology, within a national context. They wanted to establish connections and interactions with aspects of the society at large. While often quite mindful of technical data and concepts, they were overwhelmingly externalist. (Only six of the articles are internal; two are borderline cases; thirteen are clearly external.) To put the issue in another manner, the Americanists wanted to define this history of science to include much more than the account of the generation of the content of science by a community of savants. They aimed for a larger audience, the readers of national history. To them it often seemed as though the historians of science wrote only for themselves, a few philosophers, a few bookish scientists, and a few stray social scientists. What was worse, the products often were a largely self-contained, insulated history of an elite and their fascinating intellectual toys. It was all too remote from the history of most of humanity.

Perhaps the roots of entry of general historians into the history of science in America are in the organization of education in this country. In many other nations there is a sharp split at the university level between the faculties of science and of the arts. Often, students could switch from one to another only with great difficulty. In some countries (e.g. Great Britain) this split even extended to the secondary school level. Single faculties of arts and sciences are the norm in the United States. Typically, students majoring in one intellectual area are required to meet a minimum requirement in other areas. In contrast to most European nations, switching fields is not particularly difficult. At the least, the U.S. arrangement produces an intellectual smorgasbord. At best, it yields a climate both tolerant and stimulating. Although participants in the events were keenly aware of painful struggles in introducing the history of science into existing university departments or forming new, independent departments, by European standards, let us say, the growth of the field was almost incredible.

In the case of the history of the sciences in the United States, this resulted in a diverse body of researchers constituting neither a school nor a self-conscious group. Some had backgrounds in the sciences; others were escapees from conventional, overcrowded sub-specialties in American history. Some, like myself, consider the History of Science Society as their professional organization. Others are indifferent or consciously spurn the Society for different affiliations. About the only common denominator is the belief that the history of the sciences in the United States is important for understanding the national experience.

Not a common denominator but fairly pervasive is an interest in applied fields—technology, medicine, agriculture. The attitude of the Americanists towards these is quite different from that of conventional historians of science. The latter might occasionally investigate instances of apparent intersections of theory and practice. Very few would follow the many Americanists who have made these applied fields principal areas of professional concern. Disregarding the distinctions between internal and external history is very easy in the cases of technology, agriculture, and medicine. A few Americanists found themselves very much at home in these emerging professional specialties.

On the distinction between internal and external history, I cannot speak with assurance about the viewpoint of fellow Americanists. I have never polled them. My suspicion is they feel as I do that the division is an irrelevancy. In the English speaking world, the distinction was pushed in reaction to purported linkages of the two, such as those in the writings of Marxist scholas. Consequently, an intellectualist view of the history of science became fashionable in which the sciences developed largely as a result of internal logics. Forces external to the scientific field intruded at initial stages and rarely thereafter.

So far as I can recollect, we were neither intellectualist nor committed for or against any particular linkage. Rather, the concern was to present and to analyze a complete situation. We saw a total configuration rather than a split into two autonomous situations. For that reason, in the literature one can find supposedly external-minded Americanists fretting over points within the so-called internal history. On my part the holistic approach resulted in the convication that searching for specific universal linkages is futile, and cyclical patterns like Kuhn's theory of scientific revolutions are illusions. Given the incredible interpenetration and interaction of so many diverse factors in the history of science, devising a taxonomy of processes is by far the more prudent course.

To return to my second historiographic question, why in the past did most Americans in the history of science shun the study of the sciences in their own nation? It is an awkward question. Even during the years 1955-1968, the history of science was not monolithic; any answer risks an injustice to friends and colleagues. Beyond that, so much has changed since 1968 in the field that much formerly taken for granted appears archaic, even incredible. I use 1968 as a dividing year not because anything momentous happened then in the history of science as such. Just as the Americanists reflected the assumptions prevalent in the twenty years after World War II, newcomers to the history of science were reacting to war and to domestic social forces. 1968 is a good year to symbolize the shift. Not only were the young historians interested in external history, they were often anxious to push forward to very current and relevant history. And that has implications for the study of the sciences in the United States. I like to think the pendulum would have swung anyway; war and domestic turmoil simply speeded up the process. I do not mean to imply the extirpation of old ideologies and work habits. They persist, stronger in some locations than in others. In relative terms the change is substantial.

To return to the old dispensation, three major factors formerly militated against the study of the sciences in the United States placing the practitioners of that specialty in a marginal position. One was previously mentioned—the bias against so-called external history. A second factor was the time focus of most of the field. Except for students of Darwin, very few historians of science in the early years ventured past the year 1800. That clearly militated against American concerns. And if this was not enough, there was the widespread belief that Americans were indifferent to basic research (applied research did not count to the intellectualist school). (I have already said all I care to say on the indifference theme in "American Indifference to Basic Research: A Reappraisal," in George H. Daniels, ed., *Nineteenth-Century American Science, A Reappraisal* (Evanston, 1972) pp. 38-63.) Therefore, with a few minor exceptions science in the United States was trivial and unworthy of serious study.

It was futile to argue that the roots of present American stature in research were in a past well meriting investigation. A good number of historians of science apparently believed that study of the national experience in research was evading the real stuff of the history of science for easy topics or for chauvinistic reasons. Let me illustrate this by a personal anecdote. Before my *Documentary History of Science in Nineteenth Century America* appeared, one of the readers of my manuscript praised it to an eminent historian of science. Shortly afterwards at a committee meeting this eminent scholar congratulated me on the good things he had heard. After I replied with proper modesty, he asked me, "Was I doing a whitewash of American science?" Not knowing whether to be outraged or to laugh, I pompously replied in a Rankean vein that I was trying to give an accurate picture of the past in its own terms. We have come a long way since that encounter.

If we step back from the sociology of the profession to the sheer progress in our knowledge, we have indeed come a long way. When I started out, the "standard text" was Bernard Jaffe's *Men of Science in America* (1944), a good book of its kind but far from a satisfactory history. In the physical sciences there was Benjamin Franklin and those two nineteenth-century exceptions to the rule of indifference, Joseph Henry and J. Willard Gibbs. All the rest were outside the canon.

The natural sciences were in a somewhat better state. As everyone thought they knew about Darwin in America, work existed on the heralds and anti-heralds of the coming. Even some of the later preachers of the words from Down Bromley had made the secondary literature. It was all very American, being somehow connected with the grand national theme of the exploration and settlement of the land. That grand theme produced a rather uneven literature purporting to explain geology, botany, and zoology on the frontier's edge. Add a few items each on medicine, technology, and agriculture, and one has the total bibliography available, ca. 1950-1955. It was not very large. Despite a few very fine items, it was hardly impressive.

If one adds to the twenty-one pieces in this collection the roughly equal number of my hypothetical non-*Isis* rival volume plus articles for the pre-1820 period, we can talk of about sixty articles since 1955. This is by no means the entire body of such writings. Without much strain I can list an equal number of books appearing in the past two decades substantially extending our knowledge of the sciences in the American setting. These sixty books are not an exhaustive listing.

In both qualitative and quantitative terms we have come a long way in twenty years. The works of the last two decades include institutional histories, biographies, "internalist" case studies, public policy pieces, sociologically-oriented writings, and contributions to the study of applied fields. The time scope is from the seventeenth century down to the early 1960's. Splendid possibilities now exist for courses at both the undergraduate and graduate levels. Nevertheless, my enthusiasm for all this progress is tempered by the recognition of problems. Some topics are overstudied, far beyond the point of diminishing returns. Many areas of great importance are almost completely neglected. I am enough of an optimist to believe the best is yet to come.

Let me conclude with a few cautious prophesies. The most notable trend visible to me is the entrance into this area of well-trained young scholars from history of science programs. I think this presages an increase in studies of an internalist kind. Hopefully, these will now reflect the recognition that scientists are human beings in very specific environments, not disembodied spewers of data and concepts. As to the prospect of the field receiving newcomers from general American history, I am somewhat uncertain. That source will not completely dry up, but large portions of the community of American historians now march to quite different drums. I am inclined to forecast a future in which historians of science will reclaim their portion of the national history from the general historians.

I suspect that the biggest push will occur in the history of the 1900-1940 period. At the same time work in progress known to me indicates no absolute falling off in concern for the history of the sciences in the nineteenth century. Too many intriguing problems remain. Education, popular attitudes, and public policy are likely to grow as areas of research.

Less promising are the prospects for what I conceive as the greatest need in the history of the sciences in the United States. Far too many writings lack a comparative perspective. Sometimes the result is a high-minded provinciality. The authors are blithely unaware of both parallelism and of differences. In this respect the history of the sciences in the United States is following the pattern of other areas of American history where insularity is giving way to the comparative. A conscious comparative perspective will certainly enrich the field. Perhaps it will eventually merge what was a marginal specialty in the history of science into the core of the discipline. Hopefully, a comparative perspective may even attract non-Americans to contribute to what is still a growing area of research.

Nathan Reingold

Science and the Common Man in Ante-Bellum America

*By Donald Zochert**

ONE OF THE SEVERAL FORCES set in motion by the flood tide of democracy during the 1830s and 1840s was the rapid diffusion of science, along with the corollary notion that the common man—no less than the philosopher—could fasten upon it to his advantage. "Science has now left her retreats . . . her selected company of votaries, and with familiar tone begun the work of instructing the race," William Ellery Channing wrote in 1841. "Through the press, discoveries and theories, once the monopoly of philosophers, have become the property of the multitudes. . . . Science, once the greatest of distinctions, is becoming popular." So pervasive was this spread of scientific knowledge, Channing thought, that the characteristic of the age was "not the improvement of science, rapid as this is, so much as its extension to all men."[1]

The popular embrace of science was part of a paradox only half described by Channing, for even as he wrote, the concerns and methodology of orthodox science were moving beyond the grasp of the common man. The traditional orientation of science toward the natural world and the naming of its parts was being challenged by a specialization which placed theoretical considerations above simple description; the gentleman philosopher able to combine leisure with the pursuit of science (and often with considerable literary grace) was being superseded by a more disciplined class of scientists with distinct claims to professionalism.[2]

While the internal development of orthodox science has had its many annotators, the ways in which science worked upon the popular mind have not. This study is addressed to three sets of questions: (1) What were the dimensions of the science which became available to the common man of this period through the process of democratization described by Channing? (2) How did this science exercise its appeal on the popular mind? What claims were made on its behalf? What attitudes toward orthodox science and scientists did it elicit from the public? (3) What was the structure of popular science? For instance, how did it handle the tension between science and theology? In what ways can there be said to have existed a democratic exercise of science?

The diffusion of scientific knowledge, as Channing suggested, occurred chiefly

**Chicago Daily News*, Chicago, Illinois 60611.

[1] Milwaukee *Sentinel*, Aug. 17, 1841.

[2] The standard account is George H. Daniels, *American Science in the Age of Jackson* (New York: Columbia University Press, 1968).

through the press. The newspaper was both the molder and the mirror of popular attitudes toward science; it was the one agency of information able, in De Tocqueville's phrase, to "put the same thought at the same time before a thousand readers."[3] The present study is based upon a survey of more than 1,500 issues of three newspapers published at Milwaukee between 1837 and 1846.[4] It should be emphasized that this material reflects parochial interests only insofar as the data have been accumulated from publications issued at a particular location, for the pioneer newspaper itself was far from parochial. Dependent to a great extent upon articles which had been published earlier in books, journals, and other newspapers from across the country, the newspapers used here reprinted material from publications in Alabama, Connecticut, the District of Columbia, Georgia, Illinois, Iowa, Kentucky, Louisiana, Maryland, Massachusetts, Michigan, Mississippi, Missouri, New York, Ohio, Rhode Island, Tennessee, Virginia, and elsewhere in Wisconsin.[5] It would be difficult to conceive of a more catholic range of evidence about the shape and structure of popular science.

Of the traditional scientific disciplines, many of them having only recently emerged as distinct entities during the period under discussion, astronomy was clearly the most accessible and appreciated. It was accessible because one needed neither elaborate instruments nor an advanced education to pause with wonder before the workings of the heavens; it was appreciated by virtue of its long service as inspiration for poetry and folklore, and as a mute, prima-facie theological proof. So thoroughly was the subject treated in the popular press that it exhibits more clearly than the other branches of science a division of public interest inherent in them all, and demonstrates the encouragement of theoretical and sophisticated scientific inquiry.

Of the many aspects of astronomy, it was comets—representing an event of novelty in an otherwise well-ordered heaven—which exercised the strongest claim upon the popular mind. The Great Comet of 1843, for instance, answering the eschatological predictions of the millenarian William Miller and his followers, generated a public interest and enthusiasm which persisted throughout its entire passage. Appearing first in the Southern Hemisphere, the comet was visible over the United States by early March. In the same month the *Courier* published a communication on the object from a "scientific friend," presumably a local resident, who began by dismissing the Millerite apprehensions on their

[3] Alexis de Tocqueville, *Democracy in America*, ed. J. P. Mayer and Max Lerner (New York: Harper & Row, 1966), p. 489.

[4] These are the Milwaukee *Sentinel*, essentially a Democratic newspaper; the Milwaukee *Courier*, representing Whig interests; and the *American Freeman*, published first at Milwaukee and subsequently at nearby Prairieville (Waukesha), an evangelical and abolitionist newspaper. The name of the *Sentinel* occasionally varied.

[5] A total of eighty-five separate newspapers from these states are represented in the Milwaukee newspapers; in addition, the latter reprinted material from another twenty-five U.S. newspapers without place names in their titles, from several British, Canadian, and French newspapers, from a number of journals and periodicals, and from such books as Wilson's *American Ornithology*, Smith's *Illustrations of the Zoology of South Africa*, MacGillivray's *British Birds*, and Liebig's *Organic Chemistry*. I would estimate that as little as one-fourth of the material reprinted in the Milwaukee newspapers was actually credited to its source, making the range of evidence all the more impressive.

own terms: the comet was located beneath the foot of the constellation Orion, he observed. "Need we fear, under such protection?" More to the point were questions raised by the comet's sudden and unannounced appearance and by its apparent lack of a nucleus.

Even though astronomers were committed to the *concept* of periodicity for comets, the correspondent wrote, the actual periods of but few comets had been determined with precision. Perhaps the present comet was the same which had appeared in 1264, again (presumably) in 1556, and which had been expected to reappear in 1848. That it should return five years early was not unusual, for "the disturbing influence of the planets, and perhaps other causes (as a thin ether pervading all space) often throw these light bodies out of their regular course, and accelerate or retard their period of return in such manner as to defy all calculation." The apparent absence of a nucleus was less easily explained, but this circumstance, the writer contended, "is not uncommon."[6]

Despite the assurances of the *Courier*'s scientific friend, the headless comet inspired an alternative explanation. "The learned have not decided whether the phenomena [*sic*] in the heavens . . . is a comet or a zodiacal light," the same newspaper observed in a dispatch reprinted from the East. "The better—certainly the safer—opinion seems to be that which calls it a zodiacal light." Another dispatch resolved the difference of opinion with true self-sufficiency: "On recurring to Rees' Cyclopedia," it was obvious that the phenomenon was indeed the zodiacal light.[7]

The nucleus of the comet was soon observed, however, and announced in an article under the dateline of Yale College, which placed the head of the comet "close to the well known star *Baten* of the Whale [Zeta *Ceti*]," provided its declination and right ascension, and described the nucleus as "a bright central spot, enveloped in a misty haze, elongated in the direction of a [i.e., the] train."[8] Yet even this pronouncement from a distinguished seat of learning could not quiet the debate over the nature of the object and the suspicion that there was something quite out of the ordinary about it. A subsequent article from upstate New York, referring cautiously to "the streak of light," reported that northern observers generally considered the object a genuine comet—suggesting that the South and perhaps the West held a different opinion. Even the celebrated "Dr. Herschell" had some nagging doubts about the authenticity of the Great Comet; the *Sentinel* quoted him to the effect that the comet "was one of enormous magnitude, 'and if it be not one, it is some phenomenon beyond the earth's atmosphere of a nature even yet more remarkable!' "[9]

Yale finally prevailed, although the close passage of the comet to the sun provoked still more discussion of its strange nature, with reports of observations by the French astronomer Dominique-François-Jean Arago and the German Johann Franz Encke. Public discussion of the comet, at once "popular" and

[6] *Courier*, Mar. 22, 1843.
[7] *Ibid.*, Mar. 29, 1843; *Sentinel*, Mar. 29, 1843.
[8] *Sentinel*, Apr. 5, 1843.
[9] *Ibid.*, Apr. 12, 1843; May 10, 1843.

technical in its fashion, devolved eventually to literary effusion, in a fanciful fare-thee-well which suggested one element of the public interest: "Good-bye, thou strange, but not unwelcome visitor! thou peacock of the skies! thou locomotive of the empyrean! thou sweeper of dust and cobwebs from the stars! thou terror of the ignorant, but wonder and admiration of the man of science!"[10]

Observations of other comets continued to be reported upon, particularly after the passage of the Great Comet of 1843. Thus observations by W. H. Clarke of St. Mary's, Georgia, Ezra Otis Kendall of Philadelphia's Central High School, Matthew Fontaine Maury at Washington, and William Cranch Bond at Harvard found their way into the press.[11] The most scientifically interesting comet of the period, Biela's Comet (1846), and the division of its nucleus, was only summarily treated.[12] But the editors of both the *Sentinel* and the *American Freeman* personally reported a comet in the western sky in 1845. From Humboldt's work "now publishing," the *Sentinel* noticed the great number of comets in the universe: "When we estimate the length of their relative orbits, the boundaries of their perihelion, and the great likelihood of their remaining invisible to the inhabitants of the earth, by the rules of probabilities, we find they must amount to such myriads as to make the imagination pause amazed."[13]

Enthusiasms such as this are especially evident in the popular treatment of astronomy. Many phenomena of the most ordinary nature, as one newspaper remarked of a meteor observed over Liverpool, England, were "well calculated to impress the mind with the profoundest awe."[14] Meteors, eclipses, and sunspots were regularly reported upon, more to illustrate the fecundity and mystery of the natural world than its mechanics.[15] Citing William Hyde Wollaston and Fredrich Wilhelm Bessel on the number of stars believed to exist, the *Sentinel* exclaimed: "What an insignificant speck in the universe is the little group of worlds that nestle under the wing of our diminutive sun!" Another paper, quoting Herschel on the same subject, asked its readers: "Who can think of what lies beyond the telescopic view? In such a thought, is not the mind lost in sublimity and grandeur?" When Urbain Jean Joseph Leverrier's "theoretical planet"—Neptune—was finally observed, the *Sentinel* touched upon the same root of popular interest in astronomy, noting that the "only remaining discordant

[10] *Courier,* June 14, 1843; *Sentinel,* May 10, 1843.

[11] *Sentinel,* Aug. 12, 1843; *Courier,* Oct. 30, 1844; *American Freeman,* Feb. 26, 1845; *Sentinel,* Mar. 25, 1846.

[12] The entire notice of Biela's Comet came in a retrospective account of achievements in astronomy and merely called attention to the "anomalous appearance" of the comet without defining the anomaly. *Sentinel,* Nov. 25, 1846. James Challis, director of the Cambridge Observatory, considered Biela's Comet "the most singular celestial phenomena that has occurred for many years." George F. Chambers, *The Story of Comets* (Oxford: Oxford University Press, 1909), p. 91.

[13] *Sentinel,* June 10, 1845, and *American Freeman,* June 12, 1845; *Sentinel,* Nov. 19, 1845.

[14] *Sentinel,* Mar. 13, 1846.

[15] For meteors see *Sentinel,* July 4, 1837, Aug. 29, 1837, Sept. 19, 1845, July 23, 1846, and *American Freeman,* Dec. 22, 1846; for eclipses see *Sentinel,* Sept. 25, 1838, Mar. 18, 1846, and *Courier,* Apr. 16, 1846; for sunspots see *American Freeman,* Apr. 9, 1845, *Sentinel,* Mar. 19, 1845, and Mar. 11, 1846.

note has been attuned by the hand of that great astronomer, and the voice of nature is now a voice of melody."[16]

Yet even the most common events often carried the curious beyond awe and wonder to the theoretical grounds of astronomy, a confrontation encouraged, indeed precipitated, by the popular press. One local correspondent, finding a measure of public maturity in the general calm which greeted the announcement of new comets, assessed the status of cometary science. He identified the components of comets and observed that the objects have a proper motion of their own, moving "around the sun in an ellipse of so great eccentricity that they cease to be visible from the earth during some portion of their revolution." The experimentation "upon the character of their light upon the test of polarization, has shown that they shine only by reflecting the light of the sun." As for the physical nature of comets, the correspondent observed, very little is known; some show a well-defined nucleus while others do not "and are evidently a mass of gaseous vapors" of very low density.[17]

Calling attention to a sunspot then visible on the sun, one newspaper explained to its readers that sunspots were generally considered to be "portions of the solid and opaque mass of the sun seen through openings in the luminous atmosphere or phosphorescent clouds with which that body is surrounded."[18] Even as the newspapers reported on "grand meteoric explosions" and "wonderful meteors," they paraphrased von Hammer's summary of ancient accounts of meteors delivered to the French Academy of Sciences, they reprinted Dionysius Lardner on novae and variable stars, and they surveyed the progress of optics, introducing readers to the work of Jacobus Metius at Alkmaar, Joseph Fraunhofer in Bavaria, Friedrich Struve at Dorpat, and Galileo and Herschel.[19] The *American Freeman* reported on observations made with Lord Rosse's new telescope in England and carried the announcement of Johann Heinrich von Madler, director of the Dorpat Observatory, that he had discovered the center of "that stratum of stars to which our sun belongs."[20]

Not the least useful device to bring the public's attention to some of the more "scientific" aspects of science was the regular use of "interesting facts" as filler material in newspaper columns. One such column, reprinted from the New York *Tribune,* reported that the density of the planet Mercury was equal to that of lead, while that of Saturn was closer to that of cork; that Mercury's proper motion was 30 miles a second; that the sun's gravity was 27.5 times that of the earth; and that "fixed stars" were at least 18.2 trillion miles from the earth.[21] One should not underestimate the role of material such as this, no matter how commonplace or even incorrect, in encouraging popular interest and speculation about science.

16 *Sentinel,* July 8, 1843; *American Freeman,* Nov. 8, 1844; *Sentinel,* Nov. 25, 1846. For another account of the discovery of Neptune see *American Freeman,* Dec. 15, 1846.

17 *Sentinel,* June 13, 1845.

18 *American Freeman,* Mar. 19, 1845.

19 *Sentinel,* July 4, 1837 (von Hammer, identified only by surname, would appear to be Joseph Freiherr von Hammer-Purgstall, the Austrian orientalist); *American Freeman,* Nov. 20, 1844; *Courier,* Apr. 10, 1844.

20 *American Freeman,* May 29, 1845, Oct. 20, 1846.

21 *Courier,* Dec. 1, 1841.

The popular interest in astronomy was clearly served by a wide range of material both confirming the majesty and mystery of creation and attempting to explain it. These two elements—the "curious" and awe-inspiring on the one hand, the technical presentation on the other—were present in the popular discussion of the other sciences as well, although we shall treat them here in less detail.

Geology made virtually the same claim upon the public mind that astronomy did. The subject was notable not only for its utility—Wisconsin Territory included both a lead mining district and a copper region—but for the manner in which it led to a consideration of time, origin, and design. It was, in other words, both practical and philosophical, and its pursuit was encouraged by both qualities. The practical-minded were provided with "useful" geological information, especially about mineral deposits, through reports on surveys such as those conducted by Douglass Houghton in Michigan's Upper Peninsula and by David Dale Owen in the southwestern section of Wisconsin Territory.[22] The philosophically inclined could find in the newspaper a variety of information touching upon physical geology and its processes. The formation of the Western prairies was one recurrent subject of speculation in this regard; earthquakes and volcanoes were also of continuing interest.[23]

Many geological subjects were brought before the public simply for their curiosity, such as descriptions of the "exotic" California deserts, a whimsical contention that the Mississippi River flowed uphill because of the disparity between the earth's polar and equatorial diameters, and an elaborately romantic description of Western rock formations, from the journal of John C. Fremont.[24] Other articles had a more orthodox basis, among them Charles T. Jackson's account of the Aroostook Mountains, a report on alluvial deposits in the Vendee region of France, and accounts of salt-well borings in Michigan and coal resources in the eastern United States.[25] In many of the more technical articles the subject was treated with a precision undoubtedly beyond the capability of most readers, as in this excerpt from Houghton's report on his survey in Upper Michigan:

> . . . the trappean portion of the veins in the latter district, is essentially made up of strings, specks and bunches of native copper, with which more or less of the oxyds and carbonates are associated; while these portions of the veins traversing the conglomerate are characterised by the occurrence of the oxyds and carbonates, with occasional metalic and pyritous copper, or the places of all these are supplied by ores of zinc, associated with more or less calcareous matter.[26]

[22] *Ibid.*, June 9, 1841; *Sentinel*, Jan. 26, 1841. An intermittent series of articles on the territory's copper region appeared in the *Sentinel* throughout 1843.

[23] For the formation of the prairies see *Courier*, Dec. 29, 1841, and *Sentinel*, June 27, 1837, Sept. 26, 1837, Nov. 29, 1845. For reports of earthquakes see *Courier*, Nov. 13, 1844, Jan. 8, 1845, and *Sentinel*, Jan. 2, 1838, Sept. 1, 1846. For reports of volcanoes see *American Freeman*, Apr. 3, 1844, and *Sentinel*, May 14, 1839, Aug. 24, 1841, Sept. 11, 1845. The causes of earthquakes and volcanoes are discussed in the *Courier*, Oct. 6, 1841, and *Sentinel*, Nov. 10, 1845, Dec. 16, 1845.

[24] *Sentinel*, Nov. 27, 1838, and Oct. 5, 1841; *Sentinel*, Jan. 18, 1843; *Courier*, Jan. 31, 1844.

[25] *Sentinel*, Sept. 9, 1843; *Courier*, June 8, 1842; *Courier*, Apr. 3, 1841; *American Freeman*, Nov. 20, 1844.

[26] *Courier*, June 9, 1841.

The use of such terms as *accretion, scoriacenus trap, interfacted, gneiss, lethoid lava, viacular, obrine, ovoidal, prismatic, columnar,* and *syenitic granite* suggests the precision with which geology was often presented. Such language can hardly serve as an index to popular understanding of geology, however; nor should it be considered a proof of professional snobbery toward the capabilities of the public. It resulted simply from the dependence of the pioneer newspaper upon previously published material, in this case taken directly from an official and technical report prepared by a qualified geologist.

Paleontology received frequent and often quite detailed attention in the newspapers, although the fossil record was rarely related to the wider fields of geology or zoology.[27] The period's most publicized figure in the field of paleontology was Albert C. Koch, a German-born promoter whose gigantic "sea serpent," unearthed in Alabama, and reconstructed mastodon, the *Great Missourium,* were publicly displayed and fully reported upon.[28] A flourish of showmanship attended the exhibition of these objects (six musicians sat in the rib cage of the *Missourium* and entertained during the exhibition in St. Louis), and a fascination with their skeletal dimensions often turned the newspaper accounts into paragraphs of statistics.

Other reports in the field of paleontology included the recovery of apparent *Zeuglodon* bones in New York State and their submission to Ebenezer Emmons of the Albany Medical School for analysis; the discovery of a deposit of gigantic fossil bones in Randolph County, Arkansas; and similar discoveries near Natchez, Mississippi, and Hinsdale, New York.[29] The arrival in England of the remains of fossilized birds found in New Zealand, the Boston *Atlas* reported, had produced the conviction among British scientists that "the impressions[found by Edward Hitchcock] in the valley of the Connecticut are genuine, and that birds, both small and gigantic, once walked over the surface of the rock in question."[30]

Interest in the fossil record was sustained in part by the diluvialism of popular theology, which had as one of its corollaries the hope that ante-diluvial human remains would eventually be unearthed. The wish proved father to the thought when a human skeleton, alleged to weigh 1,500 pounds, was reported discovered in Franklin County, Tennessee, although one editor—less inclined to fancy—commented: "We think we see our old friend Kock walking deliberately into a druggist's shop, and asking for a corrosive sublimate!"[31]

27 An exception occurred in a review of the work of Cuvier, which described his catastrophism, commented upon the usefulness of fossils in dating rock strata, and noted his conclusion that "the species of extinct animals are more numerous than the living ones." *Sentinel,* Aug. 17, 1844.

28 For the "serpent" see *Sentinel,* June 24, 1844, and *Courier,* Sept. 24, 1845. For the mastodon see *Sentinel,* Feb. 9, 1841, and *Courier,* May 1, 1841. The best account of the elusive Dr. Koch is Ernst A. Stadler's biographical sketch in Albert C. Koch, *Journey Through a Part of the United States of North America in the Years 1844 to 1846,* trans. and ed. Ernst A. Stadler (Carbondale: Southern Illinois University Press, 1972).

29 *American Freeman,* June 12, 1845, *Sentinel,* June 2, 1845; *Courier,* May 8 and 15, 1844, and *Sentinel,* May 25, 1844; *Sentinel,* May 22, 1845; *Sentinel,* Apr. 27, 1841.

30 *Courier,* May 31, 1843.

31 *Ibid.,* Oct. 1, 1845 (quotation) and Jan. 14, 1846.

The springs of natural theology also drove a popular interest in archaeology and anthropology, although here there was less room for showmanship and wit. When the Reverend C. Foster claimed to have deciphered an inscription found in the ruins of a city on the Arabian coast, one commentator pointed quite explicitly to this element of the public's interest in ancient ruins: "It is not antiquity of these monuments, however high, which constitutes their value, it is the precious central truths of revealed religion which they record and which they have handed down from the first ages of the post-diluvian world."[32] Carlo Bonucet's excavations at Pompeii, the Neapolitan excavations at Herculaneum, and the work of Paul Emile Botta and the young Austen Henry Layard at Nineveh were reported on, as were the uncovering of Egyptian tombs and the discovery of ruins in Peru's Chachapoyes Province.[33]

So too was there an interest in remote and exotic contemporary peoples, particularly those with a tradition of having been instructed in the arts of civilization by members of a strange race, preferably white.[34] Yet there were exotic races closer to home, and it was upon the American continent that popular interest in archaeology and anthropology was concentrated.

> No one can reflect on the myriads of our species who have occupied this half of the globe—perhaps from times anterior to the flood—without longing to know something of their history; of their physical and intellectual condition; their language, manners and arts; of the revolutions through which they passed; and especially of those circumstances which caused them to disappear before the progenitors of the present red men.[35]

These ambitious desiderata of both popular and orthodox scientific interests were neither new nor capable of satisfaction, save by speculation and guesswork. But of these activities there was no end.

The central question of American antiquities, and certainly the most confusing, involved the origin of the American Indian. The difference between contemporary Indian culture and pre-Columbian Indian culture was itself indistinct. Nevertheless, there was a consensus that Western earthworks and the associated artifacts represented a culture higher than that manifested by contemporary Indians. That this culture originated in the Orient was the most common view espoused in the newspapers, and much evidence was marshalled in its support.[36] Yet contrary evidence was also produced: no less an authority than Henry Rowe Schoolcraft interpreted inscriptions found on a tabular stone unearthed at Grave Creek, West Virginia, as being related to Runic, Etruscan, or Pelasgic characters. So confused was the issue, and so imprecise the

[32] *Ibid.*, Aug. 21, 1844.

[33] *Sentinel*, Dec. 20, 1845; *Courier*, Oct. 27, 1841; *American Freeman*, Dec. 18, 1844, and *Sentinel*, Dec. 30, 1846; *Sentinel*, Oct. 16, 1838; *American Freeman*, Dec. 11, 1844, and Feb. 5, 1845.

[34] *Sentinel*, July 25, 1837, and July 16, 1845. Even here, however, credulity could be strained, as in one straightforward account from Ethiopia of "a species of the human family" the members of which lived entirely in trees and were covered with hair. "Their language seems to resemble the chattering of monkeys." *Courier*, Nov. 1, 1843.

[35] *Courier*, Nov. 2, 1842.

[36] *Ibid.*, Oct. 12, 1842; *Sentinel*, Jan. 29, 1839, and Aug. 19, 1843. See also *Courier*, Nov. 2, 1842.

arguments—even among orthodox scientists—that when contemporary Indians contended "they had seen the common Indian mound built," the implications of this observation for American antiquities went unnoticed.[37]

The popular image of botany rested in large measure upon curious facts ("The largest tree in the world is in Africa, where several negro families reside inside the trunk") and exotic plants: the aloe, thought to bloom but once a century; the compass plant of the Western prairies, which oriented itself in a polar direction; the pitcher plant of the Orient; the fabled Upas tree of Java, with its aroma of death.[38] James J. Mapes touched upon the curious behavior of flowers which followed the apparent motion of the sun and then made an observation applicable to both botany and popular science as a whole: "These facts, though we have not yet got at the reason of them, are still extremely interesting."[39]

Despite this emphasis upon the exotic and "curious," theoretical aspects of botany were not entirely ignored. When a local correspondent of the *American Freeman* reported that plants experience internal temperature variations, respirate, and have lactiferous tissue—when he found a correspondence between sap and blood, fiber and bone, pith and nerves, bark and skin—he was pursuing an orthodox and acceptable mode of scientific thought—analogy. When the same correspondent peppered his essay with references to *Dionaca muscipula, Desmodium gyrans, Myrica pennsylvanica, Tillandria,* and *Palode vaco* he was demonstrating a more than polite acquaintance with orthodox botany. But at the same time—and here he struck a chord of popular interest—the writer eventually lapsed into a catalogue of dubious curiosities: there is a tree in the Congo which yields butter; copper exists in wheat, sulfur in barley; gold is found in some plants, discreetly unnamed.[40]

As with botany, popular interest in zoology touched only incidentally upon the concerns of orthodox scientists in that field. The most sustained treatment of zoology consisted of excerpts from James J. Mapes' address before the Mechanics Institute of New York in 1845, which did treat legitimate zoological subjects: the structure of the bones and eyes of birds, the habits of parasitic birds, the feet of flies and lizards, the hydrodynamic shape of fish.[41] Yet the majority of zoological material in the popular press rested unperturbed upon a sense of the exotic: the eyes of insects, a mouse deer, pelicans in Cleveland, "tall chickens," a toad with five feet, whales, flies, spiders, flying fish, sharks, and orangutans which could "paint."[42] Even Wilson's *American Ornithology* was

[37] *Courier,* Feb. 15, 1843; May 1, 1841.

[38] *Sentinel,* Aug. 6, 1846; *Courier,* Sept. 28, 1842; *Courier,* Dec. 4, 1844; *American Freeman,* Oct. 27, 1846. The quotation is from the *Courier,* Sept. 24, 1845.

[39] *Courier,* Apr. 16, 1845.

[40] *American Freeman,* Oct. 19, 1844. For another exercise in analogy, an excerpt from Jean Chaptal's *Chemistry,* see the *Sentinel,* Sept. 19, 1837. The only other speculative question in botany involves an issue touched upon briefly in our discussion of geology: the origin of the Western prairies. The botanical aspect of the question—why were the prairies destitute of timber?—was considered equally puzzling to "philosophers as well as common men." See the *Sentinel,* Feb. 1, 1843.

[41] *Courier,* Apr. 16, 1845.

[42] *Sentinel,* Aug. 9, 1845; Nov. 7, 1845; June 10, 1843; Feb. 8, 1843; May 17, 1843; July 24, 1846; July 8, 1846; Feb. 12, 1842; Apr. 2, 1839; Sept. 25, 1838; Aug. 24, 1844.

represented, poorly, by an account of the trainability of crows; and a subject truly full of curiosity—the instincts of animals—was illustrated by anecdotes such as this: "A lady of title informed Buffon that she knew a blackbird who looked at the barometer every morning, and would not go out if it pointed wet."[43]

A strong sense of the marvelous also informed public discussion of two other aspects of science—microscopy and meteorology—although both occasionally demonstrated a more orthodox basis. Microscopy, for instance, was pursued chiefly as an object of wonder, illustrating the variety of creation and the pervasiveness of its principles, even when orthodox elements were reported upon. Among the merely curious items discussed was the microscopic examination of a diseased potato, which disclosed "animalculae with bodies like the soldier ant, and legs like the hairy garden spider." Even more bizarre was the claim that if a single drop of human blood were subjected to high enough magnification it would reveal the presence of "all the species of animals now existing on the earth, or that have existed during the different stages of creation for millions of years past."[44] By contrast, a local communication to the *American Freeman* alluded to the work of (Abraham?) Bennet and Christian Gottfried Ehrenberg; an article reprinted from the Galena (Illinois) *Jeffersonian* defined the field of microscopy, discussed Ehrenberg's work, mentioned the minuteness and reproductive capacity of microorganisms, and alluded to their role in the formation of chalk and in the discoloration of Alpine snow. Yet even the latter article swung around eventually to the sense of wonder which informed popular science as a whole:

> Thus from causes invisible to the unaided eye, the greatest changes have been and are now being wrought. The earth we tread, the air we breathe, even the composition of our own bodies is made up of infinite multitudes of living, organized animals. In the language of a distinguished philosopher, 'probably there is not an atom of the solid materials of the globe, which has not passed through the complex and wonderful laboratory of life.'[45]

Popular interest in meteorology was encouraged by the same sense of the marvelous. Thus we find recurrent reports of waterspouts, thunderclaps, Louisiana's "Upasian winds" (so-named from the poisonous Upas tree), and strange showers of pebbles, eels, flesh, and blood.[46] Explanations of these strange phenomena were occasionally attempted and sometimes took a technical turn. When a light rain fell in Italy on the night before an eruption of Mount Etna, the rain changed the color of silk umbrellas; perhaps, the account

[43] *Courier*, Aug. 16, 1843; *Sentinel*, Dec. 15, 1845.

[44] *Sentinel*, Jan. 1, 1845; June 5, 1845. The latter claim was attributed to Henry Bronson, professor of materia medica and therapeutics at Yale Medical School, "in illustration of his position that man contains within himself all the principles of the universe." The contention was "curious" enough to be reprinted in the same newspaper a month later, July 24, 1845.

[45] *American Freeman*, Nov. 8, 1844; *Courier*, Apr. 9, 1845.

[46] *Sentinel*, Sept. 17, 1845; *Courier*, May 11, 1842, and *Sentinel*, June 11, 1842; *Sentinel*, Aug. 10, 1841; *Sentinel*, Jan. 8, 1845, and *Courier*, Dec. 6, 1843; *Courier*, Mar. 20, 1844.

suggested, the rainfall was charged with "a large quantity of muriatic acid." A "shower of sulphur" during a thunderstorm in Tennessee was explained in this way by a correspondent of the Nashville *Union:* "I account for it by supposing sulphureted hydrogen decomposed by electric action—the hydrogen forming new combinations and leaving the sulphur free to come down on our astonished heads."[47]

But to the pioneer, meteorology was of necessity a matter of more than exotic curiosity: the role of climate in agricultural production and the importance of a salubrious climate in popularizing a new and unsettled region were reasons enough to pursue this aspect of science with some seriousness. Consequently it was not uncommon for the newspaper to discuss a problem in meteorology that would have engaged the mind more trained and subtle than most in a pioneer community. The *Courier,* for instance, discussed the question of whether dew rises from the ground or descends from the atmosphere, citing the opinions of James Hutton and Adam Clarke. The issue, the newspaper concluded, "is not yet decided," although those "pro and con have alternately preponderated." The same newspaper, six months later, carried an excerpt from John Finlay Johnstone's *Agricultural Chemistry,* in which the radiation of heat was explained in fine Rumford fashion and it was concluded that dew descends.[48]

Climate held the public's attention most strongly. The observations of the popular Nathaniel P. Willis on differences between the climates of Europe and America were reprinted from the New York *Mirror;* another newspaper reprinted the results of a study on how the winters from 1834 to 1837 had influenced vegetation in New England; and the *Courier* called for additional information on a theory that climates change every 666 years.[49] Such considerations led naturally to the compilation of daily meteorological data, which appeared intermittently in tabular form in the Milwaukee newspapers. These reports commonly recorded temperatures at three different hours—sunrise, noon, and sunset—and included brief remarks on general weather conditions such as precipitation, wind direction, and cloudiness.[50]

That this sort of activity could be pursued with few prerequisites was evident, as the *Sentinel* suggested in commending the prodigious meteorological data submitted to that newspaper by a sixteen-year-old youth who was both deaf and dumb.[51] The accumulation of such data could likewise serve much more than a parochial interest: it could involve the untutored observer in an exercise of theoretical science. Evidence of this lay in an appeal issued by James Pollard Espy to all "friends of Science" and reprinted in the *Courier:*

> A strong desire now pervades the scientific world to know more of the nature of storms. This desire may be gratified in the following way:

---

[47] *Courier,* Mar. 6, 1844; May 10, 1843.

[48] *Ibid.,* May 11, 1842; Nov., 2, 1842.

[49] *Sentinel,* July 4, 1837; Jan. 2, 1838; *Courier,* Feb. 12, 1845.

[50] Dr. L. C. Slye of Prairieville compiled weather data for the *American Freeman;* Robert Davis contributed similar data to the *Sentinel* and J. B. Smith to the *Courier;* a Milwaukee Lyceum committee, which included Increase A. Lapham and C. J. Lynde, submitted local meteorological information to both the *Courier* and the *Sentinel.*

[51] *Sentinel,* July 22, 1845.

> Let all persons who keep journals of the weather send their journals monthly to the Navy Department, Washington, in whose service I am now employed.
>
> Those who have barometers, thermometers, and rain gauges, are not the only persons who can be useful in the advancement of science. All that will send an account of the time of the beginning and end of rains and snows, the direction of the wind, its force, (as near as they can estimate it,) and the time of its change in storms, will do much that is particularly wanted.[52]

For those who were not inspired by the opportunity to advance theoretical science—Espy said he wished to "enable meteorology to take its rank among the exact sciences"—he also calculated his appeal in more pragmatic terms: "Who does not wish to know, when a great storm comes within disturbing influence, if that storm will pass to the north or the south of him, and at what time?" Either way, meteorology was the one discipline which energetically encouraged popular participation.[53]

Curiosity about the nature and properties of large bodies of water was evident in a number of articles dealing with hydrology or oceanology. Gradual fluctuations in the level of Lake Michigan and the other Great Lakes had an economic interest as well, because falling lake levels interfered with shipping.[54] Sudden rises and falls in lake levels—such as occurred at Toledo in the summer of 1840 and at Valletta in the Mediterranean—were inexplicable and returned the reader to the mysteries of nature: "What is the cause or whence came this mighty swelling of the waters?" the Toledo *Blade* asked biblically. An excerpt from Mary Somerville's *Connection of the Physical Sciences,* an article on the origin of lakes, and a report from the *American Journal of Science* on the hydraulic power of Niagara Falls answered in a general way to this interest.[55]

Chemistry and physics were most often presented to the public in theoretical terms; yet here, too, simple curiosity held part of the ground. Petrification was one chemical process which received rather sustained attention, although the mechanics of the process were never clearly explained. The report of a petrified human body which had been exhumed from a graveyard in Iowa was of sufficient interest that it was reprinted four times, in all three newspapers considered here, within the space of three months.[56] More conventional objects of petrification, especially wood, were periodically reported upon.[57] Atmospheric luminescence and cases in which natural gas deposits were struck while boring for wells were other phenomena which attracted attention as curiosities.[58]

[52] *Courier,* Aug. 10, 1842.

[53] Those who responded to Espy's appeal received a circular containing more detailed information, a sample form for recording observations, a modest request for subscriptions to Espy's *Philosophy of Storms,* and an emphatic reiteration of the usefulness of even the most common observations: "Let not those who have no barometers be discouraged—their journals of the winds and rains and snows will be of great importance." *Courier,* Aug. 10, 1842.

[54] *Courier,* Dec. 8, 1841, and *Sentinel,* July 4, 1837, Apr. 10, 1838, June 26, 1838, July 17, 1838, Sept. 18, 1838, Dec. 11, 1841.

[55] *Sentinel,* July 14, 1840 (Toledo); July 28, 1846 (Somerville); Feb. 10, 1844.

[56] *Courier,* Jan. 29, 1845; *Sentinel,* Feb. 4, 1845; *Sentinel,* Mar. 3, 1845; *American Freeman,* Mar. 5, 1845.

[57] *Courier,* May 1, 1841; *Sentinel,* Dec. 23, 1843; *Courier,* Mar. 13, 1844; *Sentinel,* Jan. 18, 1845. For the report of a petrified honeycomb—a "singular curiosity"—see *Sentinel,* Nov. 6, 1838.

[58] *Sentinel,* May 3, 1845; *Sentinel,* Nov. 28, 1837, and *American Freeman,* Apr. 10, 1844.

Most of the newspaper material relating to chemistry involved its application to agriculture, a flourishing field in which Justus Liebig was the unquestioned leader. So common were articles of this sort that they have not been tabulated. Here is an example:

> Plaster on Land—Professor Leibeg of Geissen, has discovered that snow and rain water always contain ammonia—hence its presence in the atmosphere. Plaster (sulphate of lime) forms this ammonia in the soil, and keeps it there to stimulate and feed vegetation, in the same manner as lime prevents the escape of the humic acid and other fertilizing grasses [gasses], from animal and vegetable manures.[59]

The more sophisticated exposition of chemistry reached the public in other ways as well, although the subject often carried a touch of the bizarre. Among the theoretical considerations discussed in the newspapers were the effect of air and steam on fire and the chemical composition of the earth's atmosphere.[60]

It was in the field of physics that popular science was at its most theoretical, however. In addition to touching upon concepts of heat and the effect of gravity,[61] extended attention was given to notions of sound and expansion and to experimentation. An article excerpted from the *Edinburgh Encyclopedia* commented upon the supposed diminution of sound during the day and cited Humboldt as its authority in explaining the phenomenon as the result of the presence of the sun, "which influences the propagation and intensity of sound, by opposing to them currents of air of different density, and partial undulations of the atmosphere produced by the heating of the different parts of the ground." A sound wave, it continued, is divided into two or more waves when it meets two layers of air of different density; thus the sound wave is "so wasted and frittered down as to be incapable of affecting the tympanum."[62] An article on the "Effects of Expansion," taken from Neil Arnott, illustrated the effect of temperature on the volume of water through the example of potholes in glaciers.[63] The effects of heat also figured in an experiment which can be simplified in this manner: (1) heat a platinum crucible until it is red hot; (2) add sulfurous acid; (3) add a few drops of water; (4) the acid "flashes off," leaving ice. This, the newspaper said, "is illustrative of the repellent power of heat radiating from bodies of high temperature, and the rapid abstraction of heat, produced by evaporation, or generally by such a change of condition, as largely increases the volume of any body."[64]

Electricity was another aspect of physics which received considerable attention. The speed and effects of lightning, the uses of electricity in stimulating plant growth, and the development of electromagnets and electric motors made the

[59] *Sentinel,* May 11, 1841.

[60] *Ibid.,* July 18, 1846; *Courier,* June 30, 1841, and Nov. 16, 1842.

[61] *Sentinel,* June 14, 1839; Aug. 26, 1846. The nature of heat also figured in the discussion of dew; see n. 48.

[62] *Courier,* Feb. 8, 1843.

[63] *Ibid.,* Nov. 16, 1842.

[64] *Sentinel,* Apr. 17, 1845. For experiments illustrating the diffraction of light, the nature of thunder, and capillary action in plants, see, respectively, *Sentinel,* Jan. 9, 1846; *Courier,* May 11, 1842; and *Courier,* June 11, 1845.

subject answer to both an interest in natural phenomena and the application of science to utilitarian tasks.[65]

The quantity of popular material published within the orthodox disciplines alone is impressive evidence of the diffusion of scientific information. Yet in addition to this, newspapers printed a large body of material which touched peripherally on science: medicine, a subject of recurring interest; agricultural chemistry and horticulture; mechanics and invention, what we would call engineering; and such enthusiasms as mesmerism and phrenology. Reports on patent frauds and hoaxes were persistent. In announcing the discovery of "concentrated essence of the sublimate spirit of steam," one newspaper commented merrily: "A person has only to put a vial of it into his pocket, and it will carry him along at the rate of fifty miles an hour, or by merely swallowing three drops when you go to bed at night, in the morning you will wake up in any part of the world you choose."[66] At this remove satire loses its sting, but we can only assume that the majority of such articles were written with tongue in cheek.

It is obvious that folklore, fable, and the "marvelous" had a vital hold on the public mind; yet the distinction between genuine science and pseudo-science was in fact sharply drawn. The *Courier,* for instance, ridiculed the reported sighting of a "sea serpent" off Cape May be pointing out that "it is not stated whether it was propelled on the principle of the 'screw' or the 'wriggle.'" The *Sentinel,* reporting on a later sighting of the same object, referred to it as "his snakeship."[67] Not even the popular fads of mesmerism, animal magnetism, and phrenology—all of which claimed some connection to orthodox science and which had the endorsement of many established scientists—wholly escaped this facility for distinction. While these subjects received extensive coverage in the press, it was often with suspended judgment and on occasion with outright cynicism. "A certain clergyman in New Hampshire, from his pulpit, recently put twenty five of his congregation to sleep at one sitting," the *Sentinel* reported under a headline which said "Animal Magnetism."[68]

On a more serious level the compilation of bits of information under such standing headlines as "Scraps of Curious Information" and "Interesting Items" has already been alluded to; if they did nothing else, these accumulations of fact and observation provided conversation pieces. They were prods to a more sustained interest in scientific matters. Articles treating the history of the sciences, such as the report on the progress of optics mentioned above or an article recounting the skill of ancient Egyptians in science and instrument making, were another source of scientific information available through the newspapers.[69] Two additional sources of popular science remain to be briefly noted.

Itinerant lecturers in the pioneer community both encouraged and reflected

[65] *Sentinel,* Oct. 27, 1840, and Sept. 30, 1843; *American Freeman,* May 29, 1845, and *Sentinel,* Aug. 10, 1844, Feb. 13, 1845, July 22, 1845; *Sentinel,* Feb. 15, 1843, and July 30, 1839.

[66] *Sentinel,* Oct. 23, 1838.

[67] *Courier,* May 22, 1844; *Sentinel,* July 1, 1846.

[68] *Sentinel,* Sept. 30, 1843.

[69] *Courier,* Apr. 3, 1844.

an interest in science and on occasion added to the public competency in the subject. In 1841 one Billard lectured at Milwaukee on the subject of astronomy; the Reverend G. R. Haswell of Ohio, bearing an endorsement from the Mechanics' Institute of Chicago, delivered himself of a series of lectures on the same subject. A Professor Porter of New York spoke on "a new system of arithmetic" in 1845, and a Mr. Bowman of Jefferson College delivered five lectures on astronomy in 1846, touching upon "the early history of the Science—proof of miracles—Stones falling from the Moon and Asteroids to the earth."[70] There were also lectures on less orthodox—although perhaps more popular—topics. Professor George C. Tew lectured on animal magnetism at Milwaukee's Presbyterian Meeting House in 1842; Professor Fauvel Gouraud's system of "Phreno-Mnemotechny" was the subject of a public lecture in 1844.[71]

Another important source of popular scientific information rested in the books available to the public through Milwaukee booksellers, or through subscription libraries. The range of titles suggests the dimensions of popular interest in scientific subjects:

> Anthon's *Greek and Roman Antiquities;* an *Agricultural Chemistry* by an unspecified author; Arago's *Popular Lectures on Astronomy;* Bacon's *Works;* Burritt's *Geography of the Heavens;* Boucharlat's *Mechanics;* Butler's *Analogy;* the *Cabinet of Natural History; Climate of the United States and its Epidemic Influences;* Comstock's works on botany, chemistry, mineralogy, natural philosophy, and physiology; Thomas Dick's popular *Celestial Scenery, Geography of the Heavens,* and *The Sidereal Heavens;* Frost's *Class Book of Nature;* the *Geological Reports of Massachusetts;* Goldsmith's *Geographical View of the World;* Grund's *Elements of Natural Philosophy;* Herschel's *Astronomy;* the *History of Insects;* Lapham's *Geographical and Topographical Description of Wisconsin;* Lardner's *Lectures on Science and Art;* biographies of Newton and Rittenhouse; Mrs. Lincoln's *Botany* and *Philosophy;* Mitchell's *Geography;* Morse's *Geography;* Newton's *Works;* Olmsted's *Natural Philosophy;* Olney's *Introduction to Geography;* Paley's *Theology;* the *Physical Condition of the Earth;* and Renwick's *Natural Philosophy.*

These works were all publicly available in Milwaukee during the period under discussion. Some, to be sure, were school texts, and the titles cited here comprise less than 5 per cent of the total number of titles publicly available in the village. Yet were we to include an equally large number of medical works, a number of popular titles on phrenology, animal magnetism, and mesmerism, a large number of school arithmetics and science series, and a sizable collection of travel narratives, which often touched upon elements of natural science, it would be evident that the intellectual climate of the pioneer village encouraged scientific reading.[72]

[70] *Sentinel,* June 22, 1841; *Courier,* May 3, 1843; *Sentinel,* Nov. 11, 1845; *Sentinel,* Nov. 16, 1846.

[71] *Courier,* Sept. 28, 1842, Oct. 16, 1844.

[72] Between 1837 and 1844 more than 1,500 separate titles were offered for sale by Milwaukee booksellers or for use through subscription libraries; after the latter date the number of titles increases significantly. Yet considering the presence of personal and private libraries, this represents only a portion of the reading material available.

Another force tending toward the encouragement and organization of scientific interests was the Lyceum, although the effects of such a group can be easily overestimated. "The organization of a Lyceum in this place is at once an interesting and an important event," observed the *Sentinel.* "It is a strong illustration of the march of mind and civilization, and an evidence that this community have done with the insane rage of making fortunes in a day."[73] The call for the establishment of a Lyceum had come on January 1, 1839, the *Sentinel*'s endorsement came a week later, and the first meeting was held shortly thereafter. But by September of the same year the *Sentinel* was wondering in print what had become of the Lyceum, and in December it printed a mock obituary of the institution. The march of mind and civilization was certainly no goose step.[74]

We have before us, nevertheless, evidence of the substantial range of scientific information placed on public view through newspapers, books, and to a lesser extent by popular lecture and discussion. As Channing said, science had indeed left its retreats. Let us now examine the working of public interest in science, the ways in which popular science exercised its appeal, and the claims made on its behalf. Among the sources of interest in science, two seem especially important: curiosity and social utility.

Simple curiosity was the datum of popular scientific interest, as the newspaper material we have just surveyed demonstrates. To be sure, the continual use of the word "curious" to describe accounts of the natural sciences was a journalistic convention, just as frontier murders were invariably "shocking" and escapes "wonderful." Yet curiosity was more than a convention. Without it there could be no interest in science; with it came an attitude toward the natural world—a stance of contemplation—which urged the observer toward the twin notions of majesty and grandeur and which fed a continuing interest in science. "The works of Nature are so wonderful—so passing strange," the editor of one Western newspaper exclaimed, "that we are often tempted to turn aside from the tedious duties of editorial life, to investigate them."[75] The *Sentinel,* reviewing Dionysius Lardner's *Popular Lectures on Science and Art,* found the first two numbers "commend themselves to all who have any love for the harmony, order, majesty and beauty of the physical creation."[76]

We shall soon look more closely at such qualities as the *Sentinel* associated with the work of Lardner, particularly harmony and order. But it is sufficient here to notice Perry Miller's observation that the acts of adoration and understanding—the natural outgrowth of an energetic curiosity—were essentially an escape from provinciality.[77] In a single intellectual exercise, the adoration of the sublime bridged the gap between the pioneer settlement and the established culture of the East; insofar as this act carried an implicit distinction between two levels of culture—one being fashioned, the other already estab-

[73] *Sentinel,* Jan. 8, 1839.

[74] *Ibid.,* Sept. 17, 1839, Dec. 24, 1839. The *Courier,* Dec. 15, 1841, carried a notice that the "Brookfield and Wauwatosa Lyceum" was soon to start its third session to discuss "moral, historic and scientific questions," but this is the only appearance of such a group.

[75] *Courier,* Apr. 9, 1845.

[76] *Sentinel,* May 19, 1845.

[77] *The Life of the Mind in America* (New York: Harcourt, Brace & World, 1965), p. 277.

lished—it was also an expression of social aspiration.

The notion of *social* utility has not been sufficiently recognized as a spur to the pursuit of science, in part because the relationship between science and utility has been so narrowly interpreted. Channing, in 1841, found that science had passed from speculation into life, becoming "an inexhaustible mechanician . . . by which nature is not only to be opened to thought, but subjected to our needs."[78] Here, in Channing's lament, is the standard definition of utility—the "improvement" of the objective order, the application of science to the needs of production, transportation, and material life. Yet there is another way in which we might understand utility, as an intangible and subjective force involving, on the one hand, the social advancement of one's self in terms of wealth, prestige, or culture, and on the other hand the social advancement of one's class, community, or nation. It is this sense of science as a potential agency of personal or community advancement which plays a crucial role in the popular appeal of science, and to adequately understand it we shall have to examine the claims made in its behalf. To begin with, we shall have to consider a question posed by Increase A. Lapham: "What possible benefit can it be to any person to become acquainted with the animals, plants, and minerals, by which he is surrounded?"[79]

Lapham was the most serious and substantial of Milwaukee's men of science. When he addressed the struggling Lyceum in 1840 on the subject of natural science he could speak as an authority: he had published (on geology) in Silliman's *American Journal of Science* at the age of seventeen, he had recently compiled a catalogue of shells and plants found in the vicinity of Milwaukee, he had corresponded with professional scientists in the East, he was working toward the publication of his ambitious and influential *Topographic Description of Wisconsin,* and he had accumulated his own cabinet of natural history objects. When he asked "What possible benefit can it be to any person to become acquainted with the animals, plants, and minerals, by which he is surrounded?" his response to that rhetorical question was not the response of an Eastern academician, delivered abstractly. Instead, it was informed by an intimate knowledge of his community and of the men who made it a community and consequently becomes an important clue to the kind of reason most likely to have motivated the members of his community toward the study and appreciation of science. Lapham suggested there were three reasons for the study of natural science.

If the question were to be considered "in the narrow light of *money making,*" he said, the pursuit of science might as well be forgotten. But in this disclaimer he was only bowing to the spirit of gentility. For in fact, he continued, when one observes the states and federal government sponsoring geological surveys, when naturalists are being assigned to government exploring expeditions, when chairs of natural history are being established at colleges and universities, "who

[78] *Sentinel,* Aug. 17, 1841. Channing complained that science "is not pursued enough for its intellectual, and contemplative uses."

[79] Increase A. Lapham, untitled lecture presented Jan. 23, 1840, before the Milwaukee Lyceum, in Increase Lapham Papers, State Historical Society of Wisconsin, Madison.

will say that these pursuits may not lead to pecuniary profit?" And for those not qualified for service with the government or universities, Lapham suggested that a knowledge of geology and mineralogy "may enable us to detect the presence of valuable ore or mineral beds, whose existence might otherwise never have been suspected."

But "there are other and higher reasons for pursuing these studies." One is pleasure, "a pure and unalloyed pleasure . . . that is seldom found anywhere else." More importantly, there is a universal need for respite from "our severer duties and studies." If this be particularly true for young persons, it behooves their elders to provide amusements "pure and rational, and which tend to their moral and intellectual advancement." "Here," he emphasized, "we have one of the most important arguments in favor of the sciences. Teach young persons to relish the pure and simple beauties of nature—excite in their bosoms an ardent and enthusiastic love of the wonderful works of the Great Creator and you have one of the surest safeguards against immorality and vice."

The notion of social utility lies lightly on Lapham's argument. In a manipulative sense it underpins his advocacy of science as a hedge against immorality and vice; in a more obvious way it promises pecuniary advantage. That science bore an application to social aspirations is also evident in Lapham's suggestion that a knowledge of plants and shells "seems now to be almost indispensable to an *accomplished* lady."[80]

Daniel Drake, addressing much the same kind of audience in Cincinnati a generation earlier, was even more explicit about the social advantages of science.

> Our lot, gentlemen, is cast in a region abundant in but few things, . . . Learning, philosophy and taste, are yet in early infancy, and the standard of excellence in literature and science is proportionately low. Hence, acquirements, which in older and more enlightened countries would scarcely raise an individual to mediocrity, will here place him in a commanding station. Those who attain to superiority in the community of which they are members, are relatively great.[81]

The objects of natural history, he concluded, not only stimulate industry and enthusiasm, but are "the springs of ambition."[82]

Such argument for the study of science came not only from professional scientists or those who bordered on professionalism; it came from the newspapers as well. In suggesting that his readers subscribe to Silliman's *Journal,* the editor of the *Sentinel* observed that "a knowledge of the natural sciences, and of the improvements in the Arts are indispensable" to men engaged in mining, as many in the western portion of the territory were. An article from the

[80] In this regard it is interesting to note the number of "floral" books designed for women, each carrying the notion of refinement. Among them were the *Lady's Book of Flowers and Poetry, The Language of Flowers, Floral Biography,* and *Flora's Lexicon,* all advertised in the Milwaukee newspapers.

[81] Daniel Drake, "Anniversary Address to the School of Literature and the Arts," Cincinnati, 1814, in Henry D. Shapiro and Zane L. Miller, eds., *Physician to the West: Selected Writings of Daniel Drake on Science and Society* (Lexington: The University Press of Kentucky, 1970), p. 59.

[82] *Ibid.*, p. 63.

Common School Assistant urged that the young mechanic "cultivate his mind, that his head may help his hands. Science will lessen and lighten his task." Another article, advocating that every farm family purchase a text on entomology, counseled young men to "unite knowledge with labor—science with practice—and the great Fountain of all knowledge will reward him a thousand fold for his well directed efforts."[83]

The newspaper argument was not restricted to promises of wealth or ease. "There is nothing," remarked one writer on science, "beneath the attention of a gentleman"—thus relating attentiveness to the natural world with the upper classes. A broader social utility underlay the argument that young men should forgo idleness, "games of chance and lounging in bed" for a course of study and self-improvement; the writer cited David Rittenhouse and James Ferguson as examples. Religion, morality, and true science, yet another writer exclaimed, are "inseparable companions" and must "exercise a controlling influence over the public mind." Science has lessons for life, the *Sentinel* contended, recalling Sir Humphry Davy's description of two adult eagles teaching fledglings to fly: "What an instructive lesson to Christian parents does this history read!"[84] The advantages of science thus could be both economically and socially elevating, returns coming to both the individual and the community at large. Science was encouraged as a spectacle, a grand and immutable teacher, a practical aid, but especially as an improver of self and station.[85]

The coin of encouragement had another side, equally telling on the popularity of science; that is, the response of the general public to science as an institution and organized aspect of contemporary life.

What is a scientific man? one newspaper asked rhetorically. And then answered:

> A scientific man will look at the thermometer to see when it is warm, at the barometer to see when it storms, at the clock to see when he is hungry, at the almanac to see how old he is, at the moon to see when it is high tide, at the stars to ascertain when he is in love. He is a scientific man, and does everything by rule but his devotions, which he doesn't do at all.[86]

Was the scientific man godless, befuddled, and bereft of common sense? That was certainly not the prevailing sentiment expressed in the popular press; the preceding, in fact, was the only instance of cynicism toward science published

[83] *Sentinel*, Feb. 2, 1841; *Sentinel*, Oct. 9, 1838; *Courier*, Aug. 27, 1845.

[84] *Sentinel*, Oct. 10, 1837; *Courier*, Mar. 6, 1844; *Sentinel*, Nov. 17, 1840; *Sentinel*, Sept. 1, 1840.

[85] The compelling thrust behind this drive for self-improvement and rephrased claim of individual worth was "the roaring flood of the new democracy," a term which Charles Beard applied to the vote for Jackson in 1824 but which Richard P. McCormick has shown to be more descriptive of the "Tippecanoe democracy" of the 1840s, the period considered here. See McCormick, "New Perspectives in Jacksonian Politics," in Seymour Martin Lipset and Richard Hofstadter, eds., *Sociology and History: Methods* (New York: Basic Books, 1968), pp. 229, 236.

[86] Milwaukee *Gazette* (merged with *Sentinel*), Feb. 3, 1846. The only other instance in which the newspapers came close to ridiculing a scientist came in a note that the wife of astronomer Dionysius Lardner had "again abandoned her husband and eloped with an officer of the army." This was placed under a headline which read: ERRATIC COMET. *Sentinel*, June 24, 1846.

in the newspapers considered here. In general, science was held to be neither sacrosanct nor inexplicable. One measure of its acceptance may be seen in the way in which it provided grist for the pun-mills of journalism, without—we may point out—malicious intent. "The report that Encke's Comet may be seen with 'good glasses,'" one newspaper reported, "is contradicted by a man who says he took *six 'good stiff glasses,'* on purpose to see it; and only *saw stars.*" Morse's fame, the same newspaper commented, "already reaches from *pole* to *pole.*" Which rock proves most fatal to rats and mice? asked another, in a series of geological conundrums. The answer: *Trap.* On what rock are loafers generally split? *Quartz.* What rock expresses the conditions of a lover? *Sienite.*[87]

The number and variety of references to established scientists in the newspapers suggests as well the value with which their names and achievements were held. These include Arago, Audubon, Buffon, Banks, Cuvier, Copernicus, Davy, De Saussure, Drake, Ehrenberg, Espy, Faraday, Franklin, Galileo, Galvani, Gay-Lussac, Hitchcock, Hutton, Herschel, Humboldt, Jenner, Kepler, Leverrier, Linnaeus, Liebig, Lyell, Leeuwenhoek, Mantell, Newton, Rittenhouse, Rosse, Struve, Silliman, Volta, and an even larger number of established scientists of lesser importance in retrospect.

The faithfulness with which the press followed surveys and expeditions connected with the pursuit of scientific objectives, as well as the establishment of new scientific institutions, also demonstrates a valued emphasis upon science. The U.S. Exploring Expedition under the command of Lieutenant Charles Wilkes was the most closely followed, with sixteen separate references to it during the course of its progress.[88] Local or regional surveys, however, were more likely to draw extended editorial comment. Houghton's survey of Michigan's Upper Penisula was commended for producing "a mass of valuable and interesting matter, which cannot but excite the earnest attention of every class of readers." Joseph N. Nicollet's account of his journey through the Upper Midwest drew this comment from the same newspaper: "To the emigrant, the farmer, the merchant and navigator—to the topographer and geographer, the botanist, the mineralogist and the geologist—to the scholar and statesman, and to the whole country, this is a work of the highest importance."[89] Activities of the Royal Academy, the British Association for the Advancement of Science, and the French Academy were regularly reported upon; many other foreign institutions and organizations were occasionally alluded to. The organization of the Smithsonian Institution was followed rather closely, although not substantively. The formation and collections of the National Institute in Washington, proposals for building a National Observatory, and the establishment of chairs in natural science at various colleges and universities were

[87] *Sentinel,* June 11, 1842; *Sentinel,* Aug. 17, 1846; *Courier,* June 15, 1842. For other examples see *Sentinel,* Dec. 31, 1839, July 20, 1846, and *Courier,* Apr. 10, 1844.

[88] *Sentinel,* July 4, 1837, Oct. 10, 1837, Apr. 17, 1838, June 12, 1838, Sept. 11, 1838 (two separate articles), Feb. 26, 1839, June 4, 1839, Dec. 10, 1839, Mar. 3, 1840, June 23, 1840, Feb. 23, 1841, May 4, 1841, Jan. 8, 1842, Aug. 17, 1842, and *Courier,* July 6, 1842.

[89] *Courier,* June 9, 1841; Mar. 27, 1841.

also the subjects of newspaper articles.[90]

This celebration of science—for so it appears, when considered with the large quantity of material on scientific subjects offered by the press—was marked by a decided patriotic fervor. When Emerson remarked that Americans had "listened too long to the courtly muses of Europe," he spoke of scholarship in general; but he reflected a concern with the achievement, or lack of achievement, within American science often expressed by American intellectuals and scientists themselves. The popular mind appears less concerned with actual achievement, however, than with an endorsement of American science per se. And it seems proper that this phenomenon is best described by the word patriotism, rather than chauvinism, for the work of European scientists and institutions was thoroughly reported upon, never with a note of envy or disparagement.

Yet this genial acknowledgment of European superiority in science held only so long as it was restricted to Europe; science in America, it was keenly felt, ought to be pursued by Americans. In all the references to foreign scientists, only Sir John Herschel, Sir Joseph Banks, and Christian Ehrenberg were recipients of eulogistic adjectives. It was much more common to find approving terms bestowed upon native or naturalized American scientists. Thus Schoolcraft was a "distinguished" interpreter, Joseph Henry a "most admirable selection" to direct the Smithsonian, Audubon a "celebrated ornithologist," and Alexander Wilson a "clever and amiable naturalist." Samuel L. Dana and Charles T. Jackson, an item from the East Coast observed, "both . . . possess and deserve the confidence of our community, as sagacious and expert analysts and reasoners . . . ."[91]

In some cases this patriotic approach to science was quite explicit. Commenting upon a survey of the Lead Region conducted by David Dale Owen and John Locke, the *Sentinel* observed:

> Contrary to the usual and somewhat censurable practice of our country in employing foreigners for scientific researches, this has been conducted by citizens—a western expedition conducted by western men; and we have some curiosity to see this production of "back-woodsmen" . . . These gentlemen will be found, we presume, to have written no worse, [than] that they *saw* what they have observed, instead of borrowing the field notes of other explorers, and depending upon them for information.[92]

When Ferdinand Hassler, a Swiss subject, died in office as superintendent of the U.S. Coastal Survey, the *Courier* expressed the hope that "an *American* can be found to fill the vacancy occasioned by his death." The appointment

[90] (Smithsonian) *Courier*, Feb. 19, 1845, and *Sentinel*, June 18, 1846, Sept. 19, 1846, and Dec. 16, 1846; (National Institute) *Sentinel*, Oct. 27, 1840, and *Courier*, Dec. 25, 1844; (National Observatory) *Courier*, Dec. 14, 1842; (chairs of natural history) *Sentinel*, Nov. 5, 1839, and Sept. 13, 1845.

[91] (Herschel) *Courier*, Mar. 5, 1845, and *American Freeman*, Oct. 6, 1846; (Banks) *Courier*, Mar. 5, 1845; (Ehrenberg) *Courier*, Apr. 9, 1845; (Schoolcraft) *Courier*, Dec. 7, 1842; (Henry) *Sentinel*, Dec. 16, 1846; (Audubon) *Sentinel*, May 10, 1843; (Wilson) *Sentinel*, Feb. 19, 1846; (Dana and Jackson) *Courier*, June 30, 1841.

[92] *Sentinel*, May 5, 1840.

of Alexander Dallas Bache to the position was subsequently announced in the *Courier,* but without comment.[93]

Not only did science make definite claims upon the popular mind, it elicited public endorsement in ways such as these. Yet in at least one instance—the natural conflict between science and religion—popular science was obliged to tread a careful path.

That the contemplation of the natural world would lead the observer almost inevitably to God—to the uncaused cause—was a central tenet of popular science; nevertheless, discoveries in paleontology and theories in geology had long since created a tension between science and religion which could not be ignored. The clear task of popular science was to reconcile the two—to assert, at the very least, that the conflict was illusory. To this task a correspondent of the *American Freeman* addressed himself in exploring "hostility against new ideas [in science] . . . merely because these truths do not agree with our preconceived ideas [in religion]." Indeed, he wrote, after citing the hostility engendered by the ideas of Jenner, Copernicus, and Galileo:

> . . . it would be difficult to name any important discovery, ever made in any branch of Natural Science, but what was at first violently opposed, as 'false in philosophy and the express word of God.' And the Bible has been appealed to almost as often to disprove Geological and Astronomical facts, as it ever was to establish any article of Christian faith . . . Nature and revelation go hand in hand, and each rightly understood becomes an unerring commentary on the other.—When the expounders of the Bible shall have learned how far its language is to be construed according to the exact letter, and how far it conformed to the state of human knowledge at the time it was delivered, no inconsistency will be found between the sacred volume and 'That elder scripture writ by God's own hand.'[94]

"Truth, brought down from on high, harmonizes with truth from below," another writer asserted, "and the Christian who refused to surrender his cherished volume to the taunts of reason, now holds it with a firmer grasp and scans the series of creations which Science has revealed."[95] Increase Lapham, in his address to the Milwaukee Lyceum, conceded that geology posed certain problems for religion, particularly for the Mosaic account of creation. But, he insisted,

> . . . when properly understood, it will be seen that the scripture no where says that the world and Adam were created at the same time. The world was created 'in the beginning' but when that beginning was we are not informed; and all that time between the first creation and the creation of man is passed over in silence by the sacred writer. Geology, so far from throwing doubts upon the Scriptures, tends to confirm their truth in many cases.[96]

[93] *Courier,* Dec. 13, 1843, Jan. 10, 1844.

[94] *American Freeman,* Oct. 5, 1844.

[95] *Sentinel,* Aug. 17, 1844.

[96] Lapham, *op. cit.* For the attempt of a Milwaukee layman to reconcile the Genesis account of creation and science—as well as for evidence of the speculative impulse—see the 1840 memorandum of James Clyman reprinted in Charles L. Camp, ed., *James Clyman, Frontiersman,* "definitive edition" (Portland, Ore.: Champoeg Press, 1960), pp. 53–54. This memorandum is discussed in Donald Zochert, "The Natural Science of an American Pioneer: A Case Study," *Transactions of the Wisconsin Academy of Sciences, Arts and Letters,* 1972, *60*: 7–15.

The evangelical criticism of science does not itself appear in the newspaper, although such rationalist argument as appears above demonstrates its presence in the community, perhaps from the pulpit. The existence of a tension between science and religion, however, is hardly at issue; what is important here is the way in which attempts were made to soften the differences, and the insistence that science was a confirmation of the existence and wisdom of God.

A second source of tension—in this case between pure and applied science—was quite simply not perceived with the same ragged and antagonistic edge as it is today. This is not to say that the division of interests within science was not noted; Channing's lament shows that it was. And it is true that the period being discussed here was preeminently one of invention and application. The scream of celebration over progress and practicality was set loose indiscriminately, by serious advances in science as well as by the most exotic ideas of the slightest of men.[97] Out of this came a body of thought stamped with optimism about the technological future of American society, not unconnected with the spirit of political and intellectual optimism which fed an interest in popular science. "The world is just on the threshold of discoveries in science and the arts, which must change the whole face and fabric of society," the *American Freeman* observed. "In this enlightened age," agreed the *Sentinel*, "discovery after discovery, and improvement after improvement, follow each other in such rapid succession, that we are prepared to believe almost everything, that may be asserted. . . ."[98]

Despite this focus of energetic utility and improvement, the more subtle pleasures of science did not suffer in their position before the public. In this regard the notion of two "utilities"—one objective and one subjective, one fueling technological advance and the other leading to cultural or social advancement—seems particularly appropriate. The conflict which subsequent observers found between pure and applied science, or even the tensions defined by such perceptive contemporaries as De Tocqueville or Channing, were simply not conceded in the scientific material and commentary shaping the popular mind through the press.

The popular science which emerged from this material was organized around the notion of the natural world as an emblem of divinity; the "sublime harmony of motion, of the physical Heavens," Daniel Drake found, was a "most beautiful emblem of the moral Heaven, where divine love, sustains and regulates the whole."[99] Here lay not only a rationale for the study of science, but an organizing principle for thinking about it. "When the human eye becomes tired with looking on the mutable things of this world," wrote a literary gentleman, "it

[97] Among the more exotic: ship pumps, flying machines, life preservers, steam plows, spark catchers, year-long clocks, Colonel Reed's puff machine (for blowing letters from Boston to New York City through an air tube), and Professor Reinagale's air-engine, which bore Faraday's endorsement as "perpetual motion of the most terrific description." For the endorsement see *Sentinel*, May 23, 1845.

[98] *American Freeman*, Oct. 12, 1844; *Sentinel*, Sept. 19, 1837.

[99] Daniel Drake, "Introductory Lecture for the Second Session of the Medical College of Ohio, November, 1821," in Shapiro and Miller, eds., *Physician to the West*, pp. 172–173.

can wing its way . . . to those mighty orbs that float in grandeur through the immeasureable sea of space, and follow in the concatenation of planets and suns, and systems, that lead to the throne of their unoriginated Creator." There, another writer remarked, one might find "that uncreated cause which regulates and upholds the universe," or—as still another wrote—that power which "regulates the unbroken music of those eternal spheres."[100]

It was upon the heavens that the ideas of order and regularity generally rested; we have already noticed the way in which Leverrier's calculations on the planet Neptune led to the conclusion that "the voice of nature is now a voice of melody." Yet one could find "a striking manifestation of the wonderful economy of Providence" just as easily in the way in which plants and animals were nourished, for instance, or, in something as commonplace as the morning dew, discover "one of those wise and beautiful adaptions, by which the whole system of things, animate and inanimate, is fitted and bound together."[101] So intense and compelling was this search for design and harmony in nature that the contemplation of these qualities came to be seen as a social, not merely an intellectual, virtue. As Samuel Young remarked,

> The great mass of the real ills of life, as well as the whole dark catalogue of imaginary woes, unquestionably result from man's ignorance and disobedience of the laws which have been impressed by the Creator upon the universe of the mind, and of matter by which he is surrounded.[102]

If the emblematic harmony of the spheres was a truism, shaping the popular approach toward science, so too were the means by which the elements of the natural world were to be identified and studied. The engines of science, to invoke Drake once again, were "acute observation, accurate comparison, judicious arrangement, and logical induction."[103] The outline of scientific laws—the design behind a regulated universe—was to be wrung from nature by a prodigious empiricism, the accumulated, perhaps endless, effort of observation and measurement. One finds reflections of this particular methodology in several instances: in the exactitude with which fossil discoveries were reported upon, for example, and in reports tabulating wind direction over a long period of time.[104] On occasion the empirical method was explicitly endorsed, as when one Milwaukee editor commended an article in Silliman's *Journal;* the author of the article, he wrote, "very properly confines himself almost entirely to an examination of the facts observed, and not being influenced by any pre-conceived theory . . . his observations may be relied upon with perfect confidence. . . . Let us gather all the facts before we begin to deduce our theories and form conjectures."[105]

[100] *Sentinel,* Jan. 7, 1840; June 13, 1845; Jan. 7, 1840.

[101] *Courier,* June 30, 1841; Nov. 2, 1842.

[102] *Courier,* Aug. 3, 1842.

[103] Drake, in Shapiro and Miller, eds., *Physician to the West,* p. 170.

[104] For the tabulation of wind direction, *Sentinel,* June 26, 1846. Meteorology was especially dependent upon the accumulated fact. "From the abundant materials which will soon be collected from all parts of the world," the *Courier* commented in regard to Espy's call for meteorological observers, "we may reasonably hope that many general principles or laws may be discovered in relation to the weather." Feb. 22, 1843.

[105] *Courier,* Feb. 8, 1843.

Despite the emphasis upon observed fact, and the feeling among many orthodox scientists that "speculation is perfectly disavowed,"[106] popular science can hardly be said to have dispensed with the speculative impulse; it formed too large a part of the public discussion of astronomy, geology, and antiquities—to cite only three disciplines—to be summarily dismissed. If there was overt hostility toward theoretical science or a bar of mob judgment which science was forced to pass, it came from a different quarter than the public reflected here.[107]

Popular science was above all a democratic prerogative. Local observers, with unstated credentials, felt no reluctance in communicating their views on scientific subjects to the newspapers, and the newspapers felt no reluctance in publishing them.[108]

Likewise, it was not uncommon for editors to exercise their own critical faculties in reviewing scientific publications. The editor of the *New England Farmer,* reviewing Liebig's *Organic Chemistry,* promised to publish excerpts from the work in future issues, "probably appending our own opinion of the correctness or inaccuracy of its leading positions."[109] (These, unfortunately, did not appear in the Milwaukee newspapers.) In like manner an unsigned review in the *American Freeman* of the Lardner edition of Arago's *Popular Lectures on Astronomy* cited page numbers and quoted passages in criticizing some of Lardner's views; "dead flies" in the "ointment of the apothecary," the reviewer termed them. Nevertheless he found Lardner to possess "the rare talent of making himself understood, even by the superficially informed," and commended the work as "a great bargain for those who have any sort of fondness for such subjects."[110]

Such criticism as this appears to have been offered with a genuine concern

[106] J. B. Stallo, professor of natural philosophy at St. John's College, quoted in Daniels, *American Science,* p. 247. The "very worst passport which a naturalist could carry about him," Stallo wrote, "is that of a metaphysician." A general discussion of this subject may be found in Richard H. Shryock, "American Indifference to Basic Science During the Nineteenth Century," *Archives Internationales d'Histoire des Sciences,* 1948, *2*: 50–65.

[107] The following excerpt from a satirical essay in the *Courier,* Mar. 20, 1844, attacks those who would declare *vox populi vox Dei*:

> The present age is eminently distinguished by its rage for mysticism and metaphysics. The reign of plain common sense seems to be at an end. Young men now, instead of being contented with a good old fashioned district school education must be sent to college to fill their heads with Latin, Greek, mathematics, chemistry, and the like useless stuff, to the utter destruction of common sense. . . . Some of these college-bred upstarts have even laid their impious hands upon that old and well-established opinion of our good fathers and mothers, that the moon is a 'green cheese.' But in spite of their fine arguments the people are not to be deceived. They can see the moon with their own eyes, and are capable of judging for themselves.

Yet the author goes on to observe that "after all the philosophers have written and disputed, there are yet differences of opinion concerning the moon," which appears to repudiate his own disdain for the philosophy of common sense and empiricism. Satire, at any rate, is an uncertain hook upon which to hang a conclusion. It would be worth examining in this regard, however, the extent to which animosity toward pure science—even toward some aspects of the geological surveys—came from politicians rather than the public at large.

[108] Some of these are: *Sentinel,* June 13, 1845; *Courier,* Aug. 10, 1842, and Mar. 22, 1843; *American Freeman,* Nov. 8, 1844.

[109] *Courier,* June 30, 1841.

[110] *American Freeman,* May 15, 1845.

for science, and not in the spirit of contention or sectarianism; it can hardly be said to have constituted a hostile "bar of opinion." The democratization of science went hand in hand with the flourishing of the democratic spirit. "Philosophers, like kings, are but men; and all men to a certain extent may become philosophers." So said Daniel Drake.[111] "People now-a-days," the *Courier* commented, as if in confirmation, "when a new and startling theory is brought up, do not wait for the wise men and the doctors to analyze and investigate it, before they can venture an opinion on its merits or demerits; but they take up the subject at once for themselves; they reason it over in their own mind, and discuss it among their neighbors."[112] If the democratic spirit was on occasion ill-used, it was also redeemed by its encouragement of the spread of science and learning.

The physical environment of the pioneer community stimulated its own sense of discipline and curiosity; the very nature of the pioneer role demanded an aggressive optimism and the conviction that neither one's character nor his station in life was unalterably fixed. All that encouraged the emergence of a strong will to individual self-improvement, of the spirit of social progress and optimism, and of the extension of public education encouraged the popularization of science as well.

Insofar as the evidence of newspapers such as those considered here provides a reading of the popular mind in general, it demonstrates in both quantity and substance a vigorous, sustained interest in science. In large measure this interest was sustained by the appeal of social utility and by the search for design and order. The popular mind looked toward orthodox science with approval; attempts were made to close rather than to widen those innate divisions between the needs and purposes of religion, utilitarianism, and science. In some areas, such as meteorology, orthodox science solicited not only the appreciation but the participation of the public; in return, the ideology of popular science served to shore up the waning vitality of Baconian observation and deduction. As Channing had so perceptively recognized, science had indeed "left her retreats . . . and with familiar tone begun the work of instructing the race."

[111] Drake, in Shapiro and Miller, eds., *Physician to the West,* p. 171.

[112] *Courier,* Oct. 5, 1842.

Science and Culture in the American Middle West

*By Walter B. Hendrickson**

IN THE NINETEENTH CENTURY, in the interior region of the United States known as the Middle West, there were founded a number of natural history societies or academies of science. The founding of these institutions was an extension of a movement that began in western civilization with the organization Academia Secretorum Naturae in Naples in 1560.[1] This initial movement was an urban phenomenon, and many other natural history societies were founded in European cities in succeeding centuries. The first scientific organization in the American colonies, the short-lived Boston Philosophical Society (1683), was patterned after those of Europe. The first permanent organization was the American Philosophical Society, founded in Philadelphia in 1769.[2] During the next seventy-five years many more societies appeared in Eastern cities where there were such other institutions as colleges, libraries, theaters, and music halls. In the cities there was also the wealth to support cultural and intellectual activities, and because of the wealth, there was the leisure to enjoy them.[3]

As this was true in the East, so it was in the Middle West.[4] In the cities of this area, as they ceased to be frontier towns and settlements, the culture of the urban centers

*Department of History, MacMurray College, Jacksonville, Illinois 62650. This study was supported by grants from the American History Research Center, the National Science Foundation, and MacMurray College. A part of it was delivered as a paper at the meeting of the History of Science Society at Indiana University in 1963.

[1] Ralph S. Bates, *Scientific Societies in the United States* (3rd ed., rev.; New York: Columbia University Press, 1965), pp. 1–2.

[2] Brooke Hindle, *The Pursuit of Science in Revolutionary America* (Chapel Hill: University of North Carolina Press, 1956), pp. 127–137.

[3] See Bates, *Scientific Societies*, pp. 28–84, *passim*, and Max Meisel, *Bibliography of American Natural History, The Pioneer Century, 1769–1865*, Vol. II (Brooklyn: The Premier Press, 1926).

[4] The Middle West was selected as a region to demonstrate that natural history societies and academies of science were manifestations of the cultural and intellectual influence of the city in American society. It provides an adequate sample of cities in one region so that fortuitous geographical conditions are eliminated and it is possible to trace the evolution of the academies through a meaningful time period. The Middle West is a region in which there was an integrated pattern of development, and by 1860 most of the region had moved from a formative stage into a period of maturation. Another consideration in choosing the Middle West was that the author is Middle Western born and now lives in the heart of the region. Its history, its geography, and its people have long been an object of his study and attention.

The culture of Middle Western cities has been discussed by Richard C. Wade, *The Urban Frontier* (Cambridge, Mass.: Harvard University Press, 1959). Other commentators are Beverly W. Bond, *The Civilization of the Old Northwest* (New York: Macmillan, 1934); Francis P. Weisenberger, "Urbanization of the Middle West: Town and Village in the Pioneer Period," *Indiana Magazine of History*, 1945, *50*:29–50. Since the early work of Arthur Meier Schlesinger in *The Rise of the City* (New York: Macmillan, 1933), a vast literature has grown up detailing the interrelations of cultural institutions in urban life.

of the older parts of the United States was reproduced. This is evident when the careers of the members of the academies of science are investigated and we note that they were also involved with other cultural institutions like those that were found in older cities. For example, in St. Louis among the founders of the Western Academy of Natural Sciences Benjamin B. Brown was a charter member of the Medical Society of Missouri; George and Theodor Engelmann, Edward Haren, Marie Le Duc, William G. Eliot, and William Weber were directors of the flourishing St. Louis Library Association. Brown was also active in a literary society or lyceum known as the Franklin Society, and he was an incorporator of the Mechanics Institute. Le Duc was a director of the St. Louis Lyceum; Eliot was a supporter of the Franklin Society. George Engelmann and others of the St. Louis German community aided in founding a German private school, and Brown and Eliot served the public schools in various capacities.[5] Similar evidence of the fact that academies of science were a part of the total culture of the cities is found in the biographies of members of other academies of science in other cities.

The development of cultural institutions, however, was not easy. One of the reasons can be found in a statement by the Reverend William G. Eliot, a Congregational minister from New England: "the grand motive which brought the vast majority of us here [to the Middle West] is not liberty of conscience, not intellectual improvement, not the desire to be good, but to better our own conditions, to make ourselves rich and influential members of society," and he warned that "the influences that are needed to advance the true interests of society must be provided by the exertion of such members of the community who are aware of the danger [to the future of the Middle West]." He urged that those "who value literature and religion, and feel . . . the importance of education and morality . . . come forward and establish institutions by which public opinion may be elevated, [and] public feeling and taste purified. . . ."[6] Perhaps all members of academies were not so analytical in their thinking, yet the fact that they supported the academies along with other cultural organizations is proof that they considered the former to be of equal significance.

INFLUENCES IN THE FOUNDING OF ACADEMIES

Important in the founding of academies of science in the Middle West was the example of similar organizations in Eastern cities. Daniel Drake, one of the promotors of the Western Museum Society of Cincinnati (1818), was stimulated by a visit in 1815 to Philadelphia, where he explored museums, attended medical lectures, and met scientific and literary leaders at meetings of the American Philosophical Society, the Academy of Natural Sciences, and in the homes of such men as Dr. Caspar Wistar, president of the American Philosophical Society.[7] Still another example is Dr. George Engelmann, a founder of both the Western Academy of Natural Sciences of St. Louis (1836) and the Academy of Science of St. Louis (1856). Engelmann's father

[5] Walter B. Hendrickson, "The Western Academy of Natural Sciences of St. Louis," *Bulletin of the Missouri Historical Society*, Jan. 1960:119.

[6] *Ibid.*, p. 115.

[7] Walter B. Hendrickson, "The Western Museum Society of Cincinnati," *The Scientific Monthly*, 1946, *53*:68. See also Emmet F. Horine, *Daniel Drake (1785–1852): Pioneer Physician of the Midwest* (Philadelphia: University of Pennsylvania Press, 1961), pp. 135 ff.

had been a member of the Senkenberg Society of Natural History of Frankfurt am Main, and as a boy of fourteen George attended meetings. When young Engelmann arrived in the United States in 1834 one of the first things he did after landing at Baltimore was to visit in Philadelphia with Thomas Nuttall and other members of the Academy of Natural Sciences. He maintained his contacts with members of the Academy, sending them specimens of plants and animals and calling on them whenever he was near Philadelphia.[8]

Thus the urge to emulate the cities of the East, which had academies of science as a part of their cultural attributes, was a strong motive behind the founding of the early scientific organizations of the Middle West. Other powerful reasons grew out of the Middle West itself. There, men of vision and enterprise saw a vast undeveloped continent with rich natural resources waiting to be tapped. Dr. Daniel Drake of Cincinnati said in "An Address to the People of the Western Country:"

> Every citizen of the western country must feel the necessity of a speedy development of our mineral resources. To find beneath our soil an adequate supply of the various minerals which are now imported [from other parts of the United States] must be regarded by all as a matter of first and greatest importance. The Managers [of the Western Museum Society] are anxious to be instrumental in the advancement of this useful work, and earnestly solicit the co-operation of the public.[9]

The Western Academy of Natural Sciences of St. Louis also issued an "Address" in which it was said, "The geological peculiarities of the country are as yet but slightly known; the immense mineral resources that it possesses also need development, for the little learned of them justifies the belief that when so developed they will create a new era in the arts and manufacturers of the West."[10] Thus the economic urge was important in the founding of the academies of science in the Middle West, perhaps to a greater degree than in other academies of science in the United States. But almost as strong was the conviction, common to members of all academies, Middle Western or Eastern or European, that educated and cultured men should be knowledgeable about science, especially natural science.

Finally, there was in the Middle West during the early part of the nineteenth century a close association between academies of science and members of the medical profession.[11] In part this was because doctors were something of natural scientists through their study of anatomy and *materia medica*; they also learned about chemistry, mineralogy, and other sciences in the course of their medical training. Moreover, most men attracted to medical careers had well-developed curiosities and were thus led to inquire into fields related to their profession.

[8] Charlotte C. Eliot, *William Greenleaf Eliot* (Boston: Houghton-Miffin, 1904), p. 35.

[9] *American Journal of Science*, 1818, *1*:203–204.

[10] *Act of Incorporation, Constitution and Bylaws of the Western Academy of Natural Sciences* (St. Louis, 1837), p. 14.

[11] After about 1870, in general, physicians ceased to be interested in the activities of the academies of science, and, in fact, in natural history, because of the development of medical societies and the concern of doctors with such scientific developments as bacteriology in their own profession. But there were many who remained members, sometimes because as citizens they supported the academies as cultural and educational institutions. Also, in the 1870s and later, with the introduction of microscopy into some of the academies—Grand Rapids and Chicago, for example—and the accompanying connections with bacteriology, they continued their active affiliation.

ANALYSIS OF THE ACADEMIES FOUNDED

As a result of these various influences thirteen academies of science were founded in the Middle West by 1860[12]:

Western Museum Society, Cincinnati . . . 1818
Western Academy of Natural Sciences, Cincinnati . . . 1835
Western Academy of Natural Sciences, St. Louis . . . 1836
Antiquarian and Natural History Society of Arkansas, Little Rock . . . 1837
Cleveland Academy of Natural Science . . . 1845
Natural History Association of Wisconsin, Milwaukee[13] . . . 1848
Louisville Natural History Society . . . 1851
Flint Scientific Institute . . . 1853
Grand Rapids Lyceum of Natural History[14] . . . 1854
Academy of Science of St. Louis . . . 1856
Chicago Academy of Sciences . . . 1856
Der Naturhistorische Verein von Wisconsin, Milwaukee . . . 1857
Society for the Advancement of Natural Sciences, Louisville . . . 1858

A study of the aspects of academy founding and the degrees of success which were common to all discloses that there were a number of other cities in the Middle West which had potentialities for academies but which did not establish them. Detroit, Michigan, Indianapolis, Indiana, Columbus, Ohio, and Madison, Wisconsin, come to mind—cities which were just as much business centers as Little Rock, Flint, or Milwaukee. No absolute reasons for the failure of these cities to have academies can be given, just as the reasons for academy founding are imprecise. But some tentative though not objectively provable factors can be discerned.

[12] Three organizations are not included in this list: the Kentucky Institute, Lexington (1824); the Iowa Historical and Geological Institute, Burlington (1843); and the New Orleans Academy of Science (1853). There is not enough about the first two beyond scanty information in newspapers, and the New Orleans Academy is geographically outside of the scope of this article. The founder of the Kentucky Institute was Constantine Rafinesque, who was for a time Professor of Botany, Natural History, and Modern Languages at Transylvania University in Lexington. The Kentucky Institute lasted for only a year or so. (See *Cincinnati Literary Gazette*, July 10, 1824.) The Burlington Institute was founded in 1843 when the town population was 2,000. It persisted and held meetings intermittently until the 1850s; a start was made on the accumulation of specimens for a museum, and a few books and government publications were acquired for a museum. (See Meisel, *Bibliography of Natural History*, Vol. II, p. 715; *Territorial Gazette* (Burlington), Dec. 28, 1844, and *Daily Telegraph* (Burlington), Feb. 1, 1854.)

[13] There is not much information about the National History Association of Wisconsin (Milwaukee), with Increase A. Lapham as the most influential member and organizer. Mentions are found in Milwaukee newspapers and local histories, but no minutes or other records exist. The Natural History Association name was also used in 1853 by a group based in Madison, and in 1855, when Lapham unsuccessfully attempted to get the Wisconsin legislature to appropriate money for the founding of a state natural history survey, the Milwaukee-based Wisconsin Natural History Association was reorganized and a state charter was granted. Lapham and some of his associates continued to hold informal meetings at which scientific subjects were discussed, and there is brief mention of the meetings as late as 1863. Walter B. Hendrickson, "The Forerunners of the Milwaukee Public Museum," *Lore* (official magazine of the Friends of the Milwaukee Public Museum), Summer 1972, 89–103.

[14] The Grand Rapids Lyceum of Natural History was reorganized as the Kent Scientific Institute in 1868. The name reflected the inclusion of the Kent Scientific Club, an organization of teen-age boys, but in effect this was simply a continuation of the Lyceum. Kent is the name of the county in which Grand Rapids lies.

In the case of Detroit the University of Michigan was located such a short distance away at Ann Arbor that there was a tendency for the cultural and intellectual interests of the older city to find an outlet there. Further, the faculty and students of the university as early as 1837 established a museum and a natural history society. In Indiana there was a well-established center of cultural and intellectual activity in New Harmony, the town which had been founded by Robert Owen and William Maclure. Columbus and Indianapolis (also Springfield, Illinois) were artificial capital cities that had been arbitrarily located in the geographical centers of their states. It was not until after the Civil War, when all three cities became prosperous railroad centers, that any of them began to develop into cultural as well as political centers.

Another question that defies an absolute answer is that of the critical size of a city for the founding and perpetuation of an academy. At the time of the founding of its natural history society in 1837, Little Rock had only 2,000 inhabitants. St. Louis in 1835 had 8,000, and Cincinnati had 9,500 in 1818. All of these cities were economic centers and had considerable wealth for the time and place. In Flint when the Scientific Institute was founded in 1853 the population was 2,000, and Cleveland, when its academy of science was founded in 1845, had 9,500 people. Since all of these organizations expired after a few years, it appears that having more than 10,000 persons in a community was necessary for success, the one exception being Grand Rapids, which had a population of 6,000 in 1854. On the other hand, Louisville in 1853 had 55,000, but the natural history organization failed. Of the cities that did maintain academies over a considerable period, in addition to Grand Rapids, Cincinnati in 1835 had 31,000; St. Louis in 1856, 80,000; Chicago in 1856, 84,000; and Milwaukee in 1857, 30,000.[15]

It is clear that there is no critical date or population necessary for founding or success. But since no academies were established among farm populations, nor any successful ones in cities less than 30,000 (except Grand Rapids!), it is safe to say that a large urban environment was necessary. Only there were there sufficient numbers of persons who could be interested in sustaining a common enterprise in natural history.

But when we have said these things we still do not explain completely why academies were founded in some population centers and not in others. Certainly there was an element of chance: in some cities there were individuals whose scientific interests were paramount in their lives; for others, good health, leisure, or financial security enabled them to devote their energies to nonvocational pursuits. Chance also entered in the matter of time and place: there might have been interested individuals, leaders with exceptional organizing abilities, or a good cultural climate, but circumstances did not bring them together at the same time. Also, the departure of key members could bring destructive changes. For example, in St. Louis the first organization founded, the Western Academy of Natural Sciences (1836), had among its leading members the doctors Benjamin B. Brown, George Engelmann, Frederick A. Wislizenus, and Henry King, but within a few years King went to Washington, D. C., Brown to California, and Wislizenus was led into personal adventures outside of St. Louis. By 1840 the Western Academy had expired, and it was not until the 1850s, when there were flourishing medical schools and other educational institutions in St. Louis, that Engelmann was able to bring together the group that became the Academy of Science of St.

[15] Population figures are from local histories, census reports, and newspaper accounts.

Louis. Similar failures of leadership were in part the cause of failure of the academies of Little Rock, Flint, and Louisville.

In addition, as we shall see, success or failure also rested in part on the ability of the academies to get funds. In the cases of both St. Louis and Chicago, sizable gifts of money or property were received so that the academies flourished. But even for these, along with the academies that were succeeded by other institutions, the ultimate financial support came from public sources. The founding and progress of the Middle Western academies of science rested, finally, on the same factors that influenced other cultural enterprises dependent on volunteerism.

Six of the cities on the above list—Cincinnati, St. Louis, Cleveland, Grand Rapids, Milwaukee, Chicago—today maintain natural history and science museums that carry on broad programs of science education, particularly directed to children. In all cases these museums are either municipally owned or receive substantial subsidies from local governments. Thus out of the nineteenth-century organizations we have been examining has come an important feature of present-day urban culture.

MEMBERSHIP OF THE ACADEMIES

The members of the academies were for the most part amateur natural scientists.[16] They carried on the tradition, so well established in Europe and America in the seventeenth and eighteenth centuries, of assembling "cabinets" or collections of natural objects—shells, rocks, minerals, plants, invertebrate fossils, bird's eggs and nests, reptiles, fish, and primitive human artifacts.[17] Essentially these men were collectors who not only brought in specimens from their own rambles through field and forest, but bought, sold, and exchanged with other naturalists. Like any other collectors they sought not only to have complete sets of common objects, but to acquire rarities and new and previously undescribed specimens. Many naturalist-collectors specialized in a particular branch of science and became quite expert in the description and identification of specimens. Others were universal scientists in the eighteenth-century sense, interested in all branches of natural history as well as natural philosophy (chemistry, physics, electricity, and so on).

In the academies of science these natural history and natural philosophy enthusiasts found associates with similar interests and could instruct one another in the wonders of the natural world. The academies often subscribed to the leading scientific publications of the United States and Europe, and the members used these libraries to supplement their own. An important feature of the academies' bookshelves was the publications of other academies which were received in exchange for their own printed transactions or proceedings.

While many members of the academies were amateurs in science, and their vocations were those of bankers, merchants, lawyers, clergymen, or doctors, there were others

[16] The statements about members, their community and scientific interests, and the academies themselves are generalizations that have grown out of an examination of minutes and proceedings, newspaper reports, and local history accounts. Over a period of some years I have visited all the academies mentioned or hunted down information about them in local libraries. Histories or partial histories of some of the academies have been published; some have been mentioned earlier, and others are noted in later footnotes.

[17] For a further discussion of this matter, see David D. Van Tassell and Michael G. Hall, eds., *Science and Society in the United States* (Homewood, Ill.: The Dorsey Press, 1966), pp. 25–30.

who were professionally interested. Some were teachers of science in schools or colleges; doctors were often teachers of chemistry, *materia medica*, or anatomy in the medical schools of the city. Others were professional geologists, engineers, botanists, or mapmakers employed by state or national governments. Through most of the early nineteenth century the lines between the amateur scientists, the advanced semi-pro's, and the fulltime scientists were never tightly drawn. The programs of the academies reflected the mixed character of the membership. If the amateur influence predominated, the museums they built were but extensions of their private cabinets, as was true of the academies of Flint, Grand Rapids, and even of Cleveland and Louisville. If there was leadership from more professionally oriented members, such as was found in the St. Louis, Chicago, and Milwaukee institutions, the specimens were better selected and more systematically arranged. But professionals and amateurs alike, they were all inquirers and investigators, even though their inquiries were more or less limited to the outer face of nature.

The abundance of new natural objects led to an emphasis on the collection and description of local natural history specimens. Much pioneer work in the description of American animals, fossils, minerals, rocks, and geological formations was done by the members of the academies. These descriptions appeared in the local publications, in the *American Journal of Science*, or in the transactions of such well-established organizations as the New York Lyceum of Natural History or the Academy of Natural Sciences of Philadelphia. Not only did the Middle Western academies collect local objects, but by exchange and gift they built respectable exhibits from other parts of the United States and even from foreign countries. The museums and collections were open to the public at certain times, and crude though some of them were, they did serve to inform and educate the people. Another activity of the organizations, or of some of their members, was the keeping of meteorological records, and there are extant long runs of such information for all parts of the Middle West.

It should be noted here that although much of the work of the academies and their members was directed toward the natural sciences of biology and paleontology, these were not their sole concern. In the large academies of St. Louis and Chicago, both of which were founded in 1856 and which had many members who were teachers in the medical schools, colleges, and other institutions, there were many whose main interest was the physical sciences of chemistry, mineralogy, mathematics, astronomy, and physics.[18] Geology was a science that had one foot in the natural sciences and the

[18] See George H. Daniels, *American Science in the Age of Jackson* (New York: Columbia University Press, 1968), pp. 25–30. Daniels points out (p. 16) that for the period 1815–1845 historians of science have made general statements without objective evidence to the effect that "(1) Natural history as opposed to the physical sciences was the dominant research interest and as such, accounted for the overwhelming bulk of American scientific work. (2) Science was dominated by interest in the practical as opposed to the merely theoretical. (3) Science was still largely a pursuit of amateurs, whose main interests or sources of livelihood, were elsewhere." In the first part of his book Daniels attempts to prove that these generalities do not hold water. His evidence is mostly deduced from considering the careers and scientific activities of men in Eastern colleges and other institutions and their contributions to medical and scientific journals. Since the Middle Western academies of science in the "Age of Jackson" were in much the same stage of development as the Eastern academies of science were in the colonial and early national period, Daniels' criticisms do not hold for them. On the contrary, the generalities which he states, with some exceptions, *do* apply to the Middle Western academies in the period before 1860.

other in the physical sciences. It is significant that the third paper to be published in the *Transactions* of the Academy of Science of St. Louis in 1856 was "Observations of Glycerin," by James Schiel, a physician practicing in the city who had been educated in Germany. Dr. Schiel described the chemical composition of glycerin, its affiliation with the alcohols, and its combining properties.

In the St. Louis, Chicago, and other academies there was no evident division between the men interested in shells, fossils, and plants and those interested in the physical sciences. The circumstances of time, place, and leadership determined the direction of scientific activities. But in most of the Middle Western academies of science from their beginnings through the 1860s and 1870s the leaders were men who were interested in the natural sciences, as will be made clear later in this article. The fact that the academies were located at or near areas where new rocks, fossils, plants, and Indian artifacts waited to be collected and identified, determined the natural history emphasis of the academies and their members.

LEADERSHIP AND AIMS OF THE ACADEMIES

In most of the Middle Western academies there was a sparkplug, an enthusiastic individual without whom the organizations would not have persisted in some cases. In Cincinnati the leader was Dr. Daniel Drake, not a specialist in any field other than medicine, but a man of catholic intellectual tastes and unbounded energy. When Drake dropped his active participation in the Western Academy of Natural Sciences, leadership fell to Robert Buchanan, a prosperous merchant and manufacturer who had a passion for horticulture, although he dabbled in a number of fields. Buchanan was not only the president of the Western Academy for many years, at various times he was also the president of the Astronomical Society and the Historical Society. He was also associated in one way or another with the Musical Fund Society, the local dental college, a medical school, Cincinnati College, and a dozen or so other community organizations.[19]

In Cleveland the civic-minded leader was Leonard Case, a sportsman whose hobby was collecting bird specimens.[20] In Grand Rapids it was John Ball, a businessman, but also an explorer, traveler, and big game hunter. In Chicago a number of men at one time or another took the lead. Among them were J. Young Scammon, banker; George C. Walker, financier; Eliphalet W. Blatchford, manufacturer; and Dr. Edmund A. Andrews, physician, all of whom gave money and time to the Chicago Academy of Sciences.

As noted above, most of the Middle Western academies had among their members one or more competent naturalists who were well educated in their special branch of science but who followed a vocation as well. Among these were John G. Anthony, an accountant in Cincinnati and a conchologist who later became curator of shells at Harvard, and Dr. Jared P. Kirtland of Cleveland, whose princpal scientific contribution was the systematic description of the birds and animals of Ohio. In Louisville

[19] Hendrickson, "The Western Museum Society of Cincinnati," p. 69; Walter B. Hendrickson, "The Western Academy of Natural Sciences of Cincinnati," *Isis*, 1946, *37*:139–140.

[20] Walter B. Hendrickson, *The Arkites and Other Pioneer Natural History Organizations of Cleveland*, The Makers of Cleveland Series, No. 1 (Cleveland: The Press of Western Reserve University, 1962).

Dr. Lunsford P. Yandell was a geologist and paleontologist.[21] In St. Louis there were Dr. George Engelmann, botantist, Dr. Benjamin F. Shumard, geologist and paleontologist, and many more.[22] The Chicago academy alone among the Middle Western organizations employed a professional scientist as secretary and curator; Robert Kennicott,[23] William Stimpson, James W. Velie, and Frank C. Baker were the most outstanding of these scientists in the nineteenth century.

Like the scientific organizations which they emulated, the academies of the Middle West stated their purpose in phrases similar to those used by the Chicago Academy of Science in its constitution of 1865: "The objects of the Society shall be the increase and diffusion of scientific knowledge by a museum, a library, the reading and publication of original papers, and other suitable means." Whereas the use of the words "increase and diffusion" reflects the strong ties between Robert Kennicott and the Smithsonian Institution, other groups used such terms as "the purposes of the Society are to promote the progress of natural science by investigating the geology, botany, and zoology of Wisconsin, the establishment of a library, the reading of original papers and by other suitable means" (the Wisconsin Natural History Association in 1860). The founders of the Flint Scientific Institute said, in 1853, that their purpose was "the improvement in scientific knowledge," and they "felt the want of books which we cannot at present command"; later their purpose was expanded to include the creation of a natural history museum. When everything else is taken into account, often the success or failure of the academies rested on the way in which they sought to achieve their purposes. We shall consider this subject under three heads: (1) the publication programs, (2) the papers read at meetings, and (3) the museums.

PUBLICATION PROGRAMS

Nearly all of the academies supported some sort of publication program. This ranged from newspaper articles written by a member of the Antiquarian and Natural History Society of Arkansas in Little Rock to the many-volumed *Transactions* of the Academy of Science of St. Louis, which were published continuously from 1856 to 1958. The *Transactions* contain the writings of naturalists who described new species of natural history objects found in the Trans-Mississippi country. Dr. George Engelmann published important taxonomical studies of the pines, morning glories, cacti, and other plant genera and families. The state geologists of Missouri, Kansas, and Illinois contributed the results of their pioneering work. The volumes all contained detailed accounts of the proceedings of the academy, which included the remarks and short papers of Engelmann and other men associated with the discovery and description of the previously unknown features of the West. By the latter part of the nineteenth century the work of the scientists who were professors at Washington University was appearing in the *Transactions*. As a publishing medium for scientific

[21] Walter B. Hendrickson, "Museums and Natural History Societies in Louisville," *The Filson Club Historical Quarterly*, 1962, *36*: 43–55.

[22] Hendrickson, "The Western Academy of Natural Sciences of St. Louis," pp. 114–129; Walter B. Hendrickson, "St. Louis Academy of Science: The Early Years," *Missouri Historical Review*, 1966, *6*:83–95.

[23] Walter B. Hendrickson, "Robert Kennicott, An Early Professional Naturalist in Illinois," *Illinois State Academy of Science Transactions*, 1970, *63*:104–106. Walter B. Hendrickson and William J. Beecher, "In the Service of Science," *Bulletin of the Chicago Academy of Sciences*, 1972, *2*:211–268.

research, however, the *Transactions* decreased in importance as national specialty journals increased in number. After the 1880s the best scientific work of the professional scientist members no longer appeared, and by the twentieth century the *Transactions* were dominated by papers written by local amateurs or graduate students from nearby institutions. But since the St. Louis publication was widely distributed (in 1900 it was exchanged for 560 scientific publications from all over the world), it was still a recognized journal, although its contents were not uniformly valuable as original research.

The Chicago Academy of Sciences also published throughout its career, but not as regularly as did the St. Louis academy. In its *Proceedings*, *Transactions*, and *Bulletin* appeared articles dealing with various aspects of the Middle West. Several papers described the collections made by Robert Kennicott in Canada and Alaska, and the work of such nationally known naturalists as J. W. Foster, W. H. Dall, F. B. Meek, S. F. Baird, and S. H. Scudder appeared in these volumes. In 1896 the academy began a series, *The Bulletin of the Chicago Natural History Survey*, in which all aspects of the natural history of the Chicago area were described by competent men, most of them connected with the colleges and universities. During the twentieth century there was a series of research papers, reflecting the interests of the several directors.

In Milwaukee, Der Naturhistorische Verein von Wisconsin, through the last part of the nineteenth century, published small annual volumes containing the annual reports of the officers and some of the papers delivered at meetings. Most of the latter were concerned in a nonprofessional way with various aspects of natural history. The academies in Grand Rapids, Cincinnati, Louisville, and Cleveland sponsored and sometimes financed the publication of monographs on local botany, geology, or ethnology which were prepared by their members.

SUBJECTS OF THE ACADEMY MEETINGS

The heart of the activities of the academies of science during the nineteenth century was the biweekly meeting. The usual order of business was the reading of the minutes followed by the report of the treasurer. After thus reviewing the past and assaying its financial condition the academy heard the librarian tell what exchanges had been received and the curator of the museum report what new acquisitions had been made. New members were voted in and communications were read. People sent the academies their observations on the weather and other natural phenomena. Sometimes they submitted natural curiosities such as strange rock formations, branches of trees on which were galls or other parasitic growths, and boards with objects driven through them by the force of the wind during tornadoes. Often plants, rocks, minerals, fossils, and Mound Builder artifacts were sent in for identification.

Then the speaker of the evening read a prepared paper, or, if no arrangements had been made, some member would bring up a topic for discussion. The papers read and the topics discussed were as varied as nature itself; at the Grand Rapids Lyceum of Natural History in the 1850s, for example, some of the subjects dealt with were these:

The Action of Water on the Surface of the Earth
The Elements of Botany

Animal Locomotion
A Trip across the Continent (by a member who had traveled in the Far West)
Classification of Genera and Species
The Phenomenon of Lightning and Thunder

In effect these were efforts at mutual education and came out of the reading and personal experience of the speakers, most of whom were amateurs. In Grand Rapids also, as elsewhere, much attention was given to local and natural history features, as these Grand Rapids papers indicate:

Changes in the Muskegon Flats
The Geology of the Grand River Valley
The Dunes and the Lake Shore
The Fish of the Grand River

In the 1870s and later, while programs on natural history continued to be given, a number were concerned with technology. For example, in 1876 Professor Strong, a teacher at the high school, lectured on photometry and the way in which the intensity of light was determined. A paper was given on electric lights (carbon arc) in 1881, and the speaker showed the materials used in them, and at various times manufacturing and extractive processes were explained. A similar change in the content of papers from natural history to technology or even to theoretical science took place in the papers given at St. Louis and Chicago. Here is a list of papers read to the Chicago Academy of Sciences in 1876:

Propagation of Food Fishes in Fresh Waters
A Partial Development of a New Formula for Conic Sections
A Law of Reversed Motion
A Monstrosity: A Double Pig (a fetus)
Ammonites
A Mode of Studying Ethnology, Institutional and Linguistic
The Repellent Force of Light as Discovered by Crookes
The So-Called Planet Vulcan
The Propagation of Shad and Salmon
Some Fishes New to Science

Some of these topics were presented by the teachers in local educational institutions. St. Louis also had such members, and they also gave papers. In Grand Rapids and other places where there were no colleges, papers were frequently given by physicians on such matters as anatomy, physiology, or public health. During the 1880s, especially in Chicago and Grand Rapids, the enthusiasm for microscopy that swept both Europe and the United States was reflected in the papers and demonstrations given at meetings. The tendency to reflect current subjects that had a scientific aspect was also illustrated by papers on the raising of food fishes. This was a live topic in the 1870s in both Chicago and Cleveland. A member of the Cleveland Academy, Dr. Theodatus Garlick, has a very good claim to having been the first in the United States to raise fish from artificially inseminated eggs.

In St. Louis, so long as Dr. Engelmann lived, the papers given at the academy meetings were largely concerned with botany and geology. Later, when teachers of physics, mathematics, and astronomy at Washington University became members, these subjects were frequently treated. So closely was the St. Louis Academy associated with the university that, during the 1880s, the headquarters were there. In both Chicago and St. Louis the academies served as common meeting places for teachers in several scientific disciplines, thus providing forums for the exchange of ideas. In academy publications they found outlets for the results of their research.

It may be concluded from the review of papers here given, that as the nineteenth century wore on, the academies were getting away from the older natural history emphasis, and that emphasis had not been essentially technical or highly scientific. It was perhaps much in the vein of the periodicals *The Scientific American* and *The Scientific Monthly*, both of which began to flourish in the 1870s and 1880s. Further, it is clear that the academies were no longer particularly concerned with scientific research, whether in field or laboratory, but were steadily becoming instruments of education in popular science. This is borne out by the offering of lecture programs illustrated by the use of stereoptican or other projectors to audiences of members' friends and their families in the 1890s and the early twentieth century. It is because of the attention given to popular education in science that when the academies needed outside financial support, municipalities offered to assume the burden of maintaining them.

MUSEUMS OF THE ACADEMIES

With greater or lesser success all the Middle Western academies of science undertook to build natural history museums. Traditionally this was a leading activity of European and older American academies. In Chicago the building venture of the Academy of Sciences was strongly influenced by the dreams of young Robert Kennicott who hoped that a Western equivalent of the Smithsonian Institution could be created. At first the academy occupied only a small space, but as the collections increased rapidly, a large aiea in the Metropolitan Block was rented. Here extensive exhibits, including the large number of specimens collected by Kennicott on his exploring expedition in Canada, were arranged. A considerable number of duplicates were sent from the Smithsonian, and some personal collections were turned over to the academy. Meanwhile the members investigated the possibilities for constructing their own building, a matter that became especially urgent after a destructive fire in 1866.

The idea that Chicago could have a great museum appealed to the businessmen and civic boosters of the growing city. They mounted a carefully planned publicity campaign that included bringing to Chicago Louis Agassiz, the Swiss-born natural scientist who was building a museum at Harvard University. Agassiz addressed an evening meeting of well-to-do citizens, and more than $50,000 was raised. This money, plus the insurance collected on the contents of the building, was used to purchase a lot and put up a museum building modeled closely on the museum in the Smithsonian Institution. The exhibits were reconstituted, and the building was opened to the public on January 1, 1868.

But tragedy struck again when the fireproof building was lost in the Great Fire of 1871. Included in the loss was the extensive collection of shells, and notes about them, made in the course of many collecting trips by William Stimpson, the director of the Chicago Academy, a promising young naturalist who had been trained by Agassiz and Augustus Gould, a leading American conchologist in Boston. At the time of the fire Stimpson was preparing a manuscript for a comprehensive work on the shells of the world, and he was completely broken by the loss of his material.

Although the building was replaced after the fire, it was put up with borrowed funds, and it was soon relinquished because the academy could not make payments as they came due. What specimens the academy had been able to replace after the fires were never adequately displayed. It was not until 1894 that the present building in Lincoln Park was built. During the time when the academy had no home, Marshall Field and other Chicago citizens provided the money for housing the natural history and anthropological exhibits that had been assembled for the Columbian Exposition (the Chicago World's Fair of 1893), and they later gave funds for the present magnificent Field Museum of Natural History. If the Chicago Academy of Sciences had had a going museum at the time of the fair, it might well have been the recipient of all that went into the Field Museum, and thus would have become the center of natural history activities in Chicago. The academy today owns an adequate building which houses a small but well-selected collection, particularly associated with the natural history of the Chicago area, and it carries on a program of popular science education. It receives support from the Chicago Park District.

In St. Louis the Academy of Science had a large and flourishing museum, but it was destroyed by fire in 1869. Efforts were made to re-establish the museum, and numerous campaigns were undertaken to obtain broad financial support. These failed to do more than provide makeshift quarters for small collections. In 1960, however, the city of Clayton (in greater St. Louis), gave the academy two large houses set in a wooded plot of ground. Here is maintained a natural history and science museum designed especially for children. In 1971 financial support of the academy's museum was approved by a tax referendum in St. Louis County.

While the academies of Chicago and St. Louis are still active after more than a hundred years, those of Grand Rapids, Cincinnati, Milwaukee, and Cleveland, primarily because of the weight of financial burdens, became moribund in the late nineteenth and early twentieth centuries. In each of these cities such collections as remained were eventually turned over to successor museums which were publicly supported.

The museums of the academies of the nineteenth century were only rudimentary combinations of research and public exhibit institutions in comparison with the museums of today. Minerals, rocks, invertebrate fossils, insects, and plants were usually arranged in glass-topped cases according to some orderly scheme. The best of the museums followed currently accepted scientific classification. Duplicates of specimens were kept in drawers in the bases of the cases. Birds and small animals were mounted in free-standing glass cases. Where there were comprehensive collections an attempt was made to arrange them in larger cases according to orders and families. Larger animals, usually few in number, were simply placed around the room in convenient places. Choice specimens of coal, crystal, and meteorites were similarly displayed.

Exhibits of shells, fossils, birds, insects, and so on were of interest to amateur collectors, but they were also the working tools of the geologists and naturalists who were employed by state and national governments to make surveys, or who were teachers of science. Since so much of the work of natural scientists in the nineteenth century was in description and classification, the museums were useful for amateurs and professionals alike. Museums were also instruments of popular education, and in a period when all school natural science was primarily taxonomical in nature, they fulfilled their purpose. By the twentieth century, after the formation of the American Association of Museums in 1906, the Chicago Academy of Sciences, the Academy of Science of St. Louis, and the successor institutions in Milwaukee, Grand Rapids, Cincinnati, and Cleveland had begun to use modern museum methods, beginning with the arrangement of habitat groups.

DECLINE AND TRANSFORMATION OF THE ACADEMIES

The Middle Western academies of science reflected some of the trends in science during the nineteenth century in America. It has been shown that the collection and description of natural objects was a major concern of scientists, amateur and professional, although the lines between them were blurred. Both were members of the academies, and the activities of the organizations themselves reflected the interests of the members. The meeting programs, the publications, and the museums were areas in which the amateur scientist, particularly the collector, made contributions which were welcomed by professionals.

On the other hand, even as the academies were being founded in the period 1830 to 1860, research was increasingly concerned with other things than taxonomy and with other fields than natural history. As the nineteenth century advanced, the academies ceased to be satisfactory agencies for study and research in science. The centers of research became universities and government agencies such as the Coast and Geodetic Survey, the Geological Survey, and the bureaus of the Department of Agriculture that were concerned with plant and animal morphology and pathology. The states, too, founded bureaus and surveys for entomology, biology, and geology. The large private universities such as Yale, Harvard, Stanford, Northwestern, and Chicago created science departments as integral parts of their academic offerings. The great state-supported universities that arose out of the Morrill Land Grant Act all stressed scientific agriculture and technology and consequently gave large support to departments of biology, chemistry, physics, and geology. The rise of graduate schools at both the private and public universities provided the staffs for government bureaus and the science departments of universities.

This takeover of research in science from the academies was accompanied by the acceleration of the trend toward specialization that had begun about the time the academies of science of the Middle West made their appearance during 1830–1860. Specialization within the academies was provided for by "departmentalizing" the membership according to special interests; at the same time the universality of science was maintained by the practice of encouraging general discussion after the presentation of papers. At first specialization and professionalism worked to the advantage of the larger academies, particularly those of Chicago and St. Louis, because the men at the local colleges and universities turned to the academies as places where they could associate with their fellows, and in whose journals they could get their research published. Because of their scientific qualifications these men were pressed into service as presidents, secretaries, and librarians. In the long run, however, the great gulf between professionalism and amateurism could not be bridged. The 1880s and 1890s were periods when professionals who staffed the universities and the government agencies formed national organizations along the lines of their specialties. They also founded professional journals for the publication of research, thus replacing the transactions of the academies. The last part of the nineteenth century and the early part of the twentieth were also periods when professional scientists formed state academies of science, where it was hoped—vainly, as it turned out—that men of all disciplines could meet on common ground.[24]

In their time and place, the academies of science of the Middle West were significant scientific and educational agencies. And, finally, even though circumstances changed, some academies disappeared, and others were transformed into public museums, their history illustrated the reliance of men in America, and in western civilization, on joining together to promote their cultural and intellectual concerns.

[24] Walter B. Hendrickson, "Alja R. Crook and the Founding of the Illinois State Academy of Science," *Ill. Acad. Sci. Trans.*, 1963, *56*:156–164.

Sources of Misconception on the Role of Science in the Nineteenth-Century American College

By Stanley M. Guralnick *

IT IS A MATTER OF DEMONSTRABLE FACT that science and the ante-bellum American college enjoyed an amicable, even a mutually profitable, relationship: the intellectual demands of science introduced to the college problems and subsequent changes which in turn led to further scientific expansion.[1] Yet the scientists, the historians of education, and even the historians of science who worship the nineteenth-century institutional precursors of the twentieth-century faith in specialized scientific education include mention of only the graduate school, the introduction of elective courses, and, for the ante-bellum period, the scientific school in their litany of exaltation—never the so-called classical liberal arts college. Indeed, on those few occasions when the college actually appears in the history of American science at all, it is summarily condemned for its failure to support science, an allegation based on the college's proven failure to subsidize research in the manner of its nineteenth-century European or twentieth-century American university cousins. Never is the question asked how, in the colleges of a society supposedly indifferent to basic research, provision was nevertheless made for a reshaping of priorities by means of which more recent scientific practice could grow. Moreover, this carelessly articulated and wholly unsubstantiated belief that the college was without relevance to contemporary science exists in conjunction with a broader historical misunderstanding of the classical college, the elements of which are so distorted and yet so pervasive that together they constitute a myth of serious proportions. Our purpose here is to examine the nature and consequences of that myth: to show how science was actually related to collegiate thought; how the historical myth of the college grew, forcing us to overlook that central relationship; and how, with the myth dispelled, the college's true role in disseminating scientific culture in nineteenth-century America can finally be seen to emerge.

By way of background it must be understood that contemporary college thought organized itself around the concept of a single required curriculum, an approach repeatedly articulated and defended in the large number of reports that issued during

* 1165F Bear Mountain Drive, Boulder, Colorado 80303. I gratefully acknowledge postdoctoral support from the Smithsonian Institution for the academic year 1969–1970, when this study was prepared.

[1] See the author's doctoral dissertation, "Science and the American College, 1828–1860" (University of Pennsylvania, 1969), published as *Science and the Ante-Bellum American College* (Philadelphia: American Philosophical Society, 1975).

the wave of convulsive evaluations of the collegiate course of study which shook the decade of the 1820s. The most widely disseminated of these, and subsequently the most frequently cited by historians, was the Yale Report of 1828. Historically though erroneously considered a defense of the older eighteenth-century curriculum with its overemphasis upon Latin and Greek at the expense of the new scientific disciplines, this document actually argued with quite different emphasis. Merely defending the concept of a defined course of study for all students, it neither attempted nor intended to impugn the value of any particular discipline. Indeed, it rather maintained quite explicitly that "the various branches of mathematics, or of history and antiquities, or of rhetoric and oratory, or natural philosophy, or astronomy, or chemistry, or mineralogy, or geology, or political economy, or mental and moral philosophy" were as central to the design of a viable collegiate program as the classical languages themselves.[2]

In essence the Report argued for an entire liberal arts curriculum and not in any way against the science that was then becoming so important an element in that curriculum. Recognizing that the content of the liberal arts courses might change, and believing that "as knowledge varies, education should vary with it,"[3] its authors prophetically remarked that if Yale ever became like the University of London with "higher and inferior courses, its scientific and practical departments, its professional, mercantile, and mechanical institutions . . . our present undergraduate courses ought

[2] *Report of the Course of Instruction in Yale College: By a Committee of the Corporation and the Academical Faculty* (New Haven: Hezekiah Howe, 1828), p. 19. This report was reprinted in 1830 and excerpted in Yale catalogues for the next two decades. It was widely known through the *American Journal of Science and Arts*, 1820, *15*:197–351, where it was printed as "Original Papers in Relation to a Course of Liberal Education." Page references are to the original published version, 1828.

Since my reading of the Yale Report is crucial to my understanding of the educational climate in which scientific instruction was to be advanced, I think it necessary to provide references to alternative interpretations. R. Freeman Butts, *The College Charts Its Course* (New York: McGraw Hill, 1939), pp. 118–125, discusses the Report in a chapter entitled "The College Controversies Take Shape: Conservative Reaction," emphasizing to a great extent Yale's reluctance to adopt an elective system. Richard J. Storr, *The Beginnings of Graduate Education in America* (Chicago: University of Chicago Press, 1952), *passim*, is similarly impressed by the Report's ability to curb university growth. Storr does, however, warn against the conclusion that the college course was unchanging, even though these changes are not his concern (pp. 5, 135). Richard Hofstadter and Wilson Smith, *American Higher Education: A Documentary History* (Chicago: University of Chicago Press, 1961), Vol. I, pp. 275–291, have condensed the Report in such a way as to emphasize its negative aspects. However, Hofstadter and C. Dewitt Hardy, in another volume, *The Development and Scope of Higher Education in the United States* (New York: Columbia University Press, 1952), pp. 15–17, warn of the "temptation to overrate its educational value," while pointing out the "elements of moderation and wisdom in the report" (p. 16n). Frederick Rudolph, *The American College and University* (New York: Random House, 1962), pp. 130–135, after dwelling upon the failure of the elective and university principles to take hold, and upon the "aristocratic" (p. 134) purposes of the Report, also introduces a discussion of the beneficial effects emanating from the 1828 tradition. George P. Schmidt, *The Liberal Arts College: A Chapter in American Cultural History* (New Brunswick, N.J.: Rutgers University Press, 1957), pp. 55–58, leans toward conservative interpretation, but adds: "The cultural values of science, it is only fair to say, were also vigorously defended." Walter P. Metzger, *Academic Freedom in the Age of the University* (New York: Columbia University Press, 1955), thinks he has found in the Report "a consistent spirit and argument: the preceptive importance of religion" (p. 5), despite the fact that religion does not enter the discussion in even the most obscure way; in fact, the word "religion" never appears. It would seem that there is still room for further interpretation of the 1828 tradition.

[3] *Report*, p. 31.

still to constitute one distinct branch of the complicated system of arrangement."[4] This insistence upon a distinct curriculum, echoed in the reports of other colleges, would define the atmosphere in which science could exist and in fact flourish. At a time when popular clamor for vocational education threatened a devaluation in the intellectual demands of science and when independent research centers were financial impossibilities, the liberal arts approach to scientific education was one to which even scientists themselves readily subscribed.

In fact, science constituted a steadily increasing proportion of the intellectual content of the prescribed program as colleges worked over the years to perfect the formula that would accomplish the liberal education of all students. The constant growth in scientific knowledge as well as in other disciplines required new courses, professors, and instructional methods if all the mental faculties were to continue to have their proper exercise. And the changes that took place in the second quarter of the nineteenth century in this context were at once startling for those who lived through them and perplexing to those who had later to consider and interpret them. The colleges upgraded the level of those sciences already taught, such as mathematics, physics, and astronomy, and added many more, such as chemistry, geology, and biology. Graduating seniors, who in 1815 could not solve algebraic equations of two unknowns or distinguish between Newton's fluxional notation and fly spots, were by 1840 dealing with ordinary differential equations, all in the course of a single generation. More professors were hired to concentrate on more narrowly defined scientific areas, and all were compensated at the same, occasionally even higher levels than their colleagues in older disciplines. Scientific apparatus at the average college increased in value from a few hundred dollars in 1825 to many thousands by mid-century, and this was concurrent with the building of laboratories and astronomical observatories to house the new acquisitions.[5]

We are not, however, concerned so much with the details of the change as with the fact that even its broadest outline usually remains hidden in an obscuring myth. The major elements of this collegiate myth are semantic distortions—the use of words in their twentieth-century connotations to describe situations in the nineteenth. The words often falsely employed to characterize the old American college are "sectarian," "classical," "aristocratic," and, most inappropriately, "anti-scientific."[6] Primarily, the college appears in twentieth-century interpretations almost as a theological seminary in which the ancient languages were used to further study of the Bible and its exegesis as a means of perpetuating a particular brand of Christianity. Without any supporting arguments, the late-nineteenth-century thesis that a warfare raged between science and

[4] *Ibid.*, p. 25.

[5] Discussions of the funds spent on scientific apparatus are included in *Science and the Ante-Bellum American College*. An example of a catalogue of instruments prepared at the end of the period is that of Wesleyan University, with a value of $7,394.20 (MS, John Johnston, "List of Apparatus in Wes. University," June 18, 1861, Wesleyan University Archives).

[6] Allan Nevins, e.g., describes the college of 1860: "it had to be wrested out of the ruts in which it had so long traveled. A revolt against it grew, compounded primarily of four elements: rejection of the tyranny of classical and theological studies, championship of science, insistence on attention to agriculture and the mechanic arts, and—most important of all—a demand for greater democracy in education." *The State Universities and Democracy* (Urbana: University of Illinois Press, 1962), p. 2. Earlier, Frederick Jackson Turner wrote that in 1830, "Emphasis was laid upon preparation for the ministry. The curricula were rather rigidly required and of the old classical type." *The United States 1830–1850: The Nation and its Sections* (New York: W. W. Norton, 1963), p. 17.

theology is allowed to stand;[7] and religion becomes a scapegoat for our failure to inspect closely other aspects of intellectual history and the college.[8] In fact, it has become almost standard procedure in general histories to include, in the limited space accorded to colleges, a recitation of their denominational affiliations, as if that information were relevant to their intellectual postures.[9] This, however, is not the reality that emerges once the ante-bellum college is examined free from the prejudices aroused by its traditional composite image. Indeed, a close look at the differences between the curricula of Baptist, Congregationalist, Presbyterian, Methodist, and Episcopalian colleges and those of independent or nonsectarian affiliations reveals, surprisingly, that there was no difference at all.

The term "denominational" is, then, probably the most ill-chosen one to be used in accounts of the American college. While the majority of colleges were associated with some sect and had some particular religious motivation at their inception, even before 1800 they could not be characterized as proselytizers. Moreover, by 1825 all the major Protestant sects had established theological seminaries to function as post-graduate schools where the real work of preparing ministers would go on; remarkably little systematic theology remained to be found in their colleges' curricula. This secularization of the college within the framework of an evangelical America, which demanded certain pious and moralistic tones from every area of social activity, was thoroughly completed in the first part of the century. A journalist writing about the denominational college in 1860 shows how nineteenth-century rhetoric could espouse a religious morality even while denying our image of the college as church-related:

> . . . they are for the most part controlled by independent Boards of Trust, owing no obligations expressed or implied to any denomination or ecclesiastical power. They are under obligations to founders, to society, and to God, to perform their sacred trust "Christe et Ecclesiae" for Christ and the Church Universal; to consecrate the Institution under their care to sound learning and Evangelical Faith, and to nothing else.[10]

---

[7] Bruce Mazlish calls attention to some assumptions that might be challenged in the preface to his edition of Andrew Dixon White, *A History of the Warfare of Science with Theology in Christendom* (New York: Free Press, 1965), p. 17.

[8] Perhaps an illustrative example from the contemporary literature would be helpful. A common procedure is to single out early scientists, preferably those unfavorably disposed to Darwin in the year 1859, and condemn them for their use of conventional religious rhetoric. Joseph L. Blau in *Men and Movements in American Philosophy* (New York: Prentice-Hall, 1952), p. 79, e.g., informs us of the eminent scientist Reverend Edward Hitchcock that "in all his work he exploited science in the service of religion." A more accurate evaluation of Hitchcock, however, shows him to have been known even throughout Europe for his scientific discoveries. His use of geological evidence to support his personal belief in the moral precepts of the revealed word was not "exploitation." Nor did he suppress scientific evidence or deny conclusions of other scientists by any but mutually shared canons of scientific criticism. And if, until his death in 1864, Hitchcock was skeptical about human evolution, his motive was the lack of paleontological evidence, a perfectly valid scientific objection. Since he had already abandoned a belief that a free interpretation of the Bible would show its historical truths to conform to physical law, it was both logical and honest for him to agree that "the real question is, not whether these hypotheses [evolution] accord with our religious views, but whether they are true." "The Law of Nature's Constancy Subordinate to the Higher Law of Change," *Bibliotheca Sacra and Biblical Repository*, 1863, *20*:524.

[9] Blau, *Men and Movements*, p. 80, is even more insistent on this point: "While they were interdenominational in their student body and invited aid from all sects, there was no attempt to be nonsectarian." I, for one, would not expect the interdenominational student bodies and donors to support such narrow enterprises.

[10] "Denominational College," *New Englander*, 1860, *69*:85.

Denominational colleges were, of course, still founded in great numbers after the splits in Presbyterianism in the late 1830s and the subsequent foundation in 1844 of the Society for the Promotion of Collegiate and Theological Education at the West. But these colleges of the West, while originally sponsored by denominational groups, were generally under the influence of the heterogeneous local community rather than any ecclesiastical body. In fact, the older Eastern colleges were the most vocal complainants against the parochial foundations that many of their own graduates tried to lay in the West. This group of hundreds of sectarian colleges founded in the antebellum period, most of which failed to survive, was a phenomenon peculiar to Western migration and does not reflect any of the activities of the colleges permanently ingrained in the Northeast.[11] There, a sharp distinction was maintained between "denominational" and "sectarian," the former being ecumenical, the latter hopelessly narrow.

Any close inspection of the established nineteenth-century college will reveal that the curriculum was entirely free of all sectarian indoctrination; and if there seems to be an excess of "religious enthusiasm" reflected in the extracurricular life of the student, it is only because the outward expression of such enthusiasm was a type of habitual behavior, essentially free of intellectual content. In fact, during the Jacksonian period only a minority of college graduates entered the ministry.[12] Those who did, of course, became products of the seminary rather than the college, where the great issues of divinity were not formally taught. The philosophical system propounded in all colleges, regardless of denominational persuasion, is more properly designated moral than theological, however neatly it may have softened those objections to which the revealed Word was subjected during an increasingly critical age. Interpretations of the rite of baptism, for instance, will not be found at Amherst or Brown, but at Newton and Andover. As the retiring president of Yale reported in 1871: "in a long acquaintance with officers of colleges controlled by various religious sects, I have discovered no spirits of proselytism and no important disagreement in regard to the meaning and essence of our common Christianity."[13]

It was, in fact, this common Christianity, as opposed to the dogma of a particular creed, that was upheld throughout the nineteenth and well into the twentieth century. As late as 1899, for instance, Arthur Twining Hadley, the first nondivine to hold the office of the Yale presidency in two centuries, was examined by the Yale Corporation to be free of heretical beliefs before assuming office. It is important to note that this pious stance of the college was a feature of American morality not confined to

[11] Donald G. Tewksbury, *The Founding of American Colleges and Universities before the Civil War* (New York: Teachers College, Columbia University, 1932), is the standard compilation on the growth and death of colleges in this period.

[12] Richard Hofstadter, *Academic Freedom in the Age of the College* (New York: Columbia University Press, 1955), p. 192, states that the number of Harvard and Yale graduates who became ministers dropped from 60% at the end of the seventeenth century to 40% in the early 1740s to 20% a hundred years later. The last figure is a very reasonable estimate for most of the older colleges during most of this period. At Bowdoin, e.g., a college traditionally recognized for its training of New England ministers, only 240 out of 1,032 graduates (1806–1850) entered the ministry. Egbert C. Smyth, *Three Discourses Upon the Religious History of Bowdoin College during the Administration of Presidents M'Keen, Appleton, & Allan* (Brunswick, Maine: J. Griffin, 1858), p. 80.

[13] Theodore D. Woolsey, *Addresses at the Inauguration of Professor Noah Porter ... As President of Yale College* (New York: Charles Scribner, 1871), p. 16.

denominational institutions but present in all of them, including the independent and state controlled. As the most comprehensive of the historians of the university suggests, "inherent differences between state and private institutions in the nineteenth century can easily be exaggerated . . . ; although state institutions de-emphasized religion somewhat more rapidly than did private colleges, the remarkable thing is how long officially sponsored religion endured at many state endowed universities."[14]

More important still is the fact that records of the daily activities of these colleges make it clear that piety within the institutions in no way hindered the development of their curricula. Despite recent historical claims, religion cannot be said to have hindered the progress of science within early-nineteenth-century collegiate institutions. Church and state were as separate in the college as they were in the street, yet in the study of political history we do not overemphasize the religious rhetoric found in the latter milieu; analogously, it is wrong to distort the study of education by assuming that rhetoric common to the expressions of religious and scientific goals indicated that instruction in these two areas was not separate and distinct.

Indeed, contemporary historians never make clear what connection or retarding influence religion is supposed to have had on either science or its progress in the college, or why that connection ceased, although there are some implicit proposals.[15] There is, for example, the suggestion that the liberalizing influence of German universities and Darwin's evolutionary hypothesis were responsible for the removal of the religious excrescence from the college. And there is the notion that scientific thought and activity are somehow incongruous with general religious sentiment, even if theology and science do not clash on specific issues. But in fact these concepts, along with others of late-nineteenth-century vintage, are unrelated to the realities of college reform earlier in the century. Scientists and science teachers were then well able to prosecute science according to the standards of that profession, without any conflict or reliance upon a particular theological system. And scientific research suffered in no way from the one view its promoters shared in common with theologians—the broad, nonmethodological belief in the "design" paradigm, according to which God created the world to serve best the ends He desired.

A convenient test of these suggestions is provided by the example of Wesleyan University, founded in the midst of the ante-bellum period. Nominally a Methodist institution, its first president, the classicist Reverend Willbur Fisk, resisted all attempts to give the Eastern college any exclusively theological flavor. He insisted that "modern literature, the natural and exact sciences, and the application of the sciences to the useful arts [are] first in importance in a useful education."[16] The school's first catalogue outlined a "partial course" for science-oriented students, and by 1836 Wesleyan

[14] Laurence R. Veysey, *The Emergence of the American University* (Chicago: University of Chicago Press, 1965), p. 112.

[15] The danger in establishing this causal connection between religion and science may be illustrated by statements appearing in two books, both published in the same year. Samuel Eliot Morison, *The Oxford History of the American People* (New York: Oxford University Press, 1965), p. 536, blames religion for retarding astronomy, while Howard Mumford Jones, *Ideas in America* (New York: Russell and Russell, 1965), p. 133, claims that religion advanced the same discipline.

[16] Willbur Fisk, *The Science of Education: An Inaugural Address delivered at the opening of the Wesleyan University in Middletown, Connecticut, Sept. 31, 1831* (New York: M'Elrath & Bangs, 1832), p. 13.

had spent over $6,000 on scientific equipment.[17] By 1840 the faculty consisted of the president and six professors, three of them in mathematics and science.[18] Most telling is the fact that there was no provision for formal theological training, as indeed there was not at any college in the East.

In light of the fact that religion in no way frustrated the introduction of scientific disciplines into the college curriculum, it is unfortunate that the spectacular progress made by science within the ante-bellum college is obscured by our interpretation of Andrew Dixon White's thesis that myth and systems of theology hindered the acceptance of scientific modes of thinking. We too easily assume that White combined this general criticism of dogmatic theology with his dislike of American denominational colleges to conclude that sectarianism in collegiate institutions prevented their promotion of science. White, however, as an ardent spokesman for universities, had as his principal complaint against denominational colleges merely the fact that their great numbers prevented the concentration of resources for advanced instruction. Fully subscribing to the contemporary belief that the college should be a center of general Christian sentiment, he did not suggest that ecclesiastical powers either inside or outside of the American college operated to suppress science.[19] The first president of the Johns Hopkins University, who watched this misinterpretation take form in the minds of others less articulate than White, declared that

> Hostility toward scientific pursuits or toward scientific instruction has never in this country been manifested to any noteworthy extent by the religious part of the community or by theological teachers. In discussions relating to the sphere of science and religion, the teachers of religion have almost been earnest in their approval of scientific research.[20]

Just as the meaning of the term "denominational" has been misconstrued by modern historians, so too has the word "classical." It is constantly placed in contradistinction to the term "scientific," a practice which obscures the fact that in the nineteenth century "classical" was often used synonomously with "the liberal arts," thus including the scientific disciplines within its definition. The term, then, was used to distinguish not the dead languages from the sciences, but liberal education from vocationalism and technical training. The usual form of the argument that we hear today—that the study of Latin and Greek kept science out of the educational program—did not apply in America, however true it may have been at Oxford. President Mark Hopkins of Williams, who noticed this impression growing even as the classical course was altered, argued in 1836:

> . . . if it be intended that improvements in the sciences are not ingrafted, as they are made, upon the scientific courses, or that new sciences are not introduced as the wants of the public demand; if it be intended that there is an adherence to things that are old because they are old, then however much ground there may have been for the charge formerly, and especially in England, from which this complaint is mostly imported, I do not think

[17] MS, Wil[l]bur Fisk, "Financial Records," Wesleyan University Archives [1836].

[18] *A Catalogue of the Officers and Students of the Wesleyan University for the Academical Year 1840–41* (Middletown: William D. Starr, 1840), p. 6.

[19] See, e.g., "The Relations of the National and State Governments to Advanced Education," *Old and New*, Oct. 1847, p. 477; *Autobiography of Andrew Dixon White* (New York: The Century Co., 1905), Vol. II, p. 527; and Mazlish, *A History of the Warfare*, p. 22.

[20] Daniel C. Gilman, "Education in America, 1776–1876," *North American Review*, 1867, p. 224.

> there is any ground for it now. It is within the memory of our older graduates that Chemistry, and Geology, and Mineralogy, and Botany, and Political Economy were either not taught at all, or scarcely at all in the college course.[21]

Hopkins' remarks are of course another elaboration of the same tradition expressed in the Yale Report of 1828.

The baseless criticism that the college curriculum bore in its own day has also been a source of historical misunderstanding. Thought to be out of spirit with an age of economic expansion, internal improvement, and increasing dependence upon the materials of the physical environment, colleges were attacked as intellectually elite and socially irrelevant, a complaint that had nothing to do with a failure to recognize science, as historians have assumed. In fact, addition of the very science that educators at first hoped would still the voices of complaint served only to quicken them. Professors of classical languages and physics were not at war with one another but were united against the Philistines. And so the opponents of classical education were, interestingly, rarely science teachers or scientists, but rather the great mass of citizens. Educators were aware that "the ordinary objections to liberal education lie not so much against this or that branch of profound or elegant learning, but against all."[22] Thus, in 1876, while reviewing the then-embryonic belief that it was the old classical curriculum with which science struggled, Daniel C. Gilman emphasized:

> . . . again, there has been very little controversy between advocates of scientific and classical culture, each party having been disposed to concede the importance of maintaining institutions in which literature and science may alike be efficiently promoted. There can be no doubt that the influence of Harvard, Yale, and the other older colleges has in this particular been powerful throughout the land.[23]

Yet even when the fusion of classical learning and science that the college promoted is understood, the subtle importation of political terminology to describe both the college and science itself still makes the facts difficult to believe. The problem is related in part to the conclusion that the college was "aristocratic," simply because less than one per cent of the college-age population partook of its liberal education[24]—a charge made without sufficiently close inspection of the usage of the term.[25] This misconception finds support in the fact that the old private Northeast colleges did achieve a

[21] Mark Hopkins, *An Inaugural Discourse, delivered at Williams College, September 15, 1836* (Troy, New York: N. Tuttle, 1836), p. 21.

[22] Martin Brewer Anderson, "End and Means of a Liberal Education," MS address, 1854, University of Rochester Archives.

[23] Gilman, "Education in America," p. 224. Compare this with the usual interpretation that science grew with stubborn opposition from classicists, as stated, e.g., by Harold Larrabee in *Naturalism and the Human Spirit*, ed. Yervant H. Krikorian (New York: Columbia University Press, 1944), p. 346.

[24] Among the nineteenth-century journals providing statistics from which such a conclusion can be drawn are *University Quarterly*, *Niles Register*, *American Almanac*, *American Quarterly Register*, and the *Quarterly Register* and *Journal of the American Education Society*. My own figure of less than 1% is derived for the year 1860 from data presented in F. A. P. Barnard, *Analysis of Some Statistics of Collegiate Education* (New York: Printed by the Trustees of Columbia College, 1870). For a group of Northeastern states having the highest college attendance rate, Barnard concludes that for that region alone 1 in 71 or 1 in 100 (depending on the calculation) of "age eligible youth" attended college. For the country as a whole it was of course a much smaller percentage (p. 14).

[25] One science professor complained that the invidious charge was also indiscriminately "brought against Science itself." Denison Olmsted, "On the Democratic Tendencies of Science," *American Journal of Education*, 1856, *1*:164–173, on p. 164.

relatively homogeneous social composition in their student bodies later in the century, the assumption being that it had always been so.[26] Actually, in the first half of the century "the sons of the poor always far out number[ed] the sons of the rich."[27] Reading the dusty records of the trustees of schools like Bowdoin, Dartmouth, Brown, or Pennsylvania, we cannot help but be struck by students' pathetic pleas to postpone payment of the few dollars needed to cover the college years, by the trials of the many students who earned their way through college by teaching in the district schools during winter recess, and by the administrations' concern with providing scholarships for the destitute. A writer at mid-century tells us:

> In point of fact, full three-fourths of the members of the country colleges are from families with small means, if not absolutely poor. Of the whole number annually graduated in the country, a large majority are of comparatively humble parentage, many of them dependent on their own resources . . . if there be any partiality in the distribution of its blessings, it is in the favor of the middling and lower classes: it is emphatically the poor man's institution.[28]

Even if we consider such subtle factors as the loss of income occasioned by four years of college as a deterrent to matriculation, we cannot conclude that the ranks of the college were closed to the poor or that the sons of what social aristocracy existed were the college youth of the day.[29] "If by aristocracy be meant a stupid and pretentious caste, founded in wealth, and birth, and an affection for European manners, no charge could be more preposterous," insisted President Eliot of Harvard; "the college is intensely American in affection, and entirely democratic in temper."[30] Recalling the first part of the century in 1899, President Hadley of Yale wrote: "differences in wealth throughout the community were less conspicuous than they are today. College education was so cheap that it fell within the reach of all. Most of the students were poor."[31] One of the most outspoken critics of the college insisted, "College education in this country is cheap enough, so cheap that no one can reasonably complain of it on that score."[32] Thus, the American invective levelled against the ante-bellum college for

[26] George E. Peterson, *The New England College in the Age of the University* (Amherst: Amherst College Press, 1964), the most penetrating publication yet to appear on an aspect of college history, deals with the college's late-nineteenth-century adjustments to electives and Progressivism.

[27] George Wyllys Benedict, *New England Educational Institutions in Relation to American Government* (Burlington, Vt.: Chauncey Goodrich, 1844), p. 44.

[28] Charles Haddock, *Collegiate Education* (Boston: T. R. Marvin, 1848), pp. 7–8. Similarly, a British historian, in a statistical study of the students at Oxford and Cambridge, has challenged some assumptions made about them, concluding that a "significant number of students were from merchant classes, artisans, and farmers." Nicholas Hans, *New Trends in Education in the Eighteenth Century* (London: Routledge & Kegan Paul, 1951), p. 41. The majority of colleges were, of course, country colleges. The term is not used here to suggest that city colleges attracted the rich. To this writer, as to most of his generation, a country college was far to be preferred to one situated in a vice-ridden city.

[29] One contemporary made this point even more clearly: "One thing is very certain, that did not the colleges exist, and exist under the same condition which they actually do exist under, none but the sons of the rich could obtain that education which is now within the reach of the poorest lad in the land, *if he but earnestly longs for it, and that too without injuring in him the wholesome feelings of being independent of the charity and patronage of others.*" (Italics mine.) G. W. Benedict, *New England Educational Institutions*, p. 44.

[30] Eliot's address in *Builders of American Universities*, ed. David Weaver (Alton, Ill.: Shurtleff College Press, 1959), Vol. I, p. 30.

[31] Hadley's address, *ibid.*, p. 45.

[32] Francis Wayland, *Thoughts on Collegiate Education*, reprinted in *American Higher Education*, ed. Hofstadter and Smith, Vol. I, p. 363. College tuition in 1831, e.g., averaged about $33 a year at Northeastern institutions. See *Quarterly Register of the American Education Society*, 1831,

its aristocratic tendencies must be considered with care. It was, after all, common to apply the same term against the public high school movement at the end of the century, when it was thought anti-democratic to keep children in school and away from honest labor so long.

The popular historical syllogism that concludes that the American college was anti-scientific because it was anti-democratic must thus be seen to contain from the first a false premise—namely, that the college was aristocratic. Logically, it ought not be necessary to consider the invalidity of the other premise which equates the anti-scientific with the anti-democratic. But since historically even illogical arguments have been persuasive, we should perhaps examine how the belief that science is democratic has functioned in the history of American science.

Spokesmen for American science in any institutional setting have at all times thought it necessary to justify their concerns in terms of the dominant democratic political tradition. Their task has been made difficult by the fact that many people consider any intellectual endeavor, including scientific inquiry, as elitist, undemocratic, and aristocratic; it has been made easy by the possibility of confusing science with socially useful technology, intentionally or not. In the context of the ante-bellum college, scientists characteristically argued for their disciplines in democratic rhetoric, and, as we have seen, the character of the college made their case seem plausible. Later, in the age of the university, the scientists and Whiggish historians who then followed in their footsteps justified their own existence by stressing the egalitarian spirit of technology and in addition by suggesting the "illiberality" of their predecessors in the supposedly "undemocratic" colleges. While not necessarily duplicitous, their emphasis mainly succeeded only in drawing our attention away from the amount of theoretical science that was actually taught in the earlier college.[33]

Interestingly, the final sundering of this dubious connection between science, education, and politics came not from historical debate but from an unexpected source. Only when Sputnik suggested that an undemocratic society can also support science did academics begin to realize that while science has its politics, and politics its science, no national political or social form is necessarily best suited to science. Thus, even if the college had been aristocratic, as it decidedly was not, that fact would still not have rendered it anti-scientific.

Such semantic errors as these have not operated alone to obscure the important position of the sciences in the old classical denominational college. Misconceptions about developments in higher education later in the century have also worked to keep the historian unaware of their precursors. An instructive example is the persistent belief that the college was transformed into the university only after intimate contact with German models. Traditionally, historians of education have been only too willing to use the notion of the superiority of a German-style university as a means to measure

---

3:298–299. Olmsted, "On the Democratic Tendencies," p. 171, claimed it was cheaper to attend Yale than the private academies.

[33] For the ante-bellum period see Olmsted, "On the Democratic Tendencies of Science." For discussion of later-nineteenth-century attempts to treat science and democracy see Laurence R. Veysey, *The Emergence of the American University* (Chicago: University of Chicago Press, 1965), pp. 62 ff., 124. For historical misunderstanding see Nevins, *State Universities*, and Frederick B. Mumford, *The Land Grant College Movement* (Columbia: University of Missouri College of Agriculture and Agricultural Experiment Station, 1940).

both the success of the American university and the failure of the earlier American college, thus leading to the belief that the improvement in science education must also have issued from that country. The idea has become so exaggerated that one is led to believe that prior to the German influence there were no lectures or laboratories in America. But the truth is that these pedagogical aids were much discussed even during the early decades of the century as they slowly emerged in existing institutional settings. There was in fact very limited knowledge of German universities in America and no widespread concern about them until the 1870s. Even the Germans themselves were annoyed at the generalizations and oversimplifications that early American travelers published after their brief Teutonic interludes in the first half of the century.[34]

Instead of insisting upon tracing a German influence that hardly existed, we might pay more attention to nineteenth-century realities. Looking for instance at the textbooks adopted in the course of the century, we find that French, English, and Scottish works were far more significant than German books, at least before 1840, if not later. The scarcity of German-trained graduates prior to 1850 should make us even more skeptical of the extent to which a German influence existed in American scientific education.

We are certainly correct in assuming that there were some people who looked to a German university as the ideal for the extension of the college's activity. Yet such an extension would have created an elitist institution in a culture not at all eager to forsake its democratic sympathies—a contradiction which was recognized even in the ante-bellum period.[35] Whatever American accounts of education in Europe may have existed then, developments in higher education in America were still quite indigenous.

A recent British historian maintains that "by the nineteenth century the curriculum and orientation of the schools were broader in America than in Europe."[36] A prominent Israeli sociologist has further argued that the German system was an eventual failure and that America achieved its twentieth-century eminence in science by the flexibility built into its own institutions.[37] Looking back, we find that as late as 1895 an important American educator writes that

> . . . an educational organization closely following the German type would not thrive in America: indeed with all its undisputed excellences, it would not meet our needs so well as the yet unsystematic, but remarkably effective, organizations that circumstances have brought into existence.[38]

The failure of America to produce German-type universities at an earlier date no doubt accounts for the concept that religious affiliations, classicism in curriculum, and elitism in student bodies were retarding factors in its slow development. But the enduring dream of a few to create a university was neither the major challenge nor the solution to the college's actual predicament, as the University of Virginia's failure to attract students for advanced instruction not associated with medical or law degrees

[34] Victor Huber, *The English Universities*, trans. Francis Newman (London: William Pickering, 1843), Vol. II, Pt. 1, p. 390n.

[35] This point is made in "Review of Dwight's Travels in the North of Germany," *Quarterly Christian Spectator*, 1829, *1*: 641.

[36] Brian Holmes, *American Criticism of American Education* (Columbus: College of Education, Ohio State University, 1956), p. 183.

[37] Joseph Ben-David, *Fundamental Research and the Universities* (Paris: Office of Economic Cooperation and Development, 1968), *passim*.

[38] Nicholas Murray Butler, in the introduction to Friedrich Paulsen, *The German Universities: Their Character and Historical Development* (New York: Macmillan, 1895), p. xiii.

had shown in the past.[39] A Jacksonian dislike for any scheme for centralization, whether in banking or in college, would leave the creation of a single national university for the production of professional scholars beyond most political comprehension, while the college of the immediate future, like that of the past, would be faced with the more urgent problem of attracting its students from a society which wanted only as little formal education as was necessary to conduct the ordinary pursuits of commercial life.[40]

It is unfortunate that the early college must constantly be criticized by comparison to its later forms. Education, by its relation to other social constructs and by its need for self-perpetuation, is inherently conservative in all societies; thus, there is never any difficulty in demonstrating that a form of the future was actively denied a place in the past. It is more valuable, however, to study the slow processes and considerations which finally contributed to change. The problems in higher education in the first half of the nineteenth century were already too numerous and complex to be characterized by something as simple-minded as a struggle between God and chemistry, Latin and physics, or college versus university education, although many modern historians have adopted this uncritical view.

Alternatively, I would suggest that the great challenge to the American college of 1828 was, as indeed it still is, to justify itself to a society which was generally uninterested in the pursuit of its "nonrelevant" programs, subscribing instead to the nineteenth-century concept of relevance as a "desire to omit all study which cannot be shown to bear immediately and directly upon making the student a better instrument for economical production."[41] That a heavy-handed conservative reaction entrenched in merely sectarian colleges prevented this spirit from overthrowing the liberal arts tradition and replacing it by a broadly scientific one is an exaggeration providing little insight.

The college did, indeed, make many attempts to extend its benefits to students other than those who, for one reason or another, subscribed to the liberal arts curriculum; yet many such attempts were unsuccessful. Columbia College, for instance, located in the prosperous commercial city of New York, found few students in the 1830s to enroll in its "Scientific and Literary Course, the whole or any part of which matriculated students may of their option attend."[42] Similar excursions at Amherst and the Univer-

[39] For attempts to form a university at Charlottesville, see Rudolph, *American College*, pp. 125–128, 213. Another indication of the prematurity of graduate scholarship may be seen from the Yale example. There, where it was possible to pursue extra-collegiate work in either the School of Science or in the Department of Philosophy and the Arts, as late as 1867 only two students were registered in the latter program. Russell H. Chittenden, *History of the Sheffield Scientific School of Yale University, 1846–1922* (New Haven: Yale University Press, 1928), Vol. I, p. 115.

[40] While giving attention to new colleges and courses, it is easy to lose sight of the fact that universal public education was not yet a reality in 1850. The college was pressed to avoid becoming the mere secondary school it had been in the eighteenth century, even while facing the problem of the opposite extreme—that of establishing graduate education. F. A. P. Barnard noted in 1867, e.g., that everyone was "occupied with the teaching of colleges—taking it apparently for granted that the course of preparatory study . . . needs no essential modification. But it is precisely at this point, as it seems to me, that modification is most necessary." Quoted from "On Early Mental Training and the Studies best fitted for it," in *The Culture Demanded by Modern Life*, by E. L. Youmans (New York: D. Appleton, 1867), p. 312.

[41] Anderson, "End and Means of a Liberal Education," *loc. cit.*

[42] MSS, Minutes of the Trustees of Columbia College; Jan. 5, 16, 1830; Jan. 16 and Feb. 3, 1836. Also James Renwick's printed circular dated Dec. 16, 1835, in Columbia College Papers, Columbia University Library.

sity of Vermont failed to awaken new interest in the college among the manual classes.

During the 1840s, in addition to the well-known examples of scientific schools at Yale and Harvard, many colleges participated in still newer activity. Amherst and Williams considered joint professorships of agriculture and technical subjects, including an experimental farm. Princeton sought a professor of applied science.[43] After failing to find a professor of agriculture, Union College settled on an analytical chemistry course; and Harvard even taught brick masonry and railway track laying in its Lawrence Scientific School.[44] Shadowy plans for a scientific education of the type that would fall between the obviously theoretical and the obviously practical were finding widespread expression in the old college. Henry Tappan, president of the University of Michigan, whose reforming voice as much saved the college as established the university, enlarged on this situation in 1851:

> This general movement of the college towards a higher position, by adding more studies to their curriculum, by endeavoring to shape themselves to more numerous classes of students, by introducing voluntary courses of study, by attempting lectures on the more advanced branches of study, and by assuming the name of University, is not a mere freak of ambitious folly, but an attempt to meet the demands of the age . . . indications of an all pervading influence which is striving in various ways to become realized. Now everything appears crude and disjointed, and sometimes even grotesque; the fused elements are running in every direction, until they find the moulds which are to give them proportion and symmetry.[45]

At the larger city institutions like Harvard, Yale, Columbia, and Pennsylvania, plans for education other than the liberal arts course shortly found "proportion and symmetry" in engineering and graduate schools, though at the smaller rural schools most of these programs fell into neglect after the War when land grant colleges assumed these functions.

I have suggested that the prosecution of science in the college is one of perhaps a number of relationships obscured in many accounts of the college. Though not every writer has been guilty of perpetuating the mistaken image of the college in its entirety, it is unfortunately still possible for one of the most quoted historians of the college to write in the 1960s that "Many a venerable campus rumbled on as though nothing of importance had happened since Caesar found all Gaul divided into three parts."[46]

What I have called the mistaken stereotype of the American college is primarily the

[43] MSS, Edward Hitchcock to Mark Hopkins, July 12, 1848; Mark Hopkins to Hon. Jos. White, Feb. 3, 1853, both in Williams College Library. Minutes of the Trustees of Princeton College, Dec. 20, 1853, Princeton University Archives.

[44] MSS, Union College: Minutes of the Trustees, July 25, 1849, and July 21, 1855. Information on Harvard derived from Charles A. Packard, "Formulae and Facts—Lawrence Scientific School. Harvard University . . .," June, 1848, in Bowdoin College Archives.

[45] Henry Tappan, *University Education* (New York: G. P. Putnam, 1851), p. 80. It is interesting to note that in this confused atmosphere the conservative *North American Review* at first advocated placing agriculture into the required classical course, but then four years later, in 1855, abandoned this hope, calling for a complete return to "those permanent studies which have been regarded by wise men, for a thousand years, as the best possible discipline for young minds," suggestive of the next age in the life of the college. "Wayland on College Education in America," Jan. 1851, *71*:60–84; and "European and American Universities," Jan. 1855, *80*:117–163, quotation from p. 151.

[46] George P. Schmidt, quoted in *A Century of Higher Education: Classical Citadel to Collegiate Colossus*, ed. William Brickman and Stanley Lehrer (New York: Society for the Advancement of Education, 1962), p. 55. Also see p. 52, where Schmidt pushes the curriculum back to Aristotle.

result of circumstances and attitudes developed in the latter half of the last century.[47] The variety of the components of this image of the college as a citadel of elite and effete ancient language study with little science detracts from the college's most fundamental character.[48] In the second quarter of the nineteenth century the college built upon its already favorable disposition to science in order to create positive relationships with the growing scientific establishment. The increase in equipment and professors, the improved quality of textbooks, and a pervasive concern for the perpetuation of the scientific spirit proceeded virtually unhindered. By mid-century there were few people who could have denied that "instruction in physical science, should hold a place, and not an inconsiderable one, in every system of education."[49]

With still no graduate schools and only a handful of popular programs beyond the liberal arts curriculum, the mid-nineteenth-century college was the major depository and dispensary of scientific expertise in education, more important than the medical or the engineering school. Since 1820 the college's own scientists had evolved from textbook editors, to textbook writers, to members of a professional scientific community. In the studies of the small group of the population that attended college, more time was devoted to science requirements than had ever been before or would ever be again after the growth of electives.[50] Thus, in 1850, when Brown's Francis Wayland argued for a free elective system while complaining that the method of universally required sciences in the classical curriculum had exhausted its possibilities, his school had four scientists on a seven-man faculty and a curriculum in which 160 out of 479

[47] In this connection it is interesting to look at a British historian's conclusions for higher education in his country undergoing similar adjustments and reactions to industrialization: "A marked distinction between 'general' education and 'technical' training became apparent. The unity of the educational ideal was severed and the social cleavage between employer of labour and employed labourers was reflected in the differentiation of social prestige between the 'classical' and 'scientific' curriculum. Classical education received a new lease of life, and divorced from science, became simply a sign of social prestige. Scientific education, narrowed down to technical skill, lost its broader emancipating appeal and was avoided by cultured families. Thus the ground for the educational setback of the nineteenth century was prepared and the reforming zeal of the eighteenth century completely forgotten." Hans, *New Trends in Education*, p. 212.

[48] R. Freeman Butts and Lawrence A. Cremin, *A History of Education in American Culture* (New York: Henry Holt, 1953), p. 179, tell us that the nineteenth-century college tradition was so caught up in scholarship that it "even argued against preparation for living in a real world as a proper goal for education." It is surprising that historians could actually apply such vague contemporary rhetoric to the past. Right or wrong, to nineteenth-century academics a college was in fact the best possible preparation for "life in the real world."

[49] "Elementary Works on Physical Sciences," *North Amer. Rev.*, Apr. 1851, *72*:358.

[50] As an example of the attention that science demanded of *all* students at this time, I cite the catalogue of Amherst College for 1852–1853. Nine professors are listed, five of whom teach science: a professor of natural theology and geology, of mathematics and natural philosophy, of chemistry and natural history, of astronomy and zoology, and of analytical and applied chemistry. The curriculum listed includes the following mathematics and science: Freshman—algebra, geometry; Sophomore—plane and spherical geometry, plane and spherical trigonometry, mensuration, surveying and navigation, anatomy and physiology, chemistry, conic sections, and analytical geometry; Junior—Practical exercises in the differential and integral calculus, mechanics, hydrostatics, pneumatics, electricity, magnetism and optics, astronomy, zoology, botany; Senior—zoology, geology. In addition to the above recitation courses there were ten courses of lectures, including geology, natural philosophy, botany and mineralogy, chemistry, psychology, anatomy and physiology, zoology, and natural theology. *Catalogue of the Officers and Students of Amherst College, for the Academical Year 1852–53* (Amherst: J. S. & C. Adams, 1852), pp. 5, 17–21.

hours of instruction were devoted to mathematics and science.[51] But although it could not be all things to all people, although it fumbled in many directions while preserving the idea of the liberal arts, although it was a visible educational enterprise at a time when the variety of alternatives was still limited, the college was neither monolithic, aristocratic, unresponsive, sectarian, nor anti-scientific.

> At once conservative and progressive in its spirit, it strives to preserve a due medium between a bigoted attachment to all that is old, and an indiscriminate passion for all that is new. Not less scientific than classical in its course of studies, it aims to engraft the science of the moderns on the wisdom of the ancients. . . . It will be modified to suit the change of circumstances; it will be enlarged to meet the growing demands of the age; but it will not be abandoned,—it cannot be abolished. It has become a fixed fact. It is the settled policy of the country.[52]

---

[51] *A Catalogue of the Officers and Students of Brown University, 1850–51* (Providence: John F. Moore, 1850), p. 15.

[52] William Seymour Tyler, *Prayer for Colleges* (New York: Society for Promotion of Theological Education at the West, 1859), pp. 59, 68. Over forty years after these thoughts were penned, Daniel C. Gilman, the first president of Johns Hopkins, America's first university, thanked the liberal arts curriculum for making the university possible. He also expressed the strong hope that in that now more sophisticated age the concept of a liberal collegiate education would not be lost. "Is it Worth While to Uphold Any Longer the Idea of Liberal Education," *Educational Review*, Feb. 1892, pp. 105–119.

The Process of Professionalization in American Science: The Emergent Period, 1820-1860

*By George H. Daniels**

RECENT WRITERS on the history of science in America generally agree that conditions underlying the pursuit of science changed drastically during the nineteenth century. By the middle of the century, the earlier pattern of gentlemanly scientific activity was rapidly becoming obsolete. The amateur was in the process of being replaced by the trained specialist—the professional who had a single-minded dedication to the interests of science. The emergence of a community of such professionals was the most significant development in nineteenth-century American science. Most of the controversies within science can be understood in terms of tensions inherent in the transition from one mode of scientific activity to another; and, as one byproduct of this paper will be to demonstrate, the necessity for a professional body to justify its existence to the public even played a part in continuing the dominance of an older philosophy of science among the first generation of professionals.

This paper seeks to develop a general framework for the study of the process of transition. The framework is, of course, a simplification, as all frameworks are, but I nevertheless think that it can help clarify a number of perplexing facts about American science in the past century.

The development of professionalization in American science can be di-

*Northwestern University. An earlier version of this paper was read at the 1965 meeting of the History of Science Society and the Society for the History of Technology, in San Francisco, 28 Dec. 1965. In the preparation of the final version, I have profited substantially from the comments of an anonymous referee for *Isis* and also from suggestions offered by Prof. John Burnham of the Ohio State University and Prof. Geraldine Joncich of the University of California, Berkeley. Also helpful were some suggestive comments in an unpublished paper kindly made available to me by the author: Robert C. Davis, "Scientists and Engineers: The Public's Image."

vided into four stages, which I shall name (1) preemption, (2) institutionalization, (3) legitimation, and (4) the attainment of professional autonomy. These stages overlap; in fact, they constantly interact. Yet, at different times, the problems of a profession can be related to one of these stages, each calls forth a particular kind of behavior from those caught up in the process, and the order in which they follow each other as the major professional interest sciences that passed through the first three stages—the "emergent period"—in is determinate.[1] I shall try to illustrate these claims by considering those the particular social conditions of mid-nineteenth-century America.[2]

Preemption

By "preemption" I refer to that stage in which a task which has customarily been performed by one group or by everybody in general comes into the exclusive possession of another particular group. This automatically occurs when the body of knowledge necessary for the task becomes esoteric, that is, when it becomes obviously unavailable to the general scholar. This is a necessary first step; for as long as it is possible to believe that "every man may contribute his mite," a secure and publicly acknowledged role cannot be developed for those who cultivate the knowledge. Since the sciences are unquestionably esoteric, it follows that there must exist some precise point at which they became so—a point that we might call "the edge of incomprehensibility." And because this point, unlike "the Edge of Objectivity," is reached only when people agree that it has been reached and begin to divide themselves into "Insiders" and "Outsiders," it is not a matter for logical analysis, but for empirical determination.

The reaction of two American commentators, each writing in 1819, to the discovery that Lavoisier's "simple and beautiful" system was not definitive—a discovery that ushered in chemistry's "edge of incomprehensibility"—is illustrative of the shock caused by the recognition that a discipline has been severed from the general stock of knowledge. John Ware, Boston physician and dabbler in chemistry, had just become aware that developments in chemistry had passed him by. His complaint was typical:

> It is impossible for those of us, who have formed our ideas of the chemical operations of nature on the principles which he [Lavoisier] taught, to turn with complacency from a theory like his to a state of science so unsettled and so obscure, as modern chemistry now is.
> . . . They have made it less captivating to the general scholar; they have lessened the interest with which it is viewed by those not immediately engaged

1 This process is concerned only with ideological factors related to the emerging self-consciousness of a scientific community. While factors such as the availability of jobs and special exemptions conferred by the society are of obvious importance, my assumption is that one test of professionalism is whether a man feels that he is a professional and is therefore able to exhibit professional behavior.

2 Since scientific communities develop in terms of a particular culture, any attempts to generalize this process beyond the United States would be hazardous. While I think that the general outlines would be analogous, the adjustments would be different in each country.

> in its pursuit, by rendering it more complicated and more difficult to be understood, and less applicable as a whole to the explication of those phenomena of the natural world with which we are most familiar.[3]

The simplicity of Lavoisier's system had been an important argument in its behalf. With its assumption (proven erroneous shortly after the turn of the century) that each element had its characteristic function in any compound, and with its sets of rigid dichotomies, it could easily be mastered with minimal study by any educated man.

Ware's point was that chemistry had in his time become so complex that only those possessing special qualifications could fully comprehend it. The specially qualified group would, as a natural consequence, begin to experience new difficulties in communicating with those outside their profession as soon as their subject became esoteric. It was the dawning awareness of this kind of alienation that Robert Hare, professional chemist, was expressing when he wrote to Benjamin Silliman:

> I was told . . . that many said they could not understand my memoir, who considered their standing such as to feel as if this were an imputation against me rather than themselves. I could not write it for those who are so ignorant, without making it too prolix and commonplace for adepts. There is our difficulty—we cannot write anything for the scientific few which will be agreeable to the ignorant many.[4]

Natural history soon began to travel the same route. Within the framework of professionalization, the significance of the natural system of classification in botany and zoology and the chemical classification of minerals was that these innovations created esoteric bodies of knowledge where before they had been parts of the general knowledge. The Linnaean system of classification was based upon obvious, easily grasped physical characteristics; the natural system, on the other hand, required the student to call upon a whole complex of knowledge to arrive at a judgment of "overall affinity."

The change from physical classification of minerals to classification based on chemical analysis was comparable to the changes in botany and zoology. Asa Gray's statement of the case for botany may therefore stand for natural history in general:

> The study of affinities is neither guesswork nor divination, but a matter of logical deduction from structure, based upon scientific principles—principles recognized and acted upon by sound botanists with considerable unanimity, although they have never been reduced to a system nor expounded in detail, so as to make them matters of elementary instruction.[5]

---

[3] John Ware, "Gorham's Chemistry," *North American Review*, 1819, *9*:134; *cf*. Denison Olmstead, "On the Present State of Chemical Science," *American Journal of Science*, 1826, *11*:352, who observed that "elder scholars" were frequently making such complaints.

[4] 28 July 1819, quoted in George P. Fisher, *Life of Benjamin Silliman, M.D., LL.D.* (2 vols., New York: Scribner, 1866), Vol. I, p. 289.

[5] [Asa Gray], "Henfrey's Botany," *Am. J. Sci.*, 1858, 2nd Ser., *24*:434.

Characteristic of the reaction of older natural historians was Chester Dewey's complaint that "the natural method takes botany from the multitude, and confines it to the learned."[6] More accurately, it was taking botany from the learned-in-general and confining it to the specially qualified.

In geology, the crucial event was the increasing use of fossil contents as guides to dating strata. Again, this was a task that required expert knowledge and wide agreement upon nomenclature.

The sum of all these changes was the creation by the 1840's of a body of esoteric knowledge called "science." As one would expect, the term "scientist" was also coined at this time to refer to those who had previously been designated "natural philosophers." William Whewell was the first to use the new term, and by the late 1840's B. A. Gould and other Americans were urging the use of this more distinctive designation.[7] In America, the term "scientific men" was in general use by 1840, and "philosophy" was beginning to take on its modern meaning. During this same period, the term "popularizer" began to be used more frequently and to take on its modern invidious connotation.[8] Popular lecturers more and more began to abandon the old effort to convey up-to-date scientific understanding, contenting themselves instead with retailing "wonders." This appeal, which was virtually absent in the first quarter of the century, had become common by the 1840's—a sure indication of the inability to communicate on other terms.[9]

The vehicle through which most of the esoteric elements were introduced was the *American Journal of Science and Arts.* Silliman, the founder and chief editor, has had his scientific competence much maligned, but at this period, he would qualify as a champion of the professional interests. Not only did he steadfastly resist every effort to make the *Journal* "more popular" or "more miscellaneous," but he and his junior editors systematically worked to introduce the new views into American chemistry, beginning with the first issue in 1818; geology, beginning in the 1820's, zoology in the 1830's, and botany and mineralogy in the 1840's.[10] Silliman, as the British geologist Robert Bakewell warned him, was "more anxious to obtain the approbation of a few scientific readers," than he was to "excite and gratify" the curious.[11] He lost his standing in the scientific community only

6 Chester Dewey to Asa Gray, 14 March 1843, quoted in A. Hunter Dupree, *Asa Gray, 1810–1888* (Cambridge, Mass.: Harvard Univ. Press, 1959), p. 102.

7 For an account of Gould's attempt to introduce the word "scientist" see his "Address upon Retiring as President of the Association," *AAAS Proceedings,* 1869, *18*:9.

8 See, e.g., "July Reviewed by September," *Atlantic Monthly,* Sept. 1860, *6*:378–383. This attack on popular science, printed under James Russell Lowell's name, was written for Lowell by W. B. Rogers. On this point see Lowell to Rogers, 1 Aug. 1860, in Emma Rogers and William T. Sedgwick, *Life and Letters of William Barton Rogers* (2 vols., Boston/New York, 1896), Vol. II, p. 37.

9 For the absence of the appeal in the first quarter of the century, see John C. Greene, "Science and the Public in the Age of Jefferson," *Isis,* 1958, *49*:25. The best example of the later type is Edward Hitchcock, "The Wonders of Science Compared with the Wonders of Romance," in *Religious Truth, Illustrated From Science* (Boston: Phillips, Sampson, 1857), esp. pp. 139–140. This was the first publication of a lecture that Hitchcock had given on several occasions during the previous decade.

10 For Silliman's determination to keep the journal scientific, see "Preface," *Am. J. Sci.,* 1829, *16*:iv–vi.

11 Robert Bakewell to Benjamin Silliman, 16 July 1834, quoted in Fisher, *Life of Benjamin Silliman . . . ,* Vol. II, p. 55.

when, at a later time, he failed to maintain the behavior appropriate for an editor.[12]

Characteristic types of conflict developed at this preemption stage. As long as a science remains in its democratic (usually fact-gathering) stage,[13] the possibility of conflict with any other profession is rather remote. The fact gatherers have no explanations of their own to offer, and their facts are equally available to all. Even those sciences that were not fact gathering in the preesoteric stage (e.g., chemistry) rarely encountered opposition, for a democratic society is predisposed to tolerate all that it can comprehend well enough to perceive as "harmless."

But once this preesoteric stage is passed, a new situation vis-à-vis the society develops. The new necessity for providing formal education prior to entry into the profession led to conflict with older, better-established groups who controlled the educational system and who had a vested interest in maintaining it against encroachment. The new possibility for a rival type of explanation led to conflict with the clergy, who had a vested interest in keeping their exclusive right to interpret natural law. The public in general, not understanding the new developments, felt vaguely threatened by them, insofar as the situation was noticed at all. Since the sciences passed into the esoteric stage at different times, we find that at one period the geologist may be held suspect, while the zoologist is considered completely respectable. Again, a later generation may have worked out a way of living with the geologist, but cannot tolerate the emergent zoologist. It follows, therefore, that all scientists will not feel public hostility or indifference with equal urgency. Some will, in fact, cling to a "public" viewpoint in opposition to the professionals of another discipline.

Within the professionalizing community itself, an inevitable conflict of generations arises at this point, quite frequently growing out of efforts to standardize nomenclature, one of the first details to be attended to in the development of a profession. Older scientists, particularly those who have achieved a reputation before the change occurs, are understandably reluctant to see their work superseded. Some react as mildly as Thomas Nuttall, who, in the same work in which he confessed his inability to cope with the natural system, announced his retirement from ornithology.[14] Others are more violent. For example, the literature throughout the late 1840's is filled with the outraged protests of Charles U. Shepard against various efforts

12 Later in his life, Silliman accepted a number of papers which were considered both unsound and unnecessarily provocative. For one good account see J. D. Dana to A. D. Bache, 6 Sept. 1851, Rhees Collection, Huntington Library.

13 A science in the fact-gathering stage is more "democratic" simply because it is more "open." Since all facts are regarded as of equal relevance, fact gathering is random and facts are welcomed from all quarters. See Thomas Kuhn, *The Structure of Scientific Revolutions* (Chicago: Univ. Chicago Press, 1962), p. 15.

14 *A Manual of the Ornithology of the United States and Canada: The Water Birds* (Boston: Hilliard, Gray, 1834), p. vi. During that same year, Nuttall's work was characterized by John Torrey as "valuable" but "far behind the science and very defective in nomenclature." Torrey attributed Nuttall's deficiencies to a lack of specialization. For Torrey's assessment, see Andrew Denny Rodgers III, *John Torrey: A Story of North American Botany* (Princeton: Princeton Univ. Press, 1942), p. 46.

to chemically reclassify (sometimes out of existence) minerals that he had previously classified on natural history grounds. Namers of zoological or botanical species can be expected to label as "species splitters" or "species lumpers" the proponents of a new taxonomy.[15] The empiric in medicine may oppose chemical speculations in physiology because they conflict with his religious beliefs, but the fact that they threaten to render his *materia medica* obsolete is at least as distasteful to him. The older generation may react by forming alliances with other groups in the society—including representatives of sciences still in an earlier stage—or they may try to enlist common cultural values on their side. One example is the alliance in the early nineteenth century among holders of a crudely empirical, antihypothetical philosophy of science appropriate to the fact-gathering stage, common sense philosophy, and Protestant theology.[16]

Institutionalization

The consciousness of being a member of a special group, which is a natural result of an awareness that one is cultivating esoteric knowledge, creates a need for regularizing relationships among colleagues and between colleagues and outsiders. The second stage of the professionalization process—institutionalization—is the structuring of behavior into established patterns that provide such means. Since the institutions are essential for the attainment of professional aims, they carry normative force. Thus, Benjamin Peirce, after writing privately that Benjamin A. Gould had "thoroughly dishonored" him in the Dudley Observatory affair and that he disagreed with him entirely, nevertheless went to Albany a few days later to deliver an impassioned defense of Gould; in this he was acting out an important institutional role.[17] The trustees, all laymen, had questioned Gould's scientific competence, thus threatening an essential part of the professional ideology which insists that only the profession itself is a competent judge of its affairs. When Asa Gray became outraged over the publication of Robert Chambers' *Explanations of the Vestiges* to the point of writing forty-one heated pages in review, he too was playing an important institutional role.[18] Chambers had acknowledged that scientists had almost universally denounced the evolutionary theory in his earlier work, *Vestiges of the Natural History of Creation* (1844), but he did not accept their judgment as conclusive. The present book was a deliberate appeal to the public over the heads of the "scientific tribunal," and it was, therefore, a cardinal sin.[19] Likewise, when Hayden and Blake refused to supply James Hall with the geological in-

15 See, e.g., John Ware, "Harlan's Fauna Americana," *N. Am. Rev.*, 1826, *22*:134–135; T. W. Harris, "Upon the Economy of Some American Species of Hispa," *Boston Journal of Natural History*, 1835, *1*:141.

16 For an account of this alliance, see my "Finalism and Positivism in Nineteenth Century American Physiological Thought," *Bulletin of the History of Medicine*, 1964, *38*:343–363.

17 Peirce to Bache, 23 Dec. 1857, Rhees Papers, Huntington Library.

18 [Asa Gray], "Explanations of the Vestiges," *N. Am. Rev.*, 1846, *62*:465–506.

19 *Ibid.*, p. 506. Gray also took a dim view of Agassiz's frequent appeals to the public. On this point, see the quotation in Dupree, *Asa Gray*, p. 229.

formation he requested, they were *failing* to play the institutional role that would have been expected of professionals; for a full and free exchange of knowledge, without regard to personal considerations, had already come to be a professional norm.[20]

The period of institutionalization of behavior is pregnant with possibilities for conflict, most of which arise over the formation and conduct of associations, which are indispensable for the maintenance of professional institutions. Professionals were generally recognized in this period by a marked caution regarding the formation of associations beyond the local level where they could be assured control. As many dangers were recognized as there were advantages; for instance, there was the ever-present fear that charlatanism would gain control and that subsequently "true" science would be subordinated to extrascientific considerations and perhaps misdirected. There is clear evidence that the failure of professional scientists to support the National Institute for the Promotion of Science in 1844 was a major cause of its collapse—even though the Institute had circulated the false notion that it had the support of the other scientific societies and of the scientific community in general.[21] Dallas Bache, to be sure, had attended the National Meeting, but only so he could be "on the ground to direct . . . the host of pseudo-savants . . . into a proper course." Joseph Henry's long-felt conviction that a "promiscuous assembly of those who call themselves men of science in this country would only end in our disgrace" was a characteristic professional reaction.[22] The heavily political, almost wholly amateur, membership of the Institute was sufficient reason for its rejection.

Bache, speaking in 1851, explained that the "most progressive" had opposed the early founding of a general scientific association, because they did not have enough strength to control it. It was wise, he said, that the first organization had been left to the geologists, who were working on common problems. Among them, "to be heard a man must have *done* something."[23]

As Bache suggested, it was a small group of working geologists who, in 1840, had founded the first truly professional association in America. It had

20 James Hall to J. P. Lesley, 10 April (1854?), Lesley Papers, American Philosophical Society. But see Robert K. Merton, "Priorities in Scientific Discovery: A Chapter of the Sociology of Science," *American Sociological Review*, 1957, *22*:635–659, for an analysis of the conflict between the need to protect one's priority and the ideal of commonalty. Merton interprets it as a conflict built into the institutional norms of science. *Cf.*, also, his "The Ambivalence of Scientists," in Norman Kaplan, ed., *Science and Society* (Chicago: Rand McNally, 1965), pp. 112–132.

21 Benjamin Silliman, Jr., to James Hall, 23 April 1844, quoted in John M. Clarke, *James Hall of Albany: Geologist and Palaeontologist, 1811–1898* (Albany, 1921), pp. 141–142.

22 Joseph Henry to Dallas Bache, 16 April 1844, 9 Aug. 1838, Henry Papers, Smithsonian Institution, as quoted in Howard S. Miller, *A Bounty for Research; the Philanthropic Support of Scientific Investigation in America, 1838–1902* (Thesis, Univ. Wisconsin, 1964), p. 14. Professor Geraldine Joncich has called to my attention an analogous situation in the National Education Association, which, in 1906, expanded its membership. The results, according to Nicholas Murray Butler, were much less fortunate than with the AAAS. The NEA "degenerated into a large popular assembly which quickly fell into the hands of a very inferior class of teachers and school officials. . . . So it passed out of the picture." Nicholas Murray Butler, *Across the Busy Years; Recollections and Reflections* (2 vols., New York: Scribner, 1935), Vol. I, p. 96.

23 "Address," *AAAS Proc.*, 1851, *6*:xliv, xlvi–xivii.

expanded its membership slowly until 1847, when the name was changed to the American Association for the Advancement of Science. By organizing in this fashion, a continuity of leadership could be maintained and the Association could be held more closely to professional lines. Leadership was exercised by a standing committee which gained an increasing amount of power. On the first standing committee were three men who had been founding members of the parent association. During the first fourteen years, Bache served on the committee eleven times, Dana seven times, Peirce and the younger Silliman six times each, and Joseph Henry five times. By 1856, when a new constitution was adopted, this committee had gained the power to exclude papers from presentation, to decide which of the papers or other proceedings should be published, to suggest topics for reports, and "to manage any other general business of the Association" between sessions. Papers judged unworthy of publication by the standing committee were not even listed by title.[24]

The Association performed its duty as arbiter of scientific matters in various other ways, assigning research projects to members, appointing investigating committees, reporting on controversial papers, and at times even ruling on questions of priority among members. This last was a questionable matter, Bache thought, but it was probably a lesser evil than avoiding a decision, for if such questions were excluded, the Association would, in effect, be driving its members to appeal to the public.[25]

Naturally, the growing role of the Association as a scientific tribunal was a source of contention, especially since it was perceived that a small group of men exercised great power. Marginal members of the profession tended to view the leaders as a clique, banded together for their private interest. There were bitter protests, in public and in private, against the "reign of scientific despotism."[26] The censorship function of the Association was often misunderstood and bitterly resented; in the case of Daniel Vaughn, who had three papers refused at the 1856 meeting, the disappointed author privately distributed a paper to the members "because original enquirers meet with so little liberality from our American Association."[27] One would-be contributor had created such a storm that a committee of the AAAS had been appointed to investigate. The committee report, signed by Dana and Peirce, was a resounding affirmation of the standing committee's decision: "The character of the paper was such—its conclusions so erroneous, and its reasoning so false—that any other action would have been wanting in fidelity to the interests of the Association and the science of the country."[28]

Despite the protests and charges of conspiracy from outsiders, the more professional, for their part, were never satisfied with the Association because it was not *entirely* in the hands of professionals. Thus, in 1860, Silliman, Jr., spoke of his amazement that "that crazy man from New York," had

24 *AAAS Proc.*, 1857, *11*:170.

25 *Ibid.*, 1851, *6*:lix, xxliv, xlv.

26 Daniel Vaughn to David A. Wells, 18 Feb. 1857, John Warner Papers, American Philosophical Society.

27 Vaughn to John Warner, 4 May 1858, Warner Papers, American Philosophical Society.

28 *AAAS Proc.*, 1855, *9*:284.

been allowed to read his "foolish speculations on the Atomic theory." Two or three such blunders would be enough to taint a whole meeting "as a single dead rat a house."[29] The "foolish speculations," however, were not listed in the Proceedings, so it was not a total catastrophe.

By the same token, the unwillingness of professional scientists to have the public judge scientific matters was not accepted gracefully by everyone. Throughout his long plagiarism controversy with Benjamin Peirce, John Warner, thinking in terms of an earlier democratic period, continually demanded that Peirce publish his explanation: "The public could and would form a just opinion." For his part, Peirce was willing to have the matter adjudicated by the AAAS, but would not submit it to an "incompetent tribunal."[30] A number of newspapers, including the *New York Times*, the Springfield *Republican*, and the Albany *Argus*, in editorial comments leaped to the unwarranted conclusion that Peirce was guilty, and his contemporary reputation suffered greatly for it. Fellow professionals, however, generally defended his refusal to submit to public judgment. John L. LeConte spoke for the professional community in general when he published an essay on the impropriety of carrying such arguments into the public press.[31] The partisans of Warner, convinced that he could not receive justice before a scientific body, continued the one-sided debate in the press.

Even though the professionals had proceeded cautiously in the formation of the Association and had maintained as much professional leadership as possible, a general scientific organization still could not completely escape problems inhering in the differential rate of development of the sciences. Periodically, suggestions were made that sessions be made public, a step that would have oriented the Association more toward the diffusion of science, rather than its advancement, which was thought by the leaders to be the proper function. Members were reminded in presidential addresses: "It is in the hope of communicating new truths and of adding something to the common stock of knowledge that we have associated ourselves."[32]

An insistence upon this distinction, which is basic to modern professional institutions, first appeared in American science in the 1840's. Earlier, scientific journals had been conceived primarily for the "conveyance and diffusion" of knowledge. Even Silliman's *Journal* contained entirely too much popular matter to suit the most advanced professionals.[33] It was not until 1849, when the *Astronomical Journal* appeared, that a journal unequivocally dedicated to "not the diffusion, but the advancement" of scientific knowledge, existed in America.[34] Similarly, the emphasis upon the distinction by the leadership of the AAAS was so new that many members did not even fully comprehend it.

[29] B. Silliman, Jr., to Dallas Bache, 17 Aug. 1860, Rhees Papers, Huntington Library.

[30] John Warner to J. P. Lesley, 26 June 1858, Warner Papers, American Philosophical Society. Peirce to John A. LeConte, 23 May 1858, LeConte Papers, American Philosophical Society.

[31] J. P. Lesley to John Warner, 26 June 1858.

[32] James Hall, *AAAS Proc.*, 1856, *10*:231; see also, A. D. Bache, *AAAS Proc.*, 1851, *5*:2.

[33] E.g., J. D. Dana to A. D. Bache, 6 Sept. 1851, Huntington Library Collection.

[34] B. A. Gould, "On the Establishment and Progress of the Astronomical Journal," *AAAS Proc.*, 1850, *3*:342. See also, J. S. Hubbard, "On the Establishment of an Astronomical Journal in the United States," *AAAS Proc.*, 1847, *3*:378.

Neither were the professionals able, in the early years, to attain satisfactory membership requirements. Under the first constitution, all those having "interest in science" were admitted indiscriminately—providing a constant threat to professional standards and especially to the basic distinction between diffusion and advancement. Efforts were made, beginning in 1858, to prescribe more rigid requirements when the standing committee recommended a distinction between "members" and "associate members"; only the members—those who had made some contribution to science—were eligible to hold office or to vote.[35] In 1868, partial success finally came when the constitution was amended to provide that members could only be admitted "upon recommendation in writing and election by a majority of the members present."[36]

Legitimation

After the rudimentary establishment of the requisite institutions of science and the formation of an association to safeguard them, the problem of legitimation becomes paramount. This is not to say that legitimation claims had not been pressed upon the public before, but at this point the need becomes much more urgent, the justification claims become patterned, and an entire institutional structure is built up beside the purely scientific one. The urgency of the need is in proportion to the extent to which the profession may be thought to come into conflict with the broader culture. In any case, some conscious efforts at legitimation are necessary for a reason that derives from the very nature of a profession as cultivator of esoteric knowledge.

The removal of a body of knowledge from the public domain at once places special obligations upon those who wish to pursue this knowledge; for in a democratic society, professionals cannot exist without a measure of public support. Since the greatest advances in a science come *after* the development of a professional consciousness, it is probably inevitable that there should be a lag between the withdrawal of knowledge from the public domain and the return to society of any positive benefits. But the frankly avowed pursuit of *pure* knowledge is a luxury that a democratic society will allow only the well-established profession. This implies that in the early stages it will be hazardous to attempt to justify scientific training and research in terms of scientific values alone. Until such time that the power of science over the environment becomes perfectly obvious, the scientist must seek some other means of contact with the relevant public. That is to say, the emergent profession has no choice but to justify its work in terms of its social purposes, and in doing so, it must appeal to general cultural values. It is this felt need for public acceptance and support of *full-time roles* that causes the incorporation of older, extrascientific elements into the institutional norms of science. These elements, although they are finally perceived as having been compromising, can only be removed *after* a secure

35 *AAAS Proc.*, 1858, *12*:296–297.

36 *Ibid.*, 1868, *17*:xi.

role is established. The transition period, therefore, is marked by an especially intense public avowal, on the part of scientists, of their adherence to these older, shared values.

In this "practical" American society, one of the most characteristic means of establishing public contact has been the making of extravagant—often irresponsible—overstatements concerning the immediate utility of research. Certainly scientists had always believed that their work was useful, but the need to *persuade* others of its utility had not previously been so strongly felt. Prominent spokesmen for every one of the sciences, including the social sciences, have made such public claims in the process of gaining legitimacy. Even scientists whose work was far removed from the "useful arts" and whose private comments indicated that their chief motivation was simply the pursuit of knowledge for its own sake—in fact, even those who protested against the necessity of making such appeals—nevertheless consciously tried to represent themselves as utilitarians. They were often led to making outrageous claims for the past importance of science, and to making promises that neither they nor their colleagues could reasonably expect to keep. Thus Silliman, in 1810, could "confidently predict" that the United States could only attain to "the pinnacle of national superiority" by training chemists at home and extending the chemical arts.[37] Elias Loomis, in 1846, testified that if only sufficient means were made available, all the laws of storms (obviously useful) would be settled within one year's time and "the subject would be well-nigh exhausted."[38] Chemists, returning in the 1840's from study with Liebig, filled the agricultural press with impossible promises of direct and easy results from soil analysis to such an extent that by 1860 a general disillusionment had come to prevail.[39] Such overstatement, although it appears to have been functionally necessary, had serious consequences. Not only has it led historians to the conclusion that American scientists were a great deal more utility-oriented than they actually were, but it often led to unfortunate conflicts with the public or with the sources of patronage when the inevitable failure to deliver occurred. Why should not state legislators, for example, who had been led to expect immediate returns in mineral wealth from geological surveys, have become disquieted by the geologist's insistence that he spend perhaps years in poring over his notes, in constructing detailed stratigraphic maps, and in wrestling over esoteric problems of nomenclature, before he began to release his findings? The result, in case after case, was the publication of a series of time-consuming and largely irrelevant "Preliminary Reports," which, it was devoutly hoped, would maintain public and legislative interest in the project until a "Final Report" embodying all the scientific results could be prepared. In some cases, the

37 "Notes to the American Edition of Henry's Chemistry," in William Henry, *An Epitome of Experimental Chemistry*, 2nd American ed. from the 5th English ed. (Boston: William Andrews, 1810), p. ii.

38 "Two Storms . . . in the Month of February, 1842," *Transactions of the American Philosophical Society*, 1846, *9*:183–184.

39 A. Hunter Dupree, *Science in the Federal Government* (Cambridge, Mass.: Harvard Univ. Press, 1957), p. 112.

endless preparation of preliminary reports not only exhausted the geologist's time, but the legislators' patience, and the final report never appeared.[40]

An even more spectacular example of the conflict that impossible claims can generate is the well-known case of the Dudley Observatory—a conflict that very nearly ripped the scientific community apart and led to a serious movement to oust the leadership of the AAAS.[41] Benjamin A. Gould and a "Scientific Council" composed of Henry, Bache, and Peirce had painted broad visions of a world-famous observatory to begin operation at Albany in August 1856—an observatory that would be the best equipped of its kind in the world and which, moreover, would utilize the astronomical clock and the telegraph to synchronize the clocks of major New York cities and the railroad lines. This function would bring a measure of scientific fame to Albany, provide a useful public service, and, so Gould indicated, through the sale of time would contribute materially to the support of the observatory. Urged on by Bache, Henry, and Peirce and their interest kept alive by the extravagant plans of Gould, the trustees had dipped into their pockets to support and enlarge an observatory which, it began to appear, might never begin operations. Understandably, two years after the appointed opening date, the trustees became disillusioned to the point that they determined to "get rid of Gould." A controversy ensued between the trustees and the scientific council, a pamphlet warfare followed, Henry withdrew in disgust, Gould was dismissed, and a lasting bitterness was the result.[42]

Despite the claims to practicality by scientists, the suspicion persisted in the public mind that scientists were in reality more interested in some kind of abstract research—some kind of "theory," possibly dangerous to established values—than they were in utilitarian results. Scientists, therefore, continued to stress the traditional claim that moral and religious aspects of science were as valuable to the public as was its utility. If they could not appeal to the ignorant part of the public in this manner, they could at least make their subject respectable to "thinking and well-educated minds," by calling attention to "the more elevated and philosophical portions" of their work.[43] To a public who believed firmly that nature was the purposeful creation of God, the study of God's works could be represented as a religious duty. Unlike the purveyors of fiction and other vanities, scientists dealt with God's own world, and by understanding the natural mechanisms of that world, they were contributing to knowledge of God's plans and, ultimately, to the moral government of the world.

In the arguments for teaching science in the schools, moral values were

40 The extreme case is that of Virginia. The survey, under W. B. Rogers, was begun in 1836 and six annual reports were published before 1841, when mounting antagonism finally caused the law creating the survey to be repealed. As late as 1854, Rogers was in Richmond urging an appropriation for preparation of a final report, but none ever appeared. See George P. Merrill, *Contributions to a History of American State Geological and Natural History Surveys*, U.S. National Museum Bulletin 109 (Washington: G.P.O., 1920), pp. 511–512.

41 Series of letters in John Warner Papers, American Philosophical Society, particularly D. A. Wells to John Warner, 25 June 1858.

42 The controversy is detailed in Miller, *A Bounty for Research*, pp. 67–81.

43 "DeCandolle's Botany," *N. Am. Rev.*, 1834, *38*:33.

stressed almost to the exclusion of any others. "The proper study of nature begets devout affections," one writer pointed out. This undoubted truth, he continued, had given rise to the common maxim "that a true naturalist cannot be a bad man."[44]

In their claims for the moral value of science, as well as in their representation of themselves as interpreters of natural law, scientists naturally risked conflict with the older, better-established theological profession. But in the American society of the mid-nineteenth century, the obvious course of denouncing their rivals as incompetent was hardly a serious possibility. Therefore, it was necessary that the very appearance of conflict be avoided; and the burden of avoiding it was on the new professionals. This consideration, I think, explains the vogue of natural theology, which not only reappeared after having been submerged in the inspirationist fervor of the first few decades of the century, but reached an all-time high in the late thirties, forties, and fifties. It is significant that during this period, unlike the late eighteenth century, the chief proponents of natural theology were scientists or apologists for science. It was not a case of theologians misusing science for their own ends, but of scientists trying to attach some of the aura of the theologian to their own profession. Theologians, in fact, generally continued to stress the inadequacy of arguments from nature. Scientists were therefore careful to insist in their public statements that what they discovered (even though allowed in the "freest and fullest liberty of investigation") could have nothing to do with God's over-all plan—the domain of the theologian. They simply discovered *how* God did things—matters that He had not deigned to reveal in the Bible—and they gave the term "law" to his customary mode of operation.[45] Their work was eminently favorable to piety, for the sciences formed a "vast storehouse" for the use of natural theology, and they "cast light upon and illustrate revelation," after it had been received by other means.[46]

The general acceptance of these claims was crucial in establishing the social role of the scientist in mid-nineteenth-century America, and there is no doubt that most of those making such claims sincerely believed them. It was not that they fabricated a belief in natural theology for the occasion; it was simply that their changing social position led them, during the crucial transition period, very near to making a monomania of this last vestige of a contact with educated nonscientists.[47] Even so, the claim that science was

[44] "The Study of Natural History," *Knickerbocker*, 1845, *25*:292.

[45] E.g., J. D. Dana, "Address," *AAAS Proc.*, 1855, *9*:1–2, 30.

[46] E.g., Edward Hitchcock, "The Relations and Consequent Mutual Duties between the Philosopher and the Theologican," *Bibliotheca Sacra*, 1853, *10*:191–192.

[47] Testifying to the special need for justification in the mid-nineteenth century is the fact that no writer of a zoology textbook in America between 1846 and 1860 neglected in his preface to make a special point of the religious value of science. None before 1846 made any such point, and only one after 1860. See Bruno A. Casile, *An Analysis of Zoology Textbooks Available for American Secondary Schools Before 1920* (Thesis, Univ. Pittsburgh, 1953), pp. 185–190. Paul L. Shank, in a similar study, *The Evolution of Natural Philosophy (Physics) Textbooks Used in American Secondary Schools Before 1880* (Thesis, Univ. Pittsburgh, 1951), pp. 19–24, found that the only texts in which religious purposes were mentioned were the two published in 1846.

the "hand-maiden of theology"—words that as professional a scientist as James D. Dana actually used[48]—was even more compromising to professional aims than was the appeal to utility. The role favored the continued dominance of a taxonomic, nonexplanatory philosophy of science better suited to an earlier stage, and it forced scientists either to impose severe limitations upon themselves or to contrive special justifications when they approached controversial areas. They had to tread carefully around any effort to explain "life" except as a vital process, or to explain the "ultimate forces" of nature—electricity, magnetism, heat, etc. All of these things involved matters beyond the reach of the senses and were consequently suspect. To be sure, these prohibitions were more often voiced than obeyed by the more professional, but even such men as Benjamin Peirce and Joseph Henry had to reject the doctrine of chance, because God's work, after all, is not a matter of chance.[49] Representatives of those sciences still in the fact-gathering stage joined with the older generation in the more advanced sciences to bitterly denounce the "seculative" tendencies of the new professionals whenever they appeared.

The role also imposed special problems of self-policing. If it is postulated that "entire harmony will be the final result of all researches in philosophy and religion," it immediately follows that if entire harmony is not the result in a given case, an error must be present.[50] Consequently, in dealing with one who has threatened the legitimation compromises on which security rests, the task of the profession is defined as the identification and exposure of an error which, it is never doubted, is present. In providing the necessary public evidence, it is important that the offender be dealt with by a recognized spokesman for the profession who will demonstrate that the offender is not a "true" scientist, but a charlatan. This explains the fervent denunciations of anything approaching evolution theory during the early part of the century; the striking thing about these denunciations is that they usually came from the scientists. Robert Chambers' *Vestiges of the Natural History of Creation* (1844) is a case in point. Of all the bad reviews this book received, the most intemperate were written by scientists. For example, the burden of Asa Gray's fifty-two-page review was that the principles used by the author of *Vestiges* were not in keeping with the best scientific thought—if they had been, the book could not possibly have contained all the religious heresies found in it.[51] Despite the fact that private statements by several leading American scientists indicate that they favored the view put forth in the book, it had no American defender. In H. D. Rogers' opinion, expressed privately to his brother, it contained the "loftiest speculative views in Astronomy and Geology and Natural History, and singularly accords with views sketched by me at times in my lectures."[52] Rogers, however, did not write a review of the book.

48 J. D. Dana, "Science and Scientific Schools," *American Journal of Education*, 1856, *2*:363. A commencement address at Yale College, August 1856.

49 *AAAS Proc.*, 1849, *2*:105.

50 Hitchcock, "The Relations . . . between the Philosopher and the Theologican," p. 191.

51 *N. Am. Rev.*, 1845, *60*:426–478.

52 H. D. Rogers to W. B. Rogers, 24 Jan. 1845. See also, John F. Frazier to S. S. Halde-

Darwin's work was, of course, an entirely different matter, and his defenders were anxious to emphasize the differences. In the first place, it commanded respectful attention even from those who opposed it, because it had been introduced by an already well-known scientist through the regular channels: a paper had been presented before a scientific society, and this had been followed by a massive presentation of the evidence for the hypothesis. Defenders and opponents within the scientific community quite generally took the line that it must be debated entirely on its scientific merits.[53] Certainly Darwin's work on purely scientific grounds was far superior to Chambers', but it is clear from reviewers' comments that the difference in reception did not simply reflect the difference in scientific merit. *The Origin of Species*, unlike Chambers' earlier work, could be debated on its scientific merits because Darwin had made it clear—as his American defenders emphasized—that the book implied nothing about such theological issues as the origin of life or the cause of evolution. One could, if he liked, hold to both Darwin and a theological view of nature. It was not that science had escaped from extraneous criteria; it was simply that one could argue, as did Asa Gray, that the extraneous criteria had been satisfied by Darwin as they had not been by Chambers. It is well known that Darwin, in order to emphasize the difference between his theories and earlier ones, had done less than justice to his predecessors. Asa Gray, his chief American defender, was so anxious to avoid raising the past that he contrived the flimsiest of excuses to S. S. Haldeman's repeated demands that Gray mention in his review an earlier work of his that had called for a reexamination of the Lamarckian hypothesis.[54] It was only by such maneuvers, calculated to emphasize the "innocent" nature of the work, that it could be made acceptable.

By a variety of compromises with professional aims the American scientific community, by the 1850's, had managed to establish a relatively secure place for itself. This security was based, as Dana suggested, on "an appreciation of the value of science, not merely for its baser purpose of turning everything into gold, but for its nobler end of opening the earlier revelation."[55] And scientists continued to justify their work in the often illusory terms of immediate practicality, and in the dangerous terms of religious value, until late in the nineteenth century. From this position of security, they were later able to rid themselves, almost, from the external controls and push their claims of autonomy.

Concerning practicality: the formula that later came to be a standard part of the value commitments of science—that all science would ultimately prove useful, but that utility was not to be a test of scientific work—had been expressed privately by many scientists in the early part of the century. But it was not until 1869 that a president of the AAAS felt free to proclaim in a presidential address that the man who based the claim of science to

---

man, 31 Jan. 1846; A. A. Gould to Haldeman, 28 Jan. 1840, 1 Dec. 1840. Haldeman Papers, Academy of Natural Sciences, Philadelphia.

[53] E.g., J. S. Newbury, "Address," *AAAS Proc.*, 1867, *16*:1.

[54] E.g., Gray to Haldeman, 30 July and 27 Sept. 1860, Haldeman Papers, Academy of Natural Sciences, Philadelphia.

[55] "Address," *AAAS Proc.*, 1855, *9*:3.

support on grounds of immediate practical utility was "no loyal follower and true friend of science." Interestingly enough, the president was Benjamin Apthorp Gould, the same man whose extravagant promises had helped to bring about the Dudley Observatory controversy fifteen years earlier.[56] Four years later, another president of the AAAS for the first time attacked the other external criterion. There was, so J. Lawrence Smith said, "Less connection between science and religion than there is between jurisprudence and astronomy, and the sooner this is understood the better it will be for both." Efforts to reconcile science and religion, of the type that had occupied the attention of the first generation of professionals, he termed a "mischievous work," and he offered the formula that later became standard: science is not inimical to religion, simply because it has nothing to do with it.[57] These dates may be taken as at least symbolic of the end of the distinction between the explicit, or public, value, and the implicit value, for after this time, as though the scientific community had been given a signal by its leaders, such statements became relatively common.

The removal of these external criteria signified that the long period of compromise involved in legitimation had ended. The leading members of the profession, at least, were finally able to justify their work in its own terms.

56 *Ibid.*, 1873, *22*:18.

57 *Ibid.*, pp. 18–19.

A Step Toward Scientific Self-Identity in the United States: The Failure of the National Institute, 1844

*By Sally Kohlstedt**

ON THE SPRING MORNING of April 1, 1844, an elaborate and well-advertised "Scientific Convention" began its proceedings in Washington, D.C. The President himself led a procession of statesmen and dignitaries to the Presbyterian Church, where the Marine Band welcomed several hundred members of the National Institute for the Promotion of Science. Elegant and patriotic orations by President John Tyler, Senator Robert J. Walker, former Secretary of War Joel R. Poinsett (who had come from South Carolina for the occasion), and other known Washington leaders set the tone on succeeding days for musical entertainment as well as scientific papers of general interest.[1] In contrast to the pomp and ceremony of the Institute meetings, another national, but decidedly smaller scientific group met in Washington only five weeks later. Its only fanfare was a brief notice in the morning newspaper.[2] The thirty-some members of the Association of American Geologists and Naturalists were obviously concerned with serious business as they listened to highly specialized papers on geology and natural history; occasional sparks flew as scientific points were debated.[3]

The relative importance of the two groups meeting in Washington that year has been distorted in historical literature, with the National Institute receiving a disaproportionate share of attention. Neither of them, however, has been sufficiently studied in detail as representatives of the tendency toward association in the United States during a period of rapid institutionalizing and reform. Throughout the 1830s private discussion and editorial opinion indicated a contemporary interest in establishing a

* Simmons College, Boston, Massachusetts 02115. The research for this study was completed while the author was Visiting Research Associate, Smithsonian Institution, Washington, D.C.

[1] *Washington National Intelligencer*, Apr. 1–4, 1844. Apparently the Institute had friends on the editorial staff, and the daily coverage was quite thorough and complimentary. Publishers W. W. Gales and Joseph Seaton of the *Intelligencer* were, with Peter Force, the publishers of Institute bulletins and circulars.

[2] *Ibid.*, May 8, 1844.

[3] *Ibid.*, May 13, 15, and 17, 1844. Also see "Abstract of the Proceedings of the Fifth Session of the Association of American Geologists and Naturalists," bound with Henry D. Rogers, *Address Delivered at the Meeting of the Association of American Geologists and Naturalists, held in Washington, May, 1844* (Washington, 1844), *passim.*

national scientific organization. In 1840 two groups emerged which aimed at filling the institutional vacuum; each had different, even antithetical, operating assumptions. The National Institute was catholic both in its conception of science and its membership policy. It wanted an on-going and direct relationship to the federal government and believed that the necessary political influences would accrue from a numerically large and geographically diverse membership. The Association of American Geologists, on the other hand, was formed by men seeking to broaden their own knowledge and to further additional work in the complex and developing field of geology. They were anxious to persuade the government to sponsor appropriate scientific research but felt that their relationship should be that of separate and distinct adviser, recommending scientific projects whose scope would be established through mutual agreement of expert and financial sponsor.

The juxtaposition of the two meetings in 1844 was not coincidental. Each organization deliberately hoped to prove itself the more viable contender in a larger effort to establish American scientific integrity and unity through a national association. The key question to be resolved in the 1844 confrontation was the role which politics and the national government should play in any institution designed to promote science.

The National Institute was only one in a series of attempts to found a society for the promotion of knowledge in the national capital, ranging from a local group in 1810 to the post-Civil War Washington Philosophical Society. Other major cities possessed active scientific societies in the early nineteenth century, but all were essentially local in actual operations.[4] The immediate predecessor of the National Institute was the Columbian Institute, whose high aspirations had faded rather quickly.[5] Founded in an atmosphere of nationalism following the War of 1812 by scientific dilettantes, the Columbian Institute was an expansion of the local Metropolitan Society, transformed to national status by Congressional charter in 1818. Its objects were chiefly utilitarian, and they included establishment of a national botanical garden for agricultural reference as well as a library and a museum. In an attempt to gain patronage, the organizers of the Institute wrote a constitution which made the President of the United States a patron; many leading men in Congress, the Army, the Judiciary, and Washington professional circles were elected either honorary or corresponding members.

During its best years in the 1820s the Columbian Institute had enjoyed the active membership of John C. Calhoun and John Quincy Adams, as well as other statesmen. But the fledgling society soon faltered. Minute books of the 1830s record with increasing frequency, "Meeting adjourned without a quorum." Troubled by a paucity of

[4] The only comprehensive survey of scientific organizations is Ralph S. Bates, *Scientific Societies in the United States* (3rd ed., Cambridge, Mass.: Massachusetts Institute of Technology Press, 1965), pp. 41–46, 67–69.

[5] The most recent account of the Columbian Institute is Madge E. Pickard, "Government and Science in the United States: Historical Backgrounds [I]," *Journal of the History of Medicine and Allied Sciences*, 1946, *1*:254–265. Pickard suggests lack of funding as a primary reason for failure of the Institute. A more detailed account is by Richard Rathbun, "The Columbian Institute for the Promotion of Arts and Sciences: A Washington Society of 1816–1838, which Established a Museum and Botanic Garden under Government Patronage," Smithsonian Institution, United States National Museum *Bulletin 101* (Washington: Smithsonian Institution, 1917). The latter is based partially on the manuscript materials in the Smithsonian Institution Archives (hereafter, SIA) which contain detailed minutes for the period from 1816 to 1834. Rathbun's notes are in the Rathbun MSS, SIA, and are a valuable reference to his undocumented article.

private and public funds, the Institute's only important achievement was an embryonic botanical garden located on five acres of enclosed swampy Mall near the Capitol, granted by Congress. Only a mineral cabinet and a miscellaneous collection housed in the Capitol survived the society. Expiration of the Institute's charter in 1838 merely ended an attempt to gain federal sponsorship for a local scientific group which aspired to wider prominence.

When it was established in 1840, the National Institute for the Promotion of Science appeared to be in part a revival of the old Columbian Institute. The new society elected Joel R. Poinsett, formerly a member of the Columbian Institute, as head of an organization whose constitution closely resembled that of the older society.[6] Even the remaining property of the defunct organization was deposited with the National Institute. Although the new organization declined to assume the name of its seeming predecessor, its goals reflected antiquated and amateur views on organization of science similar to those of the earlier body.

The second quarter of the nineteenth century was a period of transition and self-definition for American scientists. One historian has called this part of an "emergent period" of professionalism, which also inspired new or revised institutional patterns.[7] Undoubtedly influenced by the contemporary thrust toward voluntarism,[8] the men of science in America displayed an increasing self-awareness and a readiness to join the scientific and philosophical societies then emerging in all major cities and in many smaller towns.[9] In addition, they gave much thought to the desirability of a national scientific organization. A dual impetus drove these men in the late 1830s: they desired an association to represent science in America internationally and they needed an organization designed to stimulate professional development and to publish reports of significant scientific activity. Because the British Association for the Advancement of Science seemed to fulfill both functions abroad, many Americans took it as a general model.[10] Furthermore, American men of science were not oblivious to popular interest in science[11] and to the self-evident financial necessity of popular approbation of science in a democratic society.

In the same years that witnessed developing ideas of scientific organization, the national government was determining its constitutional stance toward science policy. Washington increasingly assumed responsibility for projects with specific goals and as much basic research as could be unobtrusively incorporated. The Coast Survey was the

[6] The constitution of the Columbian Institute is reproduced in Rathbun, "Columbian Institute," pp. 67–70. The constitutions of the National Institute were published separately: "Constitution and By-Laws of the National Institution for the Promotion of Science, Established at Washington, May, 1840" (Washington, 1840) and "Constitution of the National Institution. . . . amended and ordered to be printed, April, 1841" (Washington, 1841).

[7] George H. Daniels, "The Process of Professionalization in American Science: The Emergent Period, 1820–1860," *Isis*, 1967, *58*: 151–160.

[8] Arthur Schlesinger, *Paths to the Present* (New York: Macmillan Co., 1949), pp. 23–50; see esp. the quotation from William Ellery Channing, p. 32.

[9] Bates, *Scientific Societies*, pp. 41–46, 67–69.

[10] *British Association for the Advancement of Science Proceedings*, 1838, *1*:44–46. A list of foreigners who attended BAAS meetings through 1837 names 40 Americans, including Samuel Dana, Robert Hare, Joseph Henry, Elias Loomis, and Henry Darwin Rogers.

[11] Two different reflections are suggested in Stephen Goldfarb, "Science and Democracy: A History of the Cincinnati Observatory, 1842–1872," *Ohio History*, 1969, *78*: 172–223, and in Wyndham D. Miles, "Public Lectures on Chemistry in the United States," *Ambix*, 1968, *15*:130–153.

private and public funds, the Institute's only important achievement was an embryonic botanical garden located on five acres of enclosed swampy Mall near the Capitol, granted by Congress. Only a mineral cabinet and a miscellaneous collection housed in the Capitol survived the society. Expiration of the Institute's charter in 1838 merely ended an attempt to gain federal sponsorship for a local scientific group which aspired to wider prominence.

When it was established in 1840, the National Institute for the Promotion of Science appeared to be in part a revival of the old Columbian Institute. The new society elected Joel R. Poinsett, formerly a member of the Columbian Institute, as head of an organization whose constitution closely resembled that of the older society.[6] Even the remaining property of the defunct organization was deposited with the National Institute. Although the new organization declined to assume the name of its seeming predecessor, its goals reflected antiquated and amateur views on organization of science similar to those of the earlier body.

The second quarter of the nineteenth century was a period of transition and self-definition for American scientists. One historian has called this part of an "emergent period" of professionalism, which also inspired new or revised institutional patterns.[7] Undoubtedly influenced by the contemporary thrust toward voluntarism,[8] the men of science in America displayed an increasing self-awareness and a readiness to join the scientific and philosophical societies then emerging in all major cities and in many smaller towns.[9] In addition, they gave much thought to the desirability of a national scientific organization. A dual impetus drove these men in the late 1830s: they desired an association to represent science in America internationally and they needed an organization designed to stimulate professional development and to publish reports of significant scientific activity. Because the British Association for the Advancement of Science seemed to fulfill both functions abroad, many Americans took it as a general model.[10] Furthermore, American men of science were not oblivious to popular interest in science[11] and to the self-evident financial necessity of popular approbation of science in a democratic society.

In the same years that witnessed developing ideas of scientific organization, the national government was determining its constitutional stance toward science policy. Washington increasingly assumed responsibility for projects with specific goals and as much basic research as could be unobtrusively incorporated. The Coast Survey was the

[6] The constitution of the Columbian Institute is reproduced in Rathbun, "Columbian Institute," pp. 67–70. The constitutions of the National Institute were published separately: "Constitution and By-Laws of the National Institution for the Promotion of Science, Established at Washington, May, 1840" (Washington, 1840) and "Constitution of the National Institution . . . . amended and ordered to be printed, April, 1841" (Washington, 1841).

[7] George H. Daniels, "The Process of Professionalization in American Science: The Emergent Period, 1820–1860," *Isis*, 1967, *58*: 151–160.

[8] Arthur Schlesinger, *Paths to the Present* (New York: Macmillan Co., 1949), pp. 23–50; see esp. the quotation from William Ellery Channing, p. 32.

[9] Bates, *Scientific Societies*, pp. 41–46, 67–69.

[10] *British Association for the Advancement of Science Proceedings*, 1838, *1*:44–46. A list of foreigners who attended BAAS meetings through 1837 names 40 Americans, including Samuel Dana, Robert Hare, Joseph Henry, Elias Loomis, and Henry Darwin Rogers.

[11] Two different reflections are suggested in Stephen Goldfarb, "Science and Democracy: A History of the Cincinnati Observatory, 1842–1872," *Ohio History*, 1969, *78*: 172–223, and in Wyndham D. Miles, "Public Lectures on Chemistry in the United States," *Ambix*, 1968, *15*:130–153.

most prominent result of this approach. But the general sponsorship of science remained an unsettled question.[12] Aware of the inadequacy of private support for research,[13] scientists were actively seeking indirect government aid on the state level, as evidenced by the state geological surveys. Many men thought money should also be made available for such semi-popular institutions as museums, observatories, and botanical gardens, which provided research possibilities as well. The National Institute, born in the mind of a politician, raised the possibility of government support, but it also raised the specter of bureaucratic political control.

A major reason for the founding of the National Institute was Congressional acceptance of the bequest of an Englishman, James Smithson, who left a half million dollars to found "an establishment for the increase and diffusion of knowledge among men" in the United States.[14] Congress deliberated over the constitutional question for several months; the final passage of a bill to accept money with such a stated purpose seemed to indicate governmental willingness to become involved with science, broadly defined. Subsequent debate centered on the problem of how to use the money, with suggestions ranging from an astronomical observatory, an objective long sought by John Q. Adams, to a renewed effort for a national university.[15] Members of the National Institute, which had been duly established under Poinsett just in time to participate in the controversy over the best use of the Smithson bequest, turned discussion toward a national history museum. They wanted it built around a nucleus of the specimens returned from the United States Exploring Expedition under Captain Charles Wilkes and those collected by David Dale Owen's western survey.[16]

Poinsett, a Southern planter and former diplomat with an avocational interest in science—the poinsettia plant bears his name—appears to have been the prime mover behind the National Institute.[17] He was chief science advocate in the cabinet of

[12] The best discussion of the attitudes of government on the national level toward science in this period is in A. Hunter Dupree, *Science in the Federal Government: A History of Policies and Activities to 1940* (Cambridge, Mass.: Belknap Press of Harvard Univ. Press, 1957), pp. 44–79.

[13] A helpful survey of private support and some discussion of its inadequacy in helping basic research effort is found in Howard S. Miller's *Dollars for Research: Science and Its Patrons in Nineteenth-Century America* (Seattle: Univ. of Washington Press, 1970), *passim.*

[14] The most detailed accounts of the early history of the Smithsonian bequest are George Brown Goode, ed., *The Smithsonian Institution, 1846–1896, The History of Its First Half Century* (Washington, D.C., 1897) and William J. Rhees, ed., *The Smithsonian Institution, Documents Relative to Its Origin and History, 1835–1899*, 2 vols. (Washington: Smithsonian Institution, D.C., 1901).

[15] The most recent analysis of the debate over the Smithsonian bequest is by Dupree, *Science in Government*, pp. 66–90. Also see Wilcomb E. Washburn, ed., *The Great Design: Two Lectures on the Smithsonian Bequest by John Quincy Adams* (Washington: Smithsonian Institution, 1965), pp. 13–41. David Madsen, *The National University: Enduring Dream of the U.S.A.* (Detroit: Wayne State Univ. Press, 1966), pp. 57–63, adds nothing new to the earlier cited accounts of the Smithsonian Institution.

[16] David B. Tyler, *The Wilkes Expedition: The First United States Exploring Expedition, 1838–1842* (Philadelphia: American Philosophical Society, 1968), p. 387. This is an excellent compendium narrative of the voyage itself but should be supplemented for post-Expedition years by Daniel Haskell, *The United States Exploring Expedition, 1838–1842, and its Publications, 1844–1874* (New York: The New York Public Library, 1942). For Poinsett's influence see George B. Goode, "The Genesis of the United States National Museum," in *A Memorial of George Brown Goode*, Smithsonian Institution, *Annual Report for 1897*, Pt. 2 (Washington: Smithsonian Institution, 1901), pp. 98–103.

[17] Son of a Southern planter, Poinsett (1779–1851) enjoyed an excellent private education as well as four years at St. Paul's, a medical school in Edinburgh. After extensive travel in Europe and western Asia, he served in the foreign service in Latin America. Active in local politics and also

Martin Van Buren. Although Congress has passed a bill authorizing funds for an exploring expedition to the South Pacific in May 1836, various problems had prevented Secretary of the Navy Dickerson from launching the expedition before Van Buren took office. Frustrated with the project, Dickerson had quite willingly transferred the task of organizing the expedition to the enthusiastic Poinsett early in 1838. Poinsett chose Charles Wilkes to head the expedition, giving him responsibility for organizational details and interfering only to require that an experienced horticulturist be taken along to secure and bring back living specimens of plant life.[18] His concern for the future deposition of the results of the expedition came precisely at the time the Smithson bequest was clearing legal tangles in England.

Several other projects gained sponsorship as the special agent of the United States, Richard Rush, sent back optimistic letters about acquiring the bequest in early 1838.[19] While others in Washington continued the initial debate between a university or research scheme and an observatory, Poinsett worked to build support for funding a national museum to be sponsored by leading statesmen and supported by the government. In December 1838 he hinted at his hopes to the elder statesman for science in the House of Representatives, John Quincy Adams.[20] By mid-1839 he also had contacted men of science through the Academy of Natural Sciences of Philadelphia. They concurred in the need to establish a national museum and urged the venerated American Philosophical Society to agree.[21] Using a Washington-based organization as sponsor, Poinsett's unabashed goal was to found a museum of display based on the results of the Exploring Expedition; the Institute emphasized saving and exhibiting the specimens, and rarely were the research possibilities of a national cabinet even mentioned. Aided primarily by a clerk in the State Department, Francis Markoe, Jr., Poinsett created a supporting society in Washington, and by May 1840 a constitution for the National Institute had been written.[22] Poinsett provided a prestigious spark for the enterprise,

---

in the cultural life of South Carolina, he became Van Buren's Secretary of War in 1837, retiring to his plantation when Van Buren left office. Of several biographies of Poinsett, Herbert E. Putnam's *Joel Robert Poinsett: A Political Biography* (Washington: Mimeoform Press, 1935) is most adequate. The library of the Historical Society of Pennsylvania has the bulk of Poinsett's papers. *A Calendar of the Joel R. Poinsett Papers in the Henry D. Gilpin Collection*, edited by Grace E. Heilman and Bernard Levin (Philadelphia: Gilpin Library of the Historical Society of Pennsylvania, 1941) offers a survey of 613 items, many in the early 1840s (hereafter, *Poinsett Papers, Gilpin Collection*).

[18] Tyler, *Wilkes Expedition*, p. 15.

[19] Rhees, *Documents*, pp. 40–66. Also see Cyrus Adler, "The Relation of Richard Rush to the Smithsonian Institution," Smithsonian Institution *Miscellaneous Collections*, 52 (Washington: Smithsonian Institution, 1910); Rush, as special agent, became a supporter of the idea of a museum (as a popular institution) coupled with public lectures.

[20] Charles F. Adams, ed. *Memoirs of John Quincy Adams, Comprising Portions of His Diary from 1795 to 1848* (Philadelphia, 1877), Vol. X, p. 57 (Dec. 8, 1838); also see pp. 112 (Apr. 8, 1839), 462 (Apr. 14, 1841), and 464 (Apr. 17, 1841) for Adams' continuing relationship to the Institute.

[21] Samuel G. Morton to John K. Kane (corresponding secretaries of the Academy of Natural Sciences of Philadelphia and the American Philosophical Society, respectively), Philadelphia, July 23, 1839, American Philosophical Society (hereafter APS) Archives. Morton notes, "Mr. Poinsett is extremely desirous that the Institutions should agree as to the plan and arrangements of a proposed National Museum, with professors who shall perform the double office of Curators and Secretaries. The Institution must be established in Washington City, in accordance with the Smithsonian Bequest." There is no evidence of any action taken by the APS.

[22] *Bulletin of the Proceedings of the National Institution for the Promotion of Science* (hereafter *Bulletin*), 1841, *1*:3. The only history of the National Institute is Pickard, *Government and Science*, pp. 265–289, which depends entirely on printed sources. A search of the Smithsonian

but in following years the State Department clerk[23] most earnestly fanned the flame of the young Institute.

Since scientific support was essential, Poinsett and his enthusiastic co-officers attempted to develop an organization which could be a mouthpiece for all science and promote all knowledge. Thus, the constitution of the Institute was broadly inclusive. It borrowed heavily from the older philosophical societies rather than from the newer specialized scientific societies. Every learned activity was to have a place: the constitution included departments of geography, astronomy, and natural philosophy; natural history; geology and mineralogy; chemistry; the application of science to the useful arts; agriculture; American history and antiquities; and literature and the fine arts.[24] Meetings were to be monthly, with a large annual meeting. Active members were to be Washington residents; but corresponding membership was nearly open-ended, with circulars distributed throughout the United States and abroad to nearly every person known to be interested in science. The obvious goal was to develop national stature by recruiting sizeable support for the aspiring Institute. The most significant indicator of the hoped-for sponsorship by the government was the creation of a directorship consisting of the "Secretaries of the Departments of State, Treasury, War, and Navy, and the Attorney General, [and the] Postmaster General." Furthermore, "the President, Vice President, or in their absence, one of the Directors in order of seniority . . . shall preside at all meetings. . . ."[25] In actual fact, Poinsett or his second-in-command, Peter Force, generally presided over meetings attended by twenty or fewer persons.

Once the National Institute had been established, the leaders began to work toward acquisition of the Smithson grant to use to fund the museum.[26] If purpose had been

---

Institution Archives as well as its branch libraries failed to uncover Richard Rathbun's manuscript history, cited in Max Meisel, *A Bibliography of American Natural History: The Pioneer Century, 1769–1865* (New York: Premier Publishing Company, 1926), Vol. II, p. 702.

[23] Adams would later describe the hard-working Markoe as the "main pillar" of the Institute, *Memoirs*, Vol. XI, p. 539 (Mar. 2, 1844). Despite a search through the Galloway-Maxcy-Markoe Papers at the Library of Congress (LC), and the Records of the State Department at the National Archives, Washington, D.C., as well as the National Institute Collection (NI MSS), SIA, Markoe (1801–1871) remains a shadowy historical figure. The best source is 35 letters of recommendation in the State Department Files, 1849–1853, and a biographical note in the Control File, NI MSS, SIA. Apparently from Pennsylvania, Markoe was appointed Head of the Bureau of Consular Affairs in 1831; by 1851 he was listed only as a clerk and possibly left the State Department when his Bureau was technically abolished in 1855. He was a son-in-law of Virgil Maxcy. The Galloway-Maxcy-Markoe papers indicate the National Institute as his chief avocational interest, coincident with his botanical collection. The records of the Institute, SIA, suggest that he often wrote several letters a day in its behalf. Perhaps as a result of his constancy he received the second highest number of votes cast for Secretary of the Smithsonian in 1846. His failure to obtain the post of Commissioner of Patents in 1843 or the Smithsonian post apparently discouraged him and by the 1850s he was no longer an active member of the Institute. Described in the letters of recommendation in the State Department as a faithful worker in a dull and tiresome office, Markoe emerges as an earnest toiler but genuinely naïve as to the expectations of men of science. He could not therefore provide the organizational leadership necessary to encourage participation by men of science.

[24] "Constitution . . . 1840," p. 9.

[25] "Constitution . . . 1841," p. 4. Also see H. C. Williams to William Barton Rogers, Washington, May 26, 1840, in Emma Rogers, ed., *Life and Letters of William Barton Rogers* (Cambridge, Mass., 1896), Vol. I, p. 170. Williams predicted, "The Society was formed under the auspices of the Secretary of War; it will be a governmental matter . . ." although temporarily run by private individuals.

[26] During this early period a "National Cabinet" in natural history was the basic proposal under Institute discussion. Markoe anticipated that overambitious expansion might

unclear earlier, the address by Poinsett at the first annual meeting in January 1841 expressly linked the National Institute to the bequest given "for the sacred purposes of increasing and diffusing knowledge among men." Stirring patriotism by reference to de Tocqueville's assessment of American science, Poinsett then shrewdly attempted to incorporate everyone's hopes for the allotment of Smithson's money:

> There can be no doubt that the National Institution, such as we contemplate, having at its command an observatory, a Museum containing collections of all the the productions of nature, a Botanic and Zoological Garden, and the necessary apparatus for illustrating every branch of Physical Science, would attract together men of learning and students from every part of our country, would open new avenues of intelligence throughout the whole of its vast extent, and would contribute largely to disseminate among the people the truths of nature and the light of science.[27]

A bill to unite the National Institute to the Smithson bequest, however, was tabled.[28] Donations of books, specimens, and curiosities from all parts of the country indicated popular interest,[29] and by March 1841 the advance collections of the Exploring Expedition were also under Institute care.[30] Cautious about pressing Congress, the leaders waited until mid-1842 before seeking a formal charter, which, incidentally, changed the name from Institution to Institute.[31]

Despite safe passage of their charter through Congress and voluminous correspondence expressing approval of the projected museum, the Institute was experiencing internal problems. Poinsett retired to South Carolina after Van Buren left office and remained unmoved by frequent requests that he return to aid the floundering organization.[32] Markoe and J. J. Abert were evidently the most persistent workers for the Institute, despite the impressive list of officers, but their departmental posts provided little political leverage and their diligent efforts were uninspired.[33] Materials, but

---

precipitate problems and noted, "It is probable that geology & mineralogy will for some time be most prominent, simply because the Gov't has a tolerable collection already in that dept." Markoe to Bache, Washington, May 19, 1840, and Hitchcock to Markoe, Amherst, Feb. 8, 1840, NI MSS, SIA. Also see Goode, "Genesis," p. 9.

[27] "Discourse on the Objects and Importance of the National Institution for the Promotion of Science, Established at Washington, 1840, Delivered at the First Anniversary" (Washington, 1841), pp. 9, 49.

[28] Pickard, "Government and Science," p. 274.

[29] *Bulletin*, 1841, *1* and *2*, *passim*. Secretarial reports at nearly every meeting opened with an account of "Donations received," which ranged from published papers and documents to zoological specimens.

[30] *Ibid.*, *1*:48 (Feb. 8, 1841).

[31] The name change was probably precipitated by a letter from Dr. Peter DuPonceau, aging but active president of the American Philosophical Society. "I would beg leave to suggest whether it would not be advisable to make some small alteration in the name of the National Institute so that it should not bear exactly the same name as the Smithsonian but one expressive of some degree of superiority. I would recommend, for instance, that of Institute, which appears to me more dignified than that of institution which is equally applicable to a school or college as to a great national establishment for the promotion of science. My idea would be to call the national establishment the 'National Institute for the Promotion of Science,' and the Subordinate one the 'Smithsonian Institution'. . . ." DuPonceau to [Markoe], Philadelphia, Apr. 1842, NI MSS, SIA.

[32] Financial problems were evident already in late 1841 when Gouverneur Kemble wrote to Poinsett asking him to return and help reunify the organization as well as aid in getting surplus revenue from the Patent Office, if the Smithsonian bequest seemed unattainable. But Poinsett resisted, stating he was too busy with family matters. Kemble, Georgetown S. C., Jan. 12, 1842. Both letters are in *Poinsett Papers, Gilpin Collection*, pp. 152–153, 156.

[33] John James Abert (1788–1863), a West Point student for three years, was a member of the Topographical Engineers after the War of 1812. He assisted Ferdinand R. Hassler in making geodetic surveys along the Atlantic Coast and

rarely money, continued to arrive in Washington with the result that "the very liberality which in continual bounty provides for its objects . . . becomes a burden when means are wanting to give the fruits of benevolence a reception and display. . . ."[34] Simply to sustain itself the National Institute desperately needed funds, and Congress alone held an adequate purse. With firm backing by either leading politicians or scientists, the Institute should have been able to attain Congressional support; its aggressive handling of affairs, however, was shifting some politicians from neutral to hostile positions.

The Institute engineered the safekeeping of the collections of the Exploring Expedition, which were threatened with dispersion as Expedition members returned home. Storage in the damp basement of the Patent Office proved unsatisfactory, and arrangements were made to move the specimens upstairs to the more physically congenial Great Hall.[35] But Henry L. Ellsworth, the Commissioner of the Patent Office, opposed the move. He resented the Institute's presumption in using his facilities and especially this new attempt to take complete and "uncontrolled possession" of his hall designated for exhibiting mechanical patents.[36] In addition, the Whigs had gained the White House and they had less interest in the Exploring Expedition than had their predecessors. The new administration inhibited expansive hopes for preserving and publishing results.[37] Nor was there a cabinet officer under presidents William Henry Harrison and John Tyler with the enthusiasm for science of Poinsett. While not openly hostile, many politicians were simply apathetic toward science and the aspiring organization designed to promote it. Others were intent on securing the Smithson bequest for different projects, now including a national library.[38]

Additional conflicts arose when Charles Wilkes returned to Washington and found that the National Institute had assumed responsibility for his collections and was using the initial funds appropriated for the preservation of specimens to hire its own curator, not a member of the scientific corps.[39] Nonetheless, he agreed to present some results

---

topographical surveys throughout the eastern United States. From 1834 to 1861 he was chief of the Topographical Bureau, which he developed into a full-fledged corps. Adept in his own department, Abert was primarily interested in supporting the Institute as an agency to preserve the materials being collected in the western surveys. *Dictionary of American Biography* (DAB), Supplement 1 (New York: Charles Scribner's Sons, 1944), pp. 2–3. Also, William Goetzmann, *Army Exploration and the American West* (New Haven: Yale Univ. Press, 1959), pp. 9–10.

[34] *Bulletin*, *3*:332 (Dec. 28, 1843).

[35] Henry King to the Committee of the National Institute, Washington, June 19, 1841; H. L. Ellsworth to Col. Force and Col. Abert, Washington, June 21, 1841; and J. J. Abert and Peter Force to Daniel Webster, Washington, June 24, 1841, NI MSS, SIA.

[36] J. J. Abert to Peter Force, [Washington], July 10, 1841. NI MSS, SIA; Henry Ellsworth to Daniel Webster, Patent Office, Apr. 18, 1842, Benjamin Tappan MSS, LC.

[37] Charles Wilkes, "Autobiography," Wilkes MSS, LC, Vol. VI, pp. 1373–1374; VII, 1472–1473. Wilkes' comments are retrospective but candid, and he notes that the Whigs were reticent to continue the publication project and it was "in a measure a party question." Wilkes felt especially harassed because he had also faced a court martial. Perhaps it was partially as a result of this shift that the National Institute became preoccupied with politics and planned a series of lectures to coincide with the extra sessions of Congress. H. D. Williams to W. B. Rogers, Washington, Apr. 1, 1841, Rogers MSS, Massachusetts Institute of Technology Archives, Cambridge, Mass.

[38] Madge E. Pickard, "Government and Science in the United States: Historical Backgrounds [II]," *J. Hist. Med.*, 1946, *1*:467–477.

[39] Wilkes, "Autobiography," Vol. VI, p. 1395: "I had no idea that the Expedition should have all its results appropriated to the foundling National Institute. . . . I offended many of the members but this was of little consequence to me." Henry King, who had previously done work for the geological survey of Missouri, was appointed curator by the Institute.

of the Exploring Expedition in June 1842, in a series of three evening sessions under Institute auspices. But this did not signify support for Institute aspirations; it meant only an opportunity to express his forceful opinions on the work still to be done in relating the results of the Expedition.[40] Wilkes became friends with Judge Benjamin Tappan of Ohio, an avocational conchologist, who was chairman of the Joint Library Committee in Congress.[41] When an appropriation bill for publication of the results passed, Tappan saw to it that in addition to publication, the transportation, preservation, and arrangement of the materials also became the responsibility of the Joint Committee.[42] The decision effectively excluded the National Institute, for Wilkes intended to employ his own scientific corps for these tasks where possible. Friends of the corps applauded the coup; George Ord of the Academy of Natural Sciences of Philadelphia editorialized that if the Institute hoped to obtain community support, "They must give evidence that their object is really the promotion of science, and not self-aggrandizement."[43] In a similar tone Boston naturalist A. A. Gould reported triumphantly to James Hall: "As to Washington, all things are in the best possible hands. King, Conrad and the whole posse are displaced—the whole thing is taken out of the hands of the Institute—everything is under lock and key—Pickering has supreme control with power to appoint all his *subs*."[44] Entirely unintentionally the Institute had aroused the opposition of men of the Expedition, who felt threatened by the Institute's activities.

Wilkes wanted to produce significant results, and he felt that his scientific corps was the best qualified to work with the collections. Clearly the National Institute could make no significant counterclaim to such expertise, even in the arranging of the collections. In fact, the corps had returned to find that someone, perhaps Curator Henry King of Missouri, had dried pickled specimens, run them through with pins, and

[40] Tyler, *Wilkes Expedition*, pp. 373–374. Also see Wilkes' "Autobiography," Vol. VI, pp. 1388–1891, for Wilkes' own account of the abrasive political encounter at the first session. The lectures were later published with no reference to Institute sponsorship. More than 400 persons attended the first session; see Markoe to DuPonceau, Washington, June 22, [1842], NI MSS, SIA.

[41] Benjamin Tappan (1775–1857) was an antislavery Democrat of Ohio who had served as a lawyer and judge and as a Senator after 1836. In the Tappan MSS, LC, there is a small file marked "conchology" with letters from Thomas Say, Isaac Lea, and John G. Anthony, indicating that Tappan was an active field worker in the 1830s. *DAB*, Vol. IX, pp. 300–301.

[42] Wilkes clearly found Tappan his most able supporter in Congress. See Wilkes to Tappan, New York, Aug. 25, 1842, and Aug. 28, 1842, Tappan MSS, LC. Tappan was seconded by Joseph P. Couthouy, a member of the scientific corps who had retired to business in New York. Couthouy suggested that there was unanimous support "among *really* scientific men" that the publication be "confided exclusively and of right to the naturalists themselves" as the most competent for the task, adding that the Society of Natural History at Boston concurred in his opinion. Couthouy to Tappan, New York, Jan. 5, 1843, Tappan MSS, LC. Specific suggestions as to how the appropriations should be established were made by Wilkes and were closely followed in Tappan's proposals and arguments.

[43] Ord to Titian R. Peale, Philadelphia, Mar. 16, 1843, Peale MSS, Historical Society of Pennsylvania, Philadelphia.

[44] Gould to Hall, Boston, Feb. 20, 1843. Hall MSS, N.Y. State Library, Albany. Also see Dana to Redfield, Washington, Feb. 1, 1843, William Redfield MSS, Beinecke Library, Yale University (hereafter BYA). Charles Pickering (1805–1878), with an M.D. from Harvard, had settled in Philadelphia and became curator of the Academy of Natural Sciences. He was the zoologist on the Exploring Expedition and worked with Dana on returning to Washington, where he published the report on "The Races of Man and Their Geographical Distribution" (*DAB*, Vol. III, p. 562).

made reference practically impossible by losing labels.[45] Against such a display of practical ineptitude, Abert's goal of nationalism had a superficial echo:

> We digested a scheme in which we thought all persons could enter because it was national; which all parties could befriend, because it was national; to which all conditions and branches of service could contribute, because it was national; to which government might extend its patronizing hand, because it was national....[46]

However symbolic and virtuous the founders of the National Institute had conceived it to be, it was not yet of sufficient scientific calibre to attain support either from leading men of science or the general public.

When the Institute officers made an inquiry about the loss of responsibility for the potential museum materials, Judge Tappan took them on in a verbal and written debate, ridiculing the idea that the results of a government expedition should be placed in the hands of a "private corporation." He unfairly suggested that they sought to supervise publication as well as take physical responsibility for the collection. Markoe's and Abert's request for a hearing only succeeded in irritating Tappan. He sarcastically dismissed Markoe as a "clerk in one of the public offices" and castigated Abert as "the head of a bureau," who "supposed they could command government appropriations."[47] Tappan's attack was a serious blow to the public image of the Institute, but in 1842 political rhetoric alone could not undermine the society. If it could gain support for the museum enterprise from men of science, Congress might yet be persuaded to assist financially.

In contrast to the distinctive impressions being fostered by the National Institute in its monthly meetings as well as in its request to Congress, the American Association of Geologists appears academic and narrow. It had been founded in April 1840 by men from several state geological surveys who had only a long-range hope of creating a national organization with more than geological interests.[48] Earlier attempts to establish a geological society had been unsuccessful, but now a significant number of professional geologists made possible another try, for within established scientific societies in New Haven, Boston, and Philadelphia, geology was in the ascendant.[49]

[45] Haskell, *Publications*, p. 7. Miller, *Dollars for Research*, pp. 13–14. Titian R. Peale to John F. Frazer, Washington, May 15, 1844, cited in Haskell, reported the comic tragedy: "my two birds (male and female) made into one,—the legs of one put on another body,—hundreds of fine insects put in 'families' without localities, although they came from all parts of the world,—bows in one end of the room—arrows in another with their ends sawed off to make them fit into fancy stands &c.—all for the great end,—the promotion of science."

[46] Abert to Wilkes, Sept. 1843, quoted in Pickard, "Government and Science," p. 287.

[47] For details relating to the Institute's side of the case see the letters reprinted in *Reply of Col. Abert and Mr. Markoe to the Hon. Mr. Tappan, of the United States Senate* (Washington, 1843).

[48] There is no analytical study of this association—whose own published proceedings are intermittent. A guide to secondary literature is in Bates, *Scientific Societies*, p. 267, and to contemporary literature in Meisel's older but excellent *Bibliography of American Natural History*, Vol. II, pp. 680–698. Some of the manuscript records of the Association of American Geologists and Naturalists are at the Academy of Natural Sciences, Philadelphia. An indication of the initiative taken by the N.Y. Geological Survey is found in John M. Clarke's *James Hall of Albany, Geologist and Palaeontologist, 1811–1898* (Albany, 1921), pp. 101–103. Edward Hitchcock stressed the early hope for a larger organization in his *Reminiscences of Amherst College, Historical Scientific, Biographical and Autobiographical*... (Northampton, Mass. 1863), pp. 368–372.

[49] Edward Hitchcock, "First Anniversary Address," *American Journal of Science and Arts*, 1841, *41*:234; Silliman's "Address before the Association of American Geologists," *Am. J. Sci.*, 1842, *43*:236 suggests that a "new order of

Scientific necessity provided the original impetus for their meeting. In search of a common American nomenclature for Palaeozoic formations, geologists representing state surveys from Michigan to Massachusetts agreed to meet and to discuss current problems, in extension of the annual caucus of the New York state geologists.[50]

Experiencing both "sanguine hope" and "timid doubt," the group originally summoned by invitation from Lardner Vanuxem of the New York survey was small and included only participants or former participants in state geological surveys.[51] Some high expectations accompanied the inauspicious birth of the Association. Already in 1838 Amherst professor Edward Hitchcock, respected geologist and head of the Massachusetts survey, had written to fellow workers about forming a national organization.[52] He obviously intended that a series of meetings of specialists would in time lead to an organization similar to the British Association for the Advancement of Science (BAAS), an organization in which geologists in the 1830s were strikingly influential.[53]

Finding the three-day meeting in the rooms of the Franklin Institute helpful, the geologists agreed to meet again the following year in Philadelphia before the field season began. Benjamin Silliman, Sr., eminent chemist at Yale University, was unanimously elected to preside over the second meeting, and each of the original members was allowed to invite six scientific friends to participate, thus extending membership beyond state geologists.[54] In subsequent years, following the example of

---

professional talent is now called into action." Also Edward Herrick to Dana, New Haven, June 22, 1841, Dana MSS, BYA.

[50] The basic problem of uniform nomenclature in natural history was long-standing. See Jacob Bigelow to Silliman, Boston, Mar. 2, 1818, Silliman Papers, BYA. Also see M. B. Anderson, "Sketch of the Life of Prof. Chester Dewey," *Proceedings of the Fifth Anniversary of the Convocation of the State of New York* (Albany, 1869), pp. 125–129.

[51] Hall, "Presidential Address," *American Association for the Advancement of Science Proceedings*, 1856, *10*:231–232. There is some disagreement as to precisely who attended the first meeting; but all sources agree that each founding member was or had been in some capacity connected with a geological survey. A typescript in the Hall MSS, N.Y. State Library, Albany, corrects the *AAAS Proceedings*, 1848, *1*:144–156, noting that James Hall and Bela Hubbard had attended, but not Caleb Briggs, Charles T. Jackson, or Solomon Roberts. Handwritten minutes at the Academy of Natural Sciences, Philadelphia, add Charles F. Jackson of Boston; J. T. Ductel, Baltimore; James B. Rogers, University of Virginia; C. Briggs, Jr., Columbus; and William Horton, Craig Hall, N.Y. The initiative was taken by the N.Y. geologists, but Hitchcock's presidency seems to lend credence to his claim of first stimulator.

[52] Hitchcock to William B. Rogers, Apr. 4, 1838, printed in Rogers, *Life and Letters*, Vol. I, pp. 154–155; Hitchcock, *Reminiscences*, pp. 368–372. The latter volume is the best account of Hitchcock (1793–1864). A Congregational minister, he became interested in geology and produced the first surveys in Massachusetts and Vermont, as well as beginning work on the First District in New York. He embodied the mid-century American belief in the compatibility of religion and science, and in his later years as president of Amherst College wrote several papers on natural theology. At Amherst he seems to have felt especially isolated and sought the companionship of scientists at Boston and New Haven whenever possible. In the letter cited above he mentioned that he had spoken of the project to several New England men of science—including Benjamin Silliman, Charles U. Shepard, George B. Emerson, and Charles B. Adams—who endorsed the plan.

[53] The BAAS, founded in 1831, was in part an answer to the exclusive and autocratic policies of the Royal Society of London, as well as a positive attempt to unite provincial scientific societies into a truly national organization. The standard history of the BAAS is O. J. R. Howarth, *The British Association for the Advancement of Science: A Retrospect* (London: The Association, 1922). For one view of the founding see L. Pearce Williams, "The Royal Society and the Founding of the British Association for the Advancement of Science," *Notes and Records of the Royal Society*, 1961, *16*:221–233.

[54] Hitchcock to Silliman, Amherst, Apr. 26, 1840, Amherst College Archives, Amherst, Mass. He adds, "I did not expect to see you at Phila-

the BAAS, the Association of American Geologists met in various urban centers: Boston in 1842, Albany in 1843, and Washington in 1844. Meetings were held in the spring, usually in April, before the geologists began their summer field work. Because the organization was small, funds for publishing proceedings and reports were elusive. Not until the third meeting in Boston was sufficient money acquired, chiefly through the generous donation of private citizens, for publication of papers presented at the meetings.[55] Evidence of Association activity reached the public, however, through other publications, especially the *American Journal of Science*, edited by Silliman. Despite the financial frustration, the participating members found the reports, the sometimes heated discussions, and the fellowship proved valuable. As a specialized society for geologists, the Association was established and creditable by 1844; it in itself accomplished the end of members' mutual edification and acquaintance.[56] Elected presidents were men of established reputation: Hitchcock, Silliman, and two Philadelphians, Samuel George Morton and Henry Darwin Rogers.

The founders considered the third meeting of the Association a milestone. Forty scientists convened in the hall of the Boston Natural History Society to discuss geology, and among their number was Charles Lyell, distinguished uniformitarian geologist, visiting from England. A crowd of nearly five hundred persons gathered to hear Silliman's presidential address on the second evening of the meeting.[57] The city of Boston, with a citizenry already attuned to popular science through the Lowell lectures, held a reception for the members. Here Benjamin Silliman, Jr., successfully hinted to Nathan Appleton of the need to publish scientific proceedings; the resulting popular subscription for the first (and only) volume of *Reports* provided prestige to the fledgling organization.[58]

Its activities brought the Association closer to its general model, namely the peripatetic meetings, the popular evening sessions, and the committee reports on the "state of knowledge" in specific areas of science. Also in 1842 the geologists added

---

delphia as I knew that the invitation which was given by the New York State Geologists was confined to those who are engaged in the state surveys. But the business is so arranged now that I trust we may hope to have you preside over the next meeting." Also see Hitchcock to Markoe, Amherst, Nov. 13, 1840, Galloway-Maxcy-Markoe MSS, LC.

Benjamin Silliman, Sr. (1779–1864) was a Yale graduate and professor of chemistry there for nearly the entire first half of the 19th century, also building one of the finest contemporary collections in mineralogy and geology. A willing popular lecturer, he inaugurated the Lowell Institute lectures. Clearly his most important contribution was his founding of the *American Journal of Science and Arts* in 1818, which for three decades was the principal publication in the natural and physical sciences. *DAB*, Vol. IX, pp. 160–163.

[55] The result was *Reports of the First, Second, and Third Meetings of the Association of American Geologists and Naturalists at Philadelphia, in 1840 and 1841, and at Boston in 1842* (Boston, 1843). The sluggish sales of the *Reports* unfortunately only demonstrated the low demand for scientific journals and the difficulty of self-supporting publications.

[56] W. B. Rogers to J. W. Bailey, Univ. of Virginia, Oct. 22, 1843, Rogers Papers, MIT Archives. "For us such reunions of the scientific brethren as our Association of Geologists are of precious value and form the best compensation we can enjoy for the prolonged restraints of our vocation [teaching]."

[57] Silliman, Jr., to Gideon Mantell, New Haven, May 14, 1842. Silliman-Mantell MSS, BYA.

[58] *Ibid.;* Silliman, Jr., to Appleton, Yale College Cabinet, May 5, 1845. Appleton MSS, Mass. Historical Society, Boston.

Benjamin Silliman, Jr. (1816–1885), after graduating from Yale in 1831 essentially followed the pattern of his father in assuming a faculty position at Yale and in editing the *Journal.* Although he is less important than his father in the history of science in America, he did play a crucial role in the young Association of American Geologists and Naturalists.

"Naturalists" to their official title and voted to permit membership to individuals belonging to any other scientific society who applied.[59] Although the Albany meeting was less popularly attended, the opportunity to see James Hall's growing collections of mineralogy and palaeontology brought full participation by the geologists themselves.[60] Earth scientists recognized the contribution of the new group, and public attention was attracted through the migratory meetings. In contrast to the earlier, more local geological societies, the Association of American Geologists and Naturalists was both national and moderately successful. The deliberate gradualism of its development shielded it from criticisms like those being directed against the more aggressive National Institute. The Association was not without internal dissension, but the debates concerned critical matters of scientific priority and the validity of a new hypothesis or description.[61] The subsequent encounter with the National Institute would emphatically demonstrate that membership loyalty was strong and cohesive.

The objectives of the National Institute and of the Association of American Geologists seemed clearly distinct when each was formed in the spring of 1840. But as the museum and scientific center in Washington appeared possible only with nationwide approval of men of science, the leaders of the Institute looked more closely at the BAAS and began to hint of similar goals. From the beginning the geologists had been aware of the BAAS model, hoping that they might develop from a specialized to a more general scientific society. The National Institute apparently envisioned the massive crowds which thronged to hear speakers like Louis Agassiz, David Brewster, and Charles Lyell,[62] while the Association of Geologists and Naturalists believed that in the specialized section meetings and the helpful reports on research needs for science there was potential growth for science itself. Both recognized that the BAAS was becoming a spokesman for the general scientific community in England. The attempt to fulfill this role provided the chief basis for competition, for there could be only one such authority in a nation. It was further evident that the composition of the organization would affect the relationship between science and the government it advised.

Francis Markoe heard of the new Association late in 1840 and solicited its support for the National Institute, suggesting that the next annual meeting be held in Washington in 1841.[63] It was too late to change plans, but Hitchcock, as initiator of the geological society and its first president, responded by inviting Markoe to attend the

[59] Lewis C. Beck to Markoe, Rutgers College, Jan. 5, 1843, NI MSS, SIA.

[60] The proceedings for 1843 were never published separately but abstracts may be derived from the *Albany Daily Argus*, Apr. 29, May 1, 3, and 4, 1843; also in Silliman's *Journal*, Jul., Oct., 1843, *5*: 135–165 and 310–353. Redfield to Hitchcock, N.Y., Nov. 15, 1842, Redfield MSS, BYA. Henry Rogers reported to William Rogers, Albany, Apr. 30, 1843, "There prevails an excellent spirit, and there has thus far been decidedly more solid work performed than last year. . . . The attendance at our discussions is very slender, this being a city of almost no taste for such matters. . . ." Rogers, *Life and Letters*, Vol. I, p. 222.

[61] *Report*, . p173.

[62] Markoe's brother-in-law Capt. George W. Hughes, a topographical engineer for the Army, attended the BAAS meeting at Glasgow in 1840 and his enthusiastic account was reprinted in the Institute's *Bulletin*, 1840, *1*:33–42. He particularly noted that recommendations from the association concerning scientific projects "are always received with great respect and attention, and are generally adopted" by the governments.

[63] Hitchcock to Silliman, Sr., Amherst, Dec. 11, 1840, Hitchcock MSS, Amherst College Archives. On Aug. 25, 1841, Markoe sent to C. F. Jackson, as secretary, a formal invitation to meet in Washington, adding that the Institute "regards with deep interest the recent formation of the Association of American Geologists." Draft copy, NI MSS, SIA.

coming session in Philadelphia as his guest.[64] The Institute's interest in the new society was not surprising, for along with its political overtures, its corresponding secretary had been establishing scientific contacts through circulars to societies and personal letters to individuals.[65] Scientific membership was as essential for credibility as political leadership seemed to be for government patronage.

Scientists generally, including members of the Association, greeted the initial announcements of the Institute favorably, assuming from public statements that its major goal was to sponsor a museum of natural history. William Barton Rogers responded to a letter seeking his advice with an offer to participate in lectures given by the Institute.[66] Struggling to build a small cabinet at Amherst, Hitchcock was enthusiastic about the plan of starting a "great collection in Natural History at Washington, embracing the whole country and the World."[67] Benjamin Silliman assured Markoe of his good will and offered the use of his *Journal* for "occasional annunciations of the proceedings of the society."[68] From Philadelphia Alexander Dallas Bache also agreed to the need for a national collection and suggested solicitation of specimens from state geologists, diplomatic agents, and various strangers visiting Washington.[69] Long and detailed letters from Peter DuPonceau, president of the American Philosophical Society, were encouraging and replete with suggestions.[70]

By late 1842, however, the tenor of response was changing. Perhaps fostered by the rift with the scientific corps and Wilkes, by the political stance of Markoe and the Institute leadership, or by the ever-expanding aspirations of the Institute, skepticism among the scientists was increasing. Mirrored in the private correspondence relating to the topic of the Institute from 1842 through early 1844 were several fears: the United States was not yet ready for such an extensive scientific organization; the Smithson bequest must give much-needed funding for the *promotion* rather than the diffusion of knowledge; government support must in no way imply governmental

[64] Hitchcock to Markoe, Amherst, Nov. 13, 1840, Galloway-Maxcy-Markoe MSS, LC. "I am glad to find that you have got the plan matured so easily too—I trust that it will be the commencement of a noble Institution for our country."

[65] *Bulletin*, 1840, *1–3*; *passim.* Markoe sent circulars to all known local scientific and historical societies in the United States, as well as to leading foreign groups which might be interested in the exchange of specimens and publications.

[66] W. B. Rogers to H. C. Williams, Univ. of Virginia, Sept. 8, 1840, *Life and Letters*, Vol. I, p. 171. The four active Rogers brothers, James Blythe (1802–1852), William Barton (1804–1882), Henry Darwin (1808–1866), and Robert Empie (1813–1884), born to Dr. Patrick Kerr Rogers, an Irish immigrant, inherited their father's interest in science. Henry, who had worked on the New Jersey and then on the Pennsylvania state geological surveys was sought as an essential participant in the new society by the other founders. With his brother William, then at the University of Virginia, with whom much of his geological work was done, Henry helped sponsor the first Association meeting at his home base in Philadelphia. The two were interested in establishing a polytechnic school, a goal realized in the Massachusetts Institute of Technology, at which William was the first president. Both were active administrators of science and Henry's familiarity with English scientific organization led them to leadership positions in the new Association; Henry was chairman in 1843; William in 1845 and 1847. *DAB*, Vol. VIII, pp. 99–100, 115, 94–95, and 99–100.

[67] Hitchcock to Markoe, Amherst, Feb. 8, 1840, NI MSS, SIA.

[68] Silliman to Markoe, New Haven, Aug. 27, 1840, NI MSS, SIA.

[69] Bache to Markoe, [Philadelphia], May 26, 1840, NI MSS, SIA.

[70] Many affirmative letters from DuPonceau are quoted at length in the first three *Bulletins*, for as president of the American Philosophical Society his support was deemed crucial: *1*:10–13 (Nov., 1840); *2*:90–92 (May 6, 1841); *2*:204–208 (Apr. 1842). The NI MSS, however, reveal a changing attitude toward the Institute, obviously not recorded for public consumption in the *Bulletin*, but indicating the general attitudes prevalent in Philadelphia.

control over scientific projects; and scientific organization must be evolved in response to and on a pattern of need for scientific intercourse, not placed into a bureaucratic superstructure. Overriding these concerns was scientific resentment of the assumption of leadership by demi-savants and politicians.

Awareness that the Institute sought to sponsor more than a museum came when John C. Spencer, chairman of an Institute committee, issued a circular giving November 28, 1842, as the date for a meeting to discuss organizing a national scientific convention. The circular brought mixed response, but nearly all respondents suggested that the notice was too short to allow attendance.[71] James Hall also commented that two such national meetings could not be sustained and that the annual geologists' meeting was already planned for Washington in April of 1844.[72] Although Markoe claimed that this new proposal was not intended to interfere with the proposed Association meeting, suspicions were aroused.

When the "Second Circular" appeared in early 1843, the doubts were proven well-founded. Without crediting Hall, Spencer noted the disadvantage of two large meetings and "after mature reflection" assumed for the Institute the expanded role of host by "respectfully inviting to Washington, *in the name of the National Institute* the members of the American Philosophical Society . . . of the Association of American Geologists and Naturalists, and the members of all other scientific and learned societies . . ." during the first week in April 1844.[73] Knowing the plans of the Association to meet in Washington, Institute action could only be interpreted as an attempt to upstage the Association or to absorb it completely.

Earlier opposition to any organization resembling that of the BAAS had stressed the unreadiness of the United States for a general organization, noting the probability of charlatanism which thrived on naïve popular support. After attending a meeting of the BAAS in 1837, Joseph Henry had recoiled from any suggestion of a similar association in the United States, wryly commenting that "a promiscuous assembly of those who call themselves men of science in this country would only end in our disgrace."[74] And the success of the geologists did not change his thinking as to

[71] E.g., see Redfield to Spencer, N.Y., Nov. 12, 1842; S. S. Haldeman to Spencer, Franklin Institute, Philadelphia, Nov. 22, 1842, both in NI MSS, SIA.

John C. Spencer (1788–1855), a Union College graduate, was a New York Whig appointed Secretary of War by President Tyler, then Secretary of the Treasury (to May 2, 1844). Unfortunately irascible when the times called for a man of compromise, he was apparently the only consistent friend of the Institute in Tyler's cabinet and left office in March 1843. *DAB*, Vol. IX, pp. 449–450.

[72] Hall to Spencer, Albany, Nov. 28, 1842, NI MSS, SIA. Also see Markoe to Lewis C. Beck, Washington, Jan. 7, 1843, NI MSS, SIA. The Association minutes are ambiguous about the reason for choosing Washington as a place of meeting, recording only a resolution that the "secretary was requested to answer the communication from the National Institute at Washington, inviting the association to meet in that city." *Report*, p. 174.

James Hall (1811–1898), who had studied under Amos Eaton at Rensselaer Polytechnic Institute, was becoming the leading American palaeontologist. He had charge of the Fourth District of the N.Y. Geological Survey and later was titled state geologist of Iowa and Wisconsin. The best account of Hall is the biography by John C. Clarke (n. 48 above).

[73] "Second Circular," Feb. 24, 1843 (italics mine). The three circulars are reprinted in the *Bulletin*, 1843, *3*:420–425.

[74] Henry to M. De La Rive, Princeton, Nov. 12, 1841, Henry MSS, SIA. Joseph Henry (1797–1878), a self-taught physicist, was, by his work on electromagnetism, already recognized as a leading American scientist. He taught at Albany Academy and was Professor of Natural Philosophy at the College of New Jersey at Princeton from 1832 to 1846. The biography by Thomas

the expediency of a society for all branches of science, for "we have among us too few workingmen and too large a number of those who would occupy the time of the meeting in idle chatter."[75] Similarly the attempt by John C. Warren to establish such an organization in 1838 had brought a negative response from the American Philosophical Society,[76] as well as from individual scientists like John Torrey, who noted, "we can hardly get up yet an association for the foundation of science. There is indeed too much charlatanism in the country, and enough to overpower *us modest men*."[77]

The theme was maintained by another perceptive commentator on American science, Alexander Dallas Bache, who noted of the Institute leadership: "[they] are worthy men with good ideas about matters generally, but have not examined closely into the state of science in our country & its wants. I have done my best to discourage the idea of a great *scientific meeting*, to represent American science ... they want something more popular, in my view."[78]

Not surprisingly, younger men like Henry and Bache held this negative view of popular effort in conjunction with the hope for a research institution for professionals. Again Bache emphasized the limitations on professionals when he wrote:

> The organization of an institution for the promotion of science must necessarily be a work of time, and one requiring considerable expenditure. There are very few positions now attained by scientific men in the United States where there is sufficient leisure for research, and the very miscellaneous nature of the duties required of our men of science is not at all favorable to the permanent reputation either of themselves or of their country.[79]

Clearly wanting a research establishment, he noted that Smithson had been an active member of the Royal Institution of London, which sponsored basic research.[80] To men like Bache there was a danger that the Smithson bequest might be "perverted" by the establishment of a museum, which basically preserved the known, and thus destroy the possibility of an institution for the advancement of science. But this view was more sophisticated and more farsighted than that of most individual scientists. In fact, many of the geologists, older and self-educated in science, were dependent on amateurs for field work and were less overtly skeptical of semi-savants.

One of the strongest reactions to the expanding aspirations of the Institute was voiced by the previously encouraging Peter DuPonceau. Commenting on the "Second

---

Coulson, *Joseph Henry: His Life and Work* (Princeton: Princeton Univ. Press, 1950) will be well supplemented by the new, definitive edition of Henry papers currently being prepared by Nathan Reingold at the Smithsonian Institution.

[75] Henry to Bache, Aug. 9, 1838, reprinted in Nathan Reingold, ed., *Science in Nineteenth Century America* (New York: Hill and Wang, 1964), p. 88.

[76] Edward Warren, ed., *The Life of John Collins Warren, M.D., compiled chiefly from his Autobiography and Journals* (Boston, 1860), Vol. I, pp. 339–340 and Vol. II, pp. 1–2.

[77] Torrey to Henry, Nov. 9, 1838, in Reingold, ed., *Science in Nineteenth Century America*, p. 91.

[78] Bache to Loomis, Washington, Mar. 7, 1844, Loomis MSS, BYA. Alexander Dallas Bache (1806–1867), a graduate of the technical program at the Military Academy at West Point, had taken a trip to Europe to assess educational institutions but had also become familiar with scientific institutions. Much of his leisure time in Philadelphia was spent working on magnetical and meteorological observations. As a result of his activity and obvious administrative talent, he was appointed head of the U.S. Coast Survey in 1843, a post he held until his death. M. M. Odgers, *Alexander Dallas Bache: Scientist and Educator, 1806–1867* (Philadelphia: Univ. of Pennsylvania Press, 1947); Nathan Reingold, "Alexander Dallas Bache: Science and Technology in the American Idiom," *Technology and Culture*, 1970, *11*: 163–177.

[79] Bache to Markoe, Philadelphia, July 21, 1841, NI MSS, SIA.

[80] *Ibid.*, Oct. 2, 1841, and Dec. 12, 1841, NI MSS, SIA.

Circular," which specifically implied the founding of an American association like that of the BAAS, he warned that such an association must not be too closely allied to politics. Rather, he suggested that a scientific meeting should be without entangling alliances:

> . . . in Europe, even in the most despotic governments, the learned are left to themselves to settle their own matters as they think best. . . . To this preliminary meeting being held at Washington I cannot reasonably object; but that it should be held during the session of Congress in the pestilent atmosphere of politics by which they could hardly avoid being contaminated appears to be greatly objectionable. . . . In leaving these matters to them [the scientists] the Institute I am convinced will be desirous to avoid the suspicion of their wanting to dictate, and to make the Association subservient to their views. I know that it is very far from their intention, but the higher they are placed, the more they should be careful of preventing suspicions which might be generated by jealous feeling. Their being also so near, and so nearly connected with the National Government, where principal officers are placed at their head, the dangers of their being involved in the affairs of political parties who succeed each other so rapidly in this country, is not perhaps so imaginary as might be supposed.[81]

DuPonceau's intensive fear for scientific autonomy reflected a Jacksonian distrust of governmental interference; apparently this was becoming a prevalent fear among the American Philosophical Society members. George Ord of the Academy of Natural Sciences also expressed a doubt about alignment with politics, fearing that with each appropriation, "a host of vagabonds . . . will rush forward, thrust aside modest men of merit, and obtain the prize."[82]

Markoe was warned of the dissatisfactions by Peter A. Browne of Philadelphia, who outlined the pending opposition from both the Association and the Philosophical Society, suggesting that the latter was especially concerned that a meeting in Washington would "fall into the hands of some political party or be made the tools of some political aspirant."[83] Browne's own skepticism of the somewhat over-cautious conclusion may have prevented Markoe from understanding that men of science valued their autonomy and felt inherent danger in becoming (or seeming) too involved with party politics. Many sincerely believed that a system of political patronage might force ill-trained men of scientific organizations and direct research into unworthy channels.

The intensity of response by the Association of American Geologists and Naturalists might be partially explained as defensive; but, in addition, their own success simply sustained a belief that the best tactic was slow development into a broader organization. Individually and as an organization the Association members had not been unfavorable to a large national institution, perhaps a museum, located in the capital. This attitude had changed as reports from the scientific corps made it clear that control

[81] DuPonceau to Markoe, Philadelphia, Nov. 8, 1842, NI MSS, SIA. Similar commentary came from the respected Samuel L. Dana from Lowell, Mass., who wrote to Spencer on Nov. 16, 1842, suggesting that there was not yet enough *esprit de corps* among scientific men to promote peripatetic meetings. Then he added, "There is danger also partly by the attractions which the deliberations of Congress may offer that the true object of the scientific meeting will be lost in the stormy sea of politics, on which even the quiet hermit of science, will be but too ready to embark when once drawn to its shores." NI MSS, SIA.

[82] Ord to Titian Ramsey Peale, Philadelphia, Mar. 16, 1843, Peale MSS, Historical Society of Pennsylvania, Philadelphia.

[83] Browne to officers and members of the National Institute, Philadelphia, Jan. 16, 1843, NI MSS, SIA. Philadelphia was a center for dissenting opinions, as indicated also by a letter from A. D. Chaloner to Thomas K. Townsend, Philadelphia, Feb. 12, 1844, NI MSS, SIA.

under nonscientific auspices was less than desirable. The examples of scientific ineptitude, coupled with the challenge raised by the circulars to the Association meeting planned for Washington in 1844, fostered direct opposition to the National Institute.

The leading members of the Association were not slow to observe the new directions of the Institute. Hitchcock wrote to William Redfield, a New York meteorologist, "What do you say of the movement at Washington to get up a scientific meeting: Is it intended to interfere with ours? or if it be successful will it not interfere?"[84] In turn, Redfield wrote to James Dwight Dana of the prevalent misgiving that "this design may interfere, intentionally or otherwise" with the Association, which was well established and with an "organization as extensive as can well be carried out."[85] Dana's response was a scathing, explosive indictment of the scientific pretensions of the Institute. Deeply involved in Washington affairs since his return from the Expedition, he noted the conflict between the Institute and Washington leadership:

> But an end has come to all their Castle-Building. The Society is now in a bad condition and unless it receives aid from the government, of which there is little if any hope, they will soon be on their way down hill and it will be a rapid slide, with a smash to nothing, at the bottom. . . . Besides all this, there is quarreling among themselves, between the Patent Office members and the others, which augurs no good. . . . Such is the great National Institute.—But I am talking too plainly with my pen, and I wish you would burn this when you have read it.[86]

Dana's candid commentary touched the core problems of amateurism and dependence on governmental support, but he underestimated the tenacity of the Institute when he went on to predict, "There is no probability that the General Meeting of the Institute . . . will take place—none at all."[87] Naturalist Lewis C. Beck wrote directly to Markoe expressing regret that "anything should be done to interfere with this Association at least until the experiment as thus got up, should have been fairly tried," adding that time and expense would prohibit more than a single annual meeting for scientists.[88]

Because the "Second Circular" implied a mutual meeting, John Locke, incoming Association president for 1844, and William Mather came from Ohio in April.[89] Dana, from his Washington vantage point, felt that the misrepresentation in the circular was deliberate and suggested, "There has been some bad underhanded work on the part of the Institute—more particularly Markoe & Co.—intended to break up

[84] Amherst, Nov. 21, 1842, Redfield MSS, BYA.

[85] New York, n.d., Redfield MSS BYA. At this time Redfield still assumed that there was to be a joint meeting of the Institute and the Association in Washington.

[86] Dana to Redfield, Washington, Feb. 1, 1843, Redfield MSS, BYA. James Dwight Dana (1813–1895) studied at Yale under Silliman and in 1838 joined the Wilkes Expedition as mineralogist and geologist. On his return he stayed in Washington to work on the reports for publication, moving in 1844 to New Haven. There he assisted in editing Silliman's *Journal* and in 1850 became Silliman Professor of Natural History and Philosophy. *DAB*, Vol. III, pp. 55–56.

[87] *Ibid.*

[88] Lewis C. Beck to Markoe, n.p., Jan. 5, 1843, NI MSS, SIA.

[89] "Third Circular," Mar. 5, 1843. Locke to Henry Ellsworth, Cincinnati, Jan. 15, 1844, Association of American Geologists and Naturalists MSS, Academy of Natural Sciences Library, Philadelphia. Although elected president of the Association, Locke was evidently not in close contact with other members and did not realize the meetings were no longer joint. In fact, he was enthusiastic about the idea of "coming together from the various parts of our country to unite our efforts in embodying detached elements of National Science and National Honour, as regards scientific requirements."

the association."[90] Benjamin Silliman, Jr., as secretary of the Association, immediately sent out additional notices to members.[91] Concerned about the coming conflict, he observed anxiously to Redfield, "We must therefore make the greater exertion to have our meeting a good one & you must do all in your power by personal presence and papers to have the thing pay off well."[92]

Down in South Carolina Poinsett had apparently heard rumors of the dissatisfaction and suggested that the Institute meeting should be postponed until May to "catch the Geologists" who were "unwilling assistants to the project."[93] Whether motivated by political acumen or sympathy for the scientists' position, his suggestion was not implemented by the active Washington members of the Institute.

Markoe, acting essentially alone, worked desperately to make the meeting of 1844 a spectacular and persuasive national convention for science.[94] With support from science, he believed he could gain financial aid from Congress.[95] But the warnings of his correspondents were too little heeded; the final result of his intensive, independent initiative was to alienate most scientific men even before the April meeting.

The meeting was, however, elaborate and impressive to the public eye. The *National Intelligencer* boldly announced under the heading of the "Literary and Scientific Convention of the National Institute" that it would "on each successive day attract assemblages of the most enlightened, dignified and respectable character."[96] Political protocol was suggested in the order of procession: "President of the United States; Heads of Departments, Orator of the day, Senators, and officers of Senate . . . Mayor and city authorities of Washington, Alexandria and Georgetown; Presidents and Faculties of Colleges . . . Resident Members of the Institute."[97] On April 1, at 10:00 A.M., President Tyler as "Patron" of the Institute led the assembled procession

[90] Dana to A. A. Gould, Washington, Apr. 11, 1844. Gould MSS, Houghton Library, Harvard University.

[91] Silliman, Jr., to Hall, New Haven, Apr. 23, 1844, Hall MSS, N.Y. State Library. Silliman adds, "We have great need of all the force we can secure at Washington, to stem the influence which the National Institute have indirectly excited against us . . . [and] by the ambiguous working of their circular—in having led some of our members to believe that *their* meeting was identical with ours, & we have 'kindly consented' to an *united* meeting. . . ."

[92] Silliman, Jr., to Redfield, New Haven, Apr. 22, 1844, Redfield MSS, BYA; also see Hitchcock to Redfield, Amherst, Apr. 24, 1844, Redfield MSS, BYA.

[93] Poinsett to Gouverneur Kemble, Greenville, S.C., May 24, 1843, *Poinsett Papers, Gilpin Collection*, p. 171.

[94] Markoe to W. H. Prescott, Washington, Feb. 21, 1844, NI MSS, SIA; also Markoe to J. R. Poinsett, Washington, Feb. 16, 1844, *Poinsett Papers, Gilpin Collection*, p. 179. Markoe's committee for arrangements consisted of J. C. Spencer, J. R. Ingersoll, W. C. Rives, W. C. Preston, A. Lawrence, R. Walker, R. Choate, and A. D. Bache; the bulk of the work appears to have been done almost exclusively by Markoe, however.

[95] Markoe to DuPonceau, Washington, Nov. 29, 1842 (draft copy), NI MSS, SIA. Markoe responded to DuPonceau's skepticism by pointing out his reasons for political affiliation: "There has never been evidenced the slightest disposition to mix politics with our cause—& while one great object is to bring our Scientific men acquainted by ocular demonstration with what we have done we believe that the whole movement will reach upon the liberality of Congress—we want & have involved the sanction of assistance of every learned man. . . ." Also see Bache to Loomis, Washington, Mar. 7, 1844, Loomis MSS, BYA, who wrote, "To obtain money directly or indirectly Congress must be satisfied that they have the confidence of men of science. . . ."

[96] Apr. 1, 1844. The *Intelligencer* was highly sympathetic to the Institute's cause and had allowed a series of editorials to be printed in the behalf of the Institute from Dec. 6, 1843, through May 25, 1844; these were reprinted as John Carroll Brent, *Letters on the National Institute, Smithsonian Legacy, the Fine Arts, and other matters connected with the interests of the District of Columbia* (Washington, 1844).

[97] *Ibid.*

from the Treasury Building along Pennsylvania Avenue to the Presbyterian Church. There the Marine Band greeted the participants. After prayer and further musical entertainment, addresses were given by Tyler, Robert J. Walker, John W. Draper, and Elias Loomis. Although the House did not pass Joseph Ingersoll's resolution for special adjournment, many Congressmen did attend.[98] Obviously impressed with the daily sessions, the *Intelligencer* editorialized that the attendance of ladies and of savants indicated a broad base of support for the "institution which has commenced its labors so propitiously, and which promises so useful an existence."[99]

Markoe's efforts were an exercise in futility; the Institute was doomed without scientific support, and such support was not forthcoming. The number of savants noted by the *Intelligencer* was limited and select. The notebook kept by Markoe of early replies to his 115 letters requesting addresses records only 52 acceptances, of which the preponderance came from local amateurs and statesmen.[100] Some men of science did attend, including Bache, Draper, Locke, Walter R. Johnson, Mattew F. Maury, James Espy, and Loomis. Motives for participation were mixed, however, and attendance did not necessarily indicate support for the Institute. Only Locke was a geologist, and most of the others were geophysicists out of touch with the Association and eager to gain federal support for magnetic and meterological investigation. Some, as Bache, were no doubt being politic and securing a position should the meeting prove successful.[101] Declinations had come, however, from an equally, if not more impressive list: Benjamin Silliman, Sr., Samuel Dana, John LeConte, Edward Hitchcock, Peter DuPonceau, Alexis Caswell, George Englemann, and John F. Frazer. Joseph Henry, Lewis R. Gibbes, John Bachman, and Benjamin Peirce had not bothered to reply.[102] Some respondents deliberately indicated that they had chosen to avoid the Institute meeting in favor of the Association. Hitchcock wrote, "For several years I have felt bound to withhold from all other societies whatever papers I am able to prepare in order that I might have something for the Geological Association having a strong desire to see that succeed."[103]

In no official sense did the Association participate in the Institute's April meeting, although a few members did attend and even presented papers.[104] It was, in fact, an

[98] Adams, *Memoirs*, Vol. XI, p. 546 and Vol. XIII, p. 5.

[99] Apr. 4, 1844. Goode, "Genesis," pp. 107–108, claims the meeting was a "brilliant success," and thus "there was every reason to believe Congress would share the general enthusiasm and take the society under its patronage."

[100] "Replies to Circulars" journal apparently kept by Markoe in Rhees MSS, Henry Huntington Library, San Marino, Calif. In all, 43 men declined and the rest did not respond.

[101] Bache took a semi-active part in the activities, serving on the committee of arrangements and presenting a paper, but his enthusiasm was lukewarm. An adroit politician, he usually kept in contact with any scientific institution to the extent that he would be in a position of power should the institution prove successful. Discussion with Dr. Reingold clarified my thinking on Bache, since his active participation was an anomaly in the reaction of the geologists and his fellow American Philosophical Society members.

[102] "Replies to Circulars," Rhees Collection, Huntington Library.

[103] Hitchcock to Markoe, Amherst, Jan. 16, 1844; also see J. W. Bailey to Markoe, West Point, Feb. 8, 1844, and S. S. Haldeman to Markoe, near Philadelphia, Mar. 4, 1844, all in NI MSS, SIA.

[104] Most historians have taken the reports of the Institute seriously in accepting the "joint meeting" of the two, including Pickard, "Government and Science [I]," p. 280, and Dupree, *Science and Government*, p. 74. Perhaps the most serious misreading of the general scientific support is in G. B. Goode's "The First National Scientific Congress (Washington, April, 1844) and its Connection with the Organization of the American Association," *AAAS Proceedings*, 1891, *60*:39–47. Goode argues that the broad

obvious and emphatic boycott. Science did not wish affiliation with and domination by politicians and amateurs and had taken this opportunity to take a stand. Walker's disclaimer that the meeting was "not designed to impede the progress or impair the usefulness of any present or future scientific institutions or societies in any of the United States," only indicated Institute awareness of opposition but did not modify it.[105] Despite the excellent addresses by Matthew F. Maury on the Gulf Stream and Bache on scientific societies,[106] there was too little real science. DuPonceau had warned, "—our Country abounds with speeches but that is not what the world will expect. They will expect some new theory or at least some new view of science, not to come from the Institute, as such, but from the assembled savans [*sic*] whom you expect to assemble at your call."[107] But the Institute encouraged general popular addresses for a lyceum-level audience.

Dana, as member of the Expedition's scientific corps, had encouraged the Association's dissension. His future brother-in-law Benjamin Silliman, Jr., was the secretary of the Association and anxious to sustain it in the upcoming confrontation with the Institute. Together Dana and Silliman wrote to friends and fellow members, reminding them to come and make the meeting a success. Sarcastically, Dana observed,

> The Institute have been making a great stir—They called in a band of music to play during the intervals—determined to make noise one way if not another—They had some men of science here, but in general the pieces read were old & superficial. Prof. Locke and W. Mather were here and understood from the Circular sent them that the Geol. Association met at the same time. . . . Will you let the Bostonians know of these matters & urge a general turn out; we are not so easily outdone as they imagine. If we can have a general attendance, the quiet business-like style of the meetings and the real value of the subjects discussed, and original matter brought out will put them in the background, bass-drum and all.[108]

Such a negative assessment was not the public response, and Markoe and the Institute appeared pleased with their meeting. In ensuing weeks, Markoe was friendly to the Association members gathering in Washington. He invited Hitchcock to stay with him during the May meeting and offered to publish a notice in the *Intelligencer*, where he had political friends.[109]

---

scientific base of the Institute made it the most logical forerunner of the AAAS, but Goode is not able to account for the "disaster" which so rapidly undermined the Institute after 1844. By missing the significance of the Association's change in plans, a major reason for failure of the Institute has been excluded; more importantly, the negative thinking of the men of science has been glossed over.

[105] *Bulletin*, 1845, *3*:439.

[106] Bache's address is in manuscript in the Bache MSS, SIA; never published, the 90 pages are an excellent statement of a leading American scientist's view of European scientific institutions, as well as of his hopes and fears for American institutions. The address was a subtle plea for specialized science at a meeting of amateurs. Bache's position was externally ambiguous (see n. 101 above) for he hoped to direct the "host of pseudo-savants" into "safe harbors." Henry to Bache, Princeton, Apr. 16, 1844, Henry MSS, SIA; Bache to Loomis, Washington, Mar. 7, 1844, Loomis MSS, BYA.

[107] DuPonceau to Markoe, Philadelphia, Mar. 22, 1843, Galloway-Maxcy-Markoe MSS, LC.

[108] Dana to A. A. Gould, Washington, Apr. 11, 1844, Gould MSS, Houghton, Harvard.

[109] Daniel Drake to Redfield, Washington, May 1, 1844, Redfield MSS, BYA. Dana commented at the end of the letter, "Markoe is very gracious of late. He had the impudence to write a notice of the Geological Association for the papers, with the larger caption overhead *National Institute* [triple underline].—Fortunately he sent it to me for my revisal, and besides other changes, I cut off the National Institute, and advised him to sign it '*a member of the National Institute*.' He moreover took no offense or showed none rather, as that is not his nature."

The Association of American Geologists and Naturalists met for six days in May with Locke in the chair and Silliman, Jr., and O. P. Hubbard as secretaries. It was quietly successful, having, as Dana suggested, "less parade than the meeting of the Institute, without the aid of a band of music, and with more real science."[110] Thanks particularly to the energetic letter-writing campaign of Silliman and Dana, which supplemented the circulars distributed by Ellsworth as chairman of the Local Committee, more members came than were anticipated and evening sessions were required.[111] The meeting opened at the Medical College and then adjourned to the larger facilities of the Unitarian Church for the remainder of the sessions. Most of the forty-some participants were geologists, but chemists with more than passing interest in mineralogy, as well as naturalists, were also at the meetings. Locke had opened the session by indicating a scientific revolution "by which the talents of our country are to be turned [toward] physical research."[112] Perhaps as a token of the broadening interests of the Association Dr. Charles G. Page exhibited two new electromagnetic instruments he had invented to produce reciprocal motion.[113]

Following the example of the previously departing presidents and, hardly coincidentally, of the BAAS, Henry D. Rogers offered a long and thorough review of geological research in the United States. First, however, he briefly noted the history of the Association and its goals. Stressing the intellectual and geographical isolation in which geologists had formerly worked, he indicated that the intercourse within the Association meetings had accomplished a new comprehension of each member's research and theoretical views. The result was that "we are now able to take the geology of three fourths of the vast region between the Atlantic and Mississippi."[114] Clearly proud of the four annual meetings, Rogers pointed out the zeal and spirit of cooperation which marked their history as well as the quality of published and unpublished papers presented to the members. His concluding point was that while the young Association "would not presume to invite a comparison between its labors and those of similar societies of Europe," yet the objects sought were the same:

> Those illustrious assemblages which in England, Germany, Italy and Switzerland have in the last ten years done so much to quicken the march of science, so much to win for the student of nature, the once withheld respect of literary scholars, statesmen and government, and best of all so much to make science and letters what they are yet far from fully being, a true republic, all were the results of the same necessities, the same intellectual and social wants, and the same high aspirations, which drew the geologists of America, a small but an enthusiastic band, together.[115]

It was these high hopes of being *the* national scientific association which were threatened by the political maneuverings of the National Institute. Staunchly independent and seeking a competent scientific membership, the Association worked to present

[110] "Abstract of Proceedings" (1844), p. 1. Although no scientific papers were published in this volume, many were abstracted in Silliman's *Journal* in various numbers of volume *47*. Dana to Spencer F. Baird, Washington, May 19, 1844, Baird MSS, SIA.

[111] Over 40 names are given in the *Intelligencer*, May 1844, as in attendance. The geological attendance is not in itself surprising, but the list of participants is almost a compendium of who was who in American geology for the period: William W. Mather, Henry D. Rogers, William B. Rogers, Douglass Houghton, Benjamin Silliman, Jr., Edward Hitchcock, James Hall, James L. Smith, Michael Tuomey, and James D. Dana.

[112] "Abstract of the Proceedings" (1844), p. 2.

[113] *Ibid.*, p. 89.

[114] Rogers, "Address . . . 1844," p. 4.

[115] *Ibid.*, pp. 4–5.

itself as a contrast to the Institute whose membership was replete with politicians and "dabblers."[116]

To close observers, the results were obvious. Politicians already neutral or opposed to the Institute were joined by the scientists as a negative force against the aspirations of the Institute. Congress adjourned in June 1844 without taking any action, indicating that a "Memorial" signed by forty "friends of science" was not persuasive evidence of scientific support. Subsequent history substantiates J. M. Gilliss' observation in the fall of 1844 that the Institute with Markoe "taking the functions of the whole society" was pursuing a "headlong course of extravagance" and publishing all sorts of "trash." The only hope would be "If Bache and one or two other men of *known* science will take hold of it, and drive out the present managers."[117] But the antidote of scientific leadership was not applied. Apparently naïve about the nature of the subtle conflict, the Institute leaders felt arbitrarily opposed. For men of science, however, there was a less ambiguous challenge being made. Politization and nonscientific or amateur control were built into the constitution and fabric of the Institute; they wanted governmental support but institutional independence. The basic structure of the Institute and the maneuverings of "Markoe and Co." only reinforced a conviction that aggressive politicians were not primarily concerned about advancing science.

Sustained by a twenty-year charter, the Institute survived and had two temporary revivals among local membership, in 1847 and in 1854. As a falling star, it shone brightest in 1844 just as it approached destruction. Its most significant meteor was the nearly complete collection of the Exploring Expedition, which had been threatened by dispersion before Wilkes' return to Washington.[118] When the Smithsonian Institution was founded in 1846, the Institute was granted the right to name two private regents to its governing board; the new Institution would eventually assume (despite Joseph Henry's reluctance) nearly all of the projected hopes of the National Institute, whose activities to 1862 were those of a local scientific society.[119]

Although the progress of the Association of American Geologists was slow and hardly linear, in 1847 it passed a resolution to become the more broadly inclusive American Association for the Advancement of Science. Hitchcock's goal of a national organization of scientists had been realized after a decade of discussion and experience. Many of the younger professional scientists retained the skepticism of Bache and Henry in reference to a semi-popular organization, in which papers might be presented and membership allowed without stringent discrimination. Their attempt to redirect the

[116] *Bulletin*, 1845, *3*:392–415. The Institute listed over 300 resident members and more than 1,400 corresponding and honorary members. A survey of the list quickly discloses that the composition was political and social as well as scientific.

[117] Gilliss to Loomis, Washington, Oct. 18, 1844, Loomis MSS, BYA.

[118] Haskell, *Publications*, p. 7. Markoe was especially interested in international exchange and founded a systematic exchange of journals with major European societies which was later continued by S. F. Baird at the Smithsonian Institution. Markoe's position in the State Department provided him with valuable consular and diplomatic contacts, and he was especially responsive to the aborted attempts of French traveller Alexandre Vattemare to establish trans-Atlantic exchanges. See *Bulletin*, 1840, *1*:8, and also Elizabeth Revai, "Le Voyage D'Alexandre Vattemare au Canada: 1840–1841. Un aperçu des relationes culturelles franco-canadiennes, 1840–1875," *Revue d'histoire de l'Amerique française*, 1968, *22*:257–299.

[119] This is evident from the manuscript minutes of the National Institute; also see Charles F. Stansbury, *Report of the Recording Secretary of the National Institute for the Year 1850* (Washington, 1850).

AAAS in the 1850s led to frequent internal discord.[120] Resisting the Continental tendency toward specialized societies, they wanted a cohesive and exclusive fellowship of leading scientific researchers. When Congress chartered the National Academy of Sciences in 1863 they temporarily felt their goal was reached.

Reasons for the failure of the Institute in 1844 are complex and involve much more than simple rejection by Congress. Political opposition fostered by various political factions was clearly important, especially after the major shift in domestic politics in 1841. Such problems were compounded by the alienation of Commissioner Ellsworth and the returning scientific corps of the Wilkes' Expedition. Yet the intensive bid for scientific support and the cautious involvement of men like Bache at the April 1844 meeting argues that had science rallied, the National Institute would have survived and might also have secured the Smithson bequest or at least a Congressional appropriation. But the separate and well-received scientific meeting of the Association of American Geologists and Naturalists signaled the end of the National Institute, whose last eighteen years were only an epilogue.[121] Distrustful of the Institute's leadership and goals, the geologists acted, it seems, for a larger and still undefined scientific community in rejecting the attempt to deliberately politicize scientific organization.

The intensive and deliberate clash of the two meetings is only in retrospect highly symbolic. Reality dictated that the National Institute could no longer anticipate either political or scientific support. Its defeat by the still amorphous scientific community had come not from the skeptics but through the very active efforts of the Association leadership. If the specific pattern of national-level scientific organization remained undefined, the example of the Institute had at least decidedly eliminated one possibility: its negative consensus demonstrated that national associations must be built from felt needs of men of science, not on a precast popular foundation with a political, bureaucratic superstructure. The defeat of the National Institute was thus one step toward professional status by men of science who refused to be indiscriminately linked to a project with popular goals and political ambitions.

[120] The author is presently completing a study of the AAAS to 1860. Older histories are brief and not analytical; see Herman L. Fairchild, "The History of the American Association for the Advancement of Science," *Science*, Apr. 5, 1924, *59*:365–369; May 2, 1924, 385–390; May 9, 1924, 410–415, and F. R. Moulton, "The American Association for the Advancement of Science: A Brief Historical Sketch," *Science*, Sept., 1948, *108*:217–218.

[121] Joseph Henry pronounced a *post mortem* at a meeting of the Washington Philosophical Society in 1871, when he warned scientific societies of the example of the Institute: "Although it included among its members a few men of true science, [it] was under the control principally of amateurs and politicians, and therefore was unfit to discharge the duty which it claimed as one of its functions, to decide questions of a strictly scientific character. It should have been borne in mind that votes on questions in science should be weighed, not counted!"

A Dynamic Theory of Mountain Building: Henry Darwin Rogers, 1842

By Patsy A. Gerstner*

ONE OF THE IMPORTANT AMERICAN contributions to theoretical science before 1850 was the mountain elevation theory of Henry Darwin Rogers (1809–1866).[1] Rogers' theory was formulated in the 1830s and presented in a paper read to the American Association of Geologists and Naturalists in April 1842 and, in a slightly condensed form, to the British Association for the Advancement of Science in June 1842.[2] The paper dealt with the Appalachian Mountains and included structural information gathered by Rogers as director of the New Jersey State Geological Survey (1835–1837) and the Pennsylvania State Geological Survey (1836–1842) and by his brother William Rogers as director of the Virginia Survey (1836–1841). The theoretical portion of the paper, although supported by the structural evidence of the Appalachians, was conceived of as a general theory applicable to every other chain of mountains.[3]

The theory argued that mountains were built during paroxysmal upheavals of the land caused by a wave-like undulation of the crust that occurred because molten matter beneath was pulsating or moving with a wave-like motion. The structure of the Appalachians suggested the validity of this theory to Rogers, and it will be helpful to examine his argument as based on the facts as they were observed.

*Howard Dittrick Museum of Historical Medicine of the Cleveland Medical Library Association, Cleveland, Ohio 44106.

[1] Although the theory discussed here is usually thought of as the joint work of Henry and his brother William, I believe, as will be shown below, that it arose primarily out of Henry's experiences in England in the early 1830s.

[2] W. B. Rogers and H. D. Rogers, "On the Physical Structure of the Appalachian Chain, as Exemplifying the Laws which Have Regulated the Elevation of Great Mountain Chains, Generally." *Reports of the First, Second, and Third Meetings of the Association of American Geologists and Naturalists at Philadelphia in 1840 and 1841, and at Boston in 1842* (Boston: Gould, Kendall, and Lincoln, 1843), pp. 474–531. The condensed version read to the British Association was reported in the *Report of the Twelfth Meeting of the British Association for the Advancement of Science; Held at Manchester in June, 1842* (London: John Murray, 1843), pp. 40–42. It was due primarily to the urging of Charles Lyell that the condensed version was sent to the British Association. Lyell was in the United States at the time and had heard Rogers' paper. Emma Rogers, ed., *Life and Letters of William Barton Rogers*, Vol. I (Boston: Houghton Mifflin, 1896), pp. 214–215.

[3] The general applicability of the theory is implied throughout the 1842 paper and is especially clear in pp. 525–531.

The Appalachian system of mountains is composed of several main ridges, each with subsidiary ridges, that are aligned in a northeast-southwest direction. The axis (axial) planes, a term which Rogers seems to have coined, that run the length of individual folds are usually long and parallel and frequently curved, which, Rogers felt, led to a natural arrangement of groups within the chain.[4] Folding is more severe along the southeastern edge of the range, and in that area faulting, intrusive dykes, overturned folds, and other features associated with intense disturbance are met with. Folding gradually becomes less and less severe toward the northwest, until horizontal strata replace the folds almost entirely. The strata are found generally to exhibit a southeast dip. Each of these features of the Appalachians was of special concern to Rogers, but he believed the dip was the most important of all, for "upon the correct interpretation of this singular feature depends . . . the clear elucidation of whatever relates to the dynamical actions which the region has experienced." Rogers interpreted the dip as the structural result of the character of the anticlinal and synclinal folds. These, he found, were not symmetrical but exhibited a steeper dip on the northwest side of every anticline than on the southeast side, or conversely a steeper dip on the southeast side of every syncline. The apex of each fold is thus to be found in advance of the center of the fold, or, put another way, the axial plane is inclined to the perpendicular to a greater extent.

Rogers had essentially three possible explanations for the nature of the folds: he could argue that the action of a vertical force was responsible, that the action of a horizontal force was responsible, or that it was a combination of both. The first two alternatives were those commonly used by elevation theorists in the 1830s, but Rogers chose the third, because

> A merely vertical force, exerted either simultaneously or successively, along a system of parallel lines, could only produce the same number of symmetrical anticlinal arches, while again, a horizontal or tangential pressure, uncombined with an alternate upward and downward motion, at regular intervals, could not possibly result in a system of parallel folds, or axes, or lead to any change in the position of the strata beyond an imperceptible bulging of the whole tract, or else a confused rumpling and dislocation dependent on local inequalities in the thickness or resistance of the crust, in different spots.[5]

---

[4] For the theory as outlined here see Rogers, *Reports of the First, Second and Third Meetings*, pp. 480–497.

[5] *Ibid.*, pp. 507–508. Vertical force was characteristic of volcanic explanations of elevation but was given particular emphasis by the work of James Hutton, who described elevation as the result of a heat force exerted from below on the rocky mass to lift it (*Theory of the Earth with Proofs and Illustrations*, Edinburgh, 1795). John Playfair, J. L. Heim, Leopold von Buch, Charles Babbage, and Charles Lyell among others in the first half of the nineteenth century adhered to the notion of some kind of vertical force at work in continental and mountain elevation. Jean Baptiste Armand Louis Leonce Élie de Beaumont introduced the notion of a horizontal or tangential force as responsible for elevation. Such force was the result of pressures set up in the earth's crust because of a shrinking interior ("Recherches sur quelques unes des révolutions de la surface du globe, presentant différens exemples de coincidence entre le redressement des couche de certain systèmes et les changemens soudains qui ont preduit les lignes de démarcation qu'on observe entre certains étages consecutifs des terrains de sédiment," *Annales des Sciences Naturelles*, 1829, *18*:5–25 and 284–416).

For Rogers such combined vertical and horizontal forces could only result in the kind of motion seen in waves of water, that is, rising and falling coupled with a constant movement forward. The wave was not, however, a wave that originated in the crust itself, but it was the movement of molten matter on which the crust of the earth rested which then imparted the same motion to the crust. The mechanism of mountain elevation began, Rogers argued, when an accumulation of vapors and gases on the molten surface exerted an upward vertical pressure on the rock crust. The pressure became strong enough to cause the crust to break into a series of long, parallel rents in successive stages. This rupture occurred primarily along the southeastern side of the Appalachians, as clearly evidenced by the abundance of faults, intrusive dykes, close folds, and overturned folds that characterize that part of the Appalachians. The sudden release of pressure on the fluid surface set the fluid into motion, causing it to rise in a great billow and carry the crust with it. Gravity acting on the rising lava would "engender a violent undulation of its whole contiguous surface, so that wave would succeed wave in regular and parallel order, flattening and expanding as they advanced, and imparting a corresponding billowy motion to the overlying strata." As the crust was thrown into parallel flexures by this force the whole crustal area was pushed forward so that "the flexures formed by the waves, would steepen the advanced side of each wave, precisely as the wind, acting on the billows of the ocean, forces forward their crests, and imparts a steeper slope to their leeward sides." The newly formed folds in the crust were frozen by the intrusion of the molten material. The folding and inversion of strata was less severe in the northwest, and this was because it was farther from the center of disturbance and because, Rogers thought, the faulted rock of the southeast served as a kind of barrier to the more violent waves and prevented them from progressing toward the northwest with great strength.

Rogers did not believe that folding had occurred simultaneously throughout the Appalachians or as the result of uniform force. This was suggested by varying kinds of folds that he observed, the arrangement of axes into groups, and the appearance of both curved and straight axes. Variations in magnitude and momentum and in the ratio of tangential to undulatory movement accounted for the different kinds of folds and for the arrangement of axes in groups. Because each curving axis plane was flanked with a straight one, Rogers argued that the straight one was formed first and then served as a deflector to waves that followed, causing them to curve. Thus, successive stages of folding had occurred. These successive stages and varying intensities were not to be associated with different epochs of elevation, however, but were instead variations within one period of mountain building as wave succeeded wave.

Rogers believed that the movement of the earth during mountain building was analogous to its motion during an earthquake and that both resulted when fluid beneath the crust pulsated.[6] He did not, however, envision earthquakes

[6]Rogers, *Reports of the First, Second and Third Meetings*, p. 517. Rogers expanded his comments on earthquakes and their cause in 1843 in a paper read to the American Association of Geologists

as the cause of mountain building so much as he believed both to be manifestations of a single dynamic process—the pulsating lava. This theory of a pulsating lava was obviously intended to be more than a theory of mountain building and earthquakes, since it was applied as early as 1842 to the phenomena of drift and coal formation. This is the same year in which the mountain-building theory was read; in fact, these further extensions of the theory were elaborated on at the same meeting of the American Association of Geologists and Naturalists.

Glacial drift, the polished and striated surfaces often observed on rocks, and large glacial erratics were major geological problems in the early nineteenth century and were dealt with in many ways. Water, floating icebergs, or some combination of both were commonly accepted causes. At the 1842 meeting of the American Association there was a lively discussion on the whole subject, during which Rogers discussed a theory offered by Charles Lyell and others that a sudden deluge of water caused by the volcanic melting of polar ice might go far to explain the phenomena. "But," Rogers said, "if we suppose that this wave was accompanied by an earthquake, rocking, or wave-like motion of the bed of the ocean, the whole mass of turned-up strata would be shoved violently from north to south, and at every heaving of the earth, a mass of water would be thrown forward, like the rolling of a tremendous surf."[7] Noting that scratches on summits of the Appalachians are oriented in a north-south direction, as is the case in New England and other localities, he pointed to the additional fact that scratches found at the bottom of valleys show "all the local deflections which a body of moving waters would encounter." He concluded that the scratches were made not by the water itself but by "the friction of the overlying stratum of drift itself, urged into rapid motion from the north by one or more sudden inundations." The continuous breaking of the sea waves on land, "abrading and dispersing the fragmentary matter during repeated oscillations of the crust," accounted for the widespread occurrence of coarse mechanical strata, such as the conglomerates, in the Appalachians.[8]

Rogers referred again to this theory of drift in 1843, and in 1844 he presented a detailed but essentially unchanged account to the American Association of Geologists and Naturalists.[9] There he dealt more explicitly with several possible

---

and Naturalists: "Abstract of the Proceedings of the Fourth Session of the Association of American Geologists and Naturalists," *American Journal of Science*, 1843, *45*:341–347.

[7] [Proceedings of the American Association of Geologists and Naturalists, 1842], *Reports of the First, Second, and Third Meetings*, p. 48.

[8] *Ibid.*, pp. 72–73.

[9] "Abstract of the Proceedings of the Fourth Session of the Association of American Geologists and Naturalists" (1843), p. 347. H. D. Rogers, *Address Delivered at the Meeting of the Association of American Geologists and Naturalists, Held in Washington May, 1844. With an Abstract of the Proceedings of Their Meeting* (New York/London: Willey and Putnam, 1844), pp. 3–58. The address also appeared in *Am. J. Sci.*, 1844, *47*:137–160 and 247–278. In 1845, in a paper read to the Boston Society of Natural History, Rogers applied the theory to the Richmond Boulder Train, a well-known train of glacial boulders. H. D. Rogers and W. B. Rogers, "An Account of Two Remarkable Trains of Angular Erratic Blocks in Berkshire, Mass., with an Attempt at an Explanation of the Phenomena," *Boston Journal of Natural History*, 1845–1847, *5*:310–330.

arguments against the theory on the origin of drift and related phenomena, principally the argument that water could not create a force strong enough to account for the features and especially not for the presence of the very large erratics. Pointing out that recent mathematical studies by the Englishman William Hopkins had showed that a current moving at only 20 miles per hour could move a block weighing 320 tons and that such a current would result from a paroxysmal elevation of 100 to 200 feet, it was easy for Rogers to argue that the paroxysms resulting from the undulations of the molten interior were more than adequate to account for the phenomena.[10]

The application of the theory to coal formation was done in a more formal manner, as Rogers presented a fully developed paper on the origin of the Appalachian coal strata to the assembled geologists in 1842.[11] Most geologists believed either that coal was formed *in situ* or that the plant material from which it was formed had drifted into the locality.[12] Rogers felt that these theories failed to account satisfactorily for certain critical factors, including the wide distribution of a clay that was found under the coal and was characterized by the presence of stigmaria (a common coal plant of the genus *Sigillaria*), the occurrence of laminated slates above the coal with an entirely different vegetable content, and the relative infrequency of tree trunks and large branches in the coal itself.[13] To account for these factors Rogers proposed that the coal area had existed as an "extensive flat, bordering a continent, and forming the shores of an ocean, or some vast bay," that the coal vegetation grew on these coasts, which slowly subsided, and that the area occasionally experienced gradual upward movements which alternated "with great paroxysmal displacement of the level, caused by those mighty pulsations of the crust we call earthquakes." During quiet periods of subsidence the coast was surrounded by large marshes in which stigmaria grew abundantly. The decay of the stigmaria, with the addition of other plants carried along the coastline by the natural movement of water, provided the material from which the coal was formed. In this situation of formation no large branches or trunks were to be expected.

During the paroxysmal upheavals the sea first drained the water away from the marsh, churning the marsh mud as it did so, and deposited small plant pieces over the entire surface. This formed the laminated slates. The sea in its violent return to the area rolled far inland, knocking over everything before it, and severely abraded the land material. In its retreat the material, including the larger branches and tree trunks, was distributed over the area. Alternately, then, the rolling sea formed fine and coarse layers above the coal. Once the earthquake was over, the marsh and sea settled into quietness. The material the water still held in suspension was "only the most finely divided sedimentary

[10] Rogers, "Address Delivered at the Meeting of the American Association of Geologists and Naturalists. . .," *Am. J. Sci.*, 1844, *47*: 274.

[11] H. D. Rogers, "An Inquiry into the Origin of the Appalachian Coal Strata, Bituminous and Anthracitic," *Reports of the First, Second, and Third Meetings*, pp. 433–474.

[12] For a summary of thoughts on coal formation among Rogers' contemporaries see his paper, *ibid.*, esp. pp. 460–463, or William Buckland, "Address Delivered on the Anniversary, Feb. 19th [1841]," *Proceedings of the Geological Society of London*, 1838–1842, *3*:487–492.

[13] Rogers, "An Inquiry into the Origin," pp. 463–466.

matter, and the most buoyant of the uptorn vegetation, that is to say, the argillaceous particles of the fire-clay, and the naturally floating hollow stems of the *Stigmariae*." As these settled they formed the stigmaria-filled clay commonly found just below every coal seam.

The dynamic process of coal formation, like that of drift, was independent of mountain elevation; that is, Rogers did not indicate in any way that mountain elevation must occur whenever the fluid interior is set in motion. The various kinds of coal found in the Appalachian area were, however, related to the elevating process. Rogers found that the types of coal exhibited a regular gradation from southeast to northwest through the Appalachians, showing a progressive increase in their volatile matter, that is, moving steadily from anthracite in the southeast to bituminous in the northwest. He believed that the bituminous elements were driven off by "intensely heated steam and gaseous matter, emitted through the crust of the earth, . . . during the undulation and permanent bending of the strata."[14] Since the focus of the forces elevating the Appalachian chain occurred in the southeast, it was in that direction that the coal was most "debituminized."

It is clear that Rogers' theory of mountain elevation was only one part of a much more extensive view of geological dynamics that had taken form in his mind sometime prior to 1842. Certainly the structure of the mountains Rogers studied during his survey of Pennsylvania encouraged him in his theory, but the history of Rogers' interest in geology suggests that the theory was not solely the outcome of a line of inductive reasoning based on observation alone but rather that it represented a particular combination of ideas that Rogers had encountered as early as 1832.[15] Although his theory as a whole was unique among elevation theories, most of the basic assumptions, including the existence of an undulating fluid, had been discussed in England either with regard to elevation or to the cause of earthquakes. Rogers was in an especially good position to be influenced by these English theories and to draw from them the ingredients of his own theory.

Rogers' interest in geology actually originated in contacts with English geologists in 1832 and 1833. Prior to that time he showed no interest in geology and devoted most of his time to chemistry.[16] In 1829 he was elected to the

[14] *Ibid.*, p. 473. Rogers noted that this gradation was first observed in 1837 when he mentioned it in a public lecture. He communicated the information to the American Association of Geologists and Naturalists in 1840.

[15] Although Rogers, as director of the Pennsylvania Survey, published six annual reports between 1836 and 1842, they are lacking in evidence concerning the evolution of the theory in Rogers' mind. As early as the second report (1838) Rogers called the mountains "undulations," but not until the fourth report (1840) did he suggest that he had a theory in mind when he called the dips and inversions an important clue to the elevation of the mountains. He also noted in this report that there was both a vertical and a horizontal force at work in elevation and that he would discuss the cause in the final report. The final report was not to be published until 1858, however, and the 1842 paper was therefore the first public treatment of the theory.

[16] The principal biographers of Rogers are W. S. W. Ruschenberger, "A Sketch of the Life of Robert E. Rogers, M.D., LL.D., with Biographical Notices of His Father and Brothers," *Proceedings of the American Philosophical Society*, 1885, *23*:104–146; J. W. Gregory, Henry Darwin Rogers, *An Address to the Glasgow University Geological Society 20th January 1916* (Glasgow: James MacLehose and Sons, 1916). Emma Rogers, ed., *Life and Letters of William Barton Rogers*, Vol. I, is also very helpful.

Chair of Chemistry and Natural Philosophy at Dickinson College, in Carlisle, Pennsylvania, but he was not especially happy at Carlisle and in 1831 left to spend the winter in New York. He was no doubt in a frame of mind to seek new challenges. In 1828 Rogers had heard a lecture by the reformer Frances Wright and was much taken with her ideas.[17] While in New York in 1831 he met Robert Dale Owen, son of the great English reformer, and was again taken with the ideas of reform. He decided to accompany Owen to England.[18]

Once in England, however, Rogers' interest in reform steadily declined as opportunities for scientific activity grew. On December 14, 1832, he wrote to his uncle James Rogers that, in addition to the other benefits of his trip, "I shall have fine chances for making myself a geologist by the free access I may have to the London Geological Society's superb museum. I was introduced personally to several of the members, De la Beche, Lyell, Babbage and others." In March 1833 he reported to William that he frequently attended small gatherings held by George Greenough, then president of the Geological Society.[19] That he was, as the above indicates, an instant success in the English community of geologists is further evidenced by his election as Fellow of the Geological Society of London on May 1, 1833—the first American to be so honored.

Rogers and Henry de la Beche became particularly good friends during this period and de la Beche had a strong influence on Rogers' career in geological surveying. De la Beche was just then beginning a mammoth geological survey of England, and he invited Rogers to visit his survey in Devonshire and Cornwall. This opportunity, Henry wrote, "will fit me to do the like at home."[20] It seems reasonable to assume that Rogers' work on the surveys of New Jersey and Pennsylvania was an outgrowth of this friendship. De la Beche and other British geologists certainly led him to an early interest in the dynamics of mountain elevation, for this was one of the most talked about problems in geology throughout the 1830s in England. A look at the array of opinions on elevation at that time will give some idea about what Rogers might have encountered during his stay in England.

In 1834 George Bellas Greenough, whose gatherings Henry had so frequently attended a year earlier, summarized the current ideas on elevation:

> The assigned *causes of elevation* are exceedingly various. One author raises the bottom of the sea by earthquakes; another, by subterranean fire; another, by aqueous

---

[17] *Life and Letters*, p. 69. Frances Wright was a humanitarian much in the same utopian vein as Robert Owen. Although an Englishwoman, she devoted most of her efforts to the United States and is best remembered for the Nashoba experiment, an unsuccessful effort to free, educate, and relocate slaves.

[18] *Ibid.*, pp. 91–92. In a letter to me Dr. John Rodgers of Yale University suggested that it was Rogers' already proven skill as a lecturer that made the reformers interested in him and especially eager to have him join them in Europe, since the purpose of the trip was to give lectures to workers.

[19] *Life and Letters*, pp. 97–98, 105.

[20] *Ibid.*, p. 106.

vapour; another by the contact of water with the metallic bases of the earths and alkalis. Heim ascribes it to gas; Playfair, to expansive force acting from beneath; Necker de Saussure connects it with magnetism; Wrede, with a slow continuous change in the position of the axis of the earth; Leslie figured to himself a stratum of concentrated atmospheric air under the ocean, to be applied, I suppose, to the same purpose.[21]

Greenough divided the theories into two classes, the first being one of explosive forces, the second of sustaining forces. Explosive forces were those of volcanic action, while the sustaining forces were those manifested in the expansive character of solids, fluids, and aeriform substances, or the elastic power of subterranean fires.

Each class was dependent on the concept of a central heat, and while Greenough divided all theories based on the presence of heat or molten lava in the earth into these two classes, two greater divisions among elevation theorists can be made. On the one hand there were those whose theories rested on the assumption of a continuing source of heat within the earth; then there were those who believed that the earth was molten when formed but was gradually cooling and solidifying. The leading proponent of this latter view was the French geologist Élie de Beaumont, who suggested, in 1829, that the cooling interior caused horizontal stresses in the already cool and hard crust which made the crust crumple and formed mountains.[22] Both because the action was sudden and because the earth would never again be in exactly the same stage of cooling, Élie de Beaumont and his followers were termed catastrophists. Adherents to the view of a continuing heat in the earth tended to think of elevation as a slow process and one that could be repeated; they were therefore uniformitarians. Much more than the others, uniformitarians tended to see mountain elevation as no more than a part of continental elevation in general. The direction of force in their view was vertical; that is, it came from below the crust.

It is possible to see in this brief overview of elevation theories just what kind of information Rogers might have brought back to America in the summer of 1833. If he had not heard many or all of these ideas while in England, he certainly would have been made aware of them through continuing contacts with his English friends.

Although Rogers' good friend de la Beche was a strong supporter of Élie de Beaumont's theory, Rogers was not apparently impressed by this argument and aligned himself with those who accepted a continuing heat in the earth. Although he was like the catastrophists in believing in the paroxysmal nature of mountain elevation and in considering mountain elevation separately rather than as a part of continental elevation, he was like the uniformitarians in that nothing in his work suggests that he thought the processes he described were phenomena limited in time. Indeed, inherent in his idea of coal formation,

[21] George Bellas Greenough, "Address Delivered at the Anniversary Meeting of the Geological Society of London on the 21st of Febraary 1834," *Proc. Geol. Soc. Lond.*, 1833–1834, 2:60–62.
[22] Élie de Beaumont, "Recherches sur quelques unes des revolutions."

earthquakes, elevation, and drift is the thought that these processes, all caused by the undulating fluid, had occurred at different times.[23] Rogers particularly admired Charles Babbage, and Babbage was a uniformitarian, a proponent of continuing heat in the earth, and very much interested in the processes of elevation.[24] As early as 1834 Babbage publicly discussed the possibilities that a sea of molten lava existed beneath the crust, that earthquakes caused mountain elevation, and that such earthquakes occurred when heated gases within the earth created such a strong force on the crust that the rock was torn apart.[25] Although Babbage did not wholly accept any of these possibilities, the tone of his discussion suggests that he took them quite seriously. Although his ideas were presented a year after Rogers left England, they may have been discussed earlier. If so, Babbage was the starting point in Rogers' reasoning.

That elevation was the result of earthquakes was discussed not only by Babbage but by others, as Greenough pointed out in his summary. While this interest was usually directed toward finding a simple cause-and-effect relation between the earthquake and elevation, Rogers was interested in knowing the most basic cause of the earthquake. It was this concern that gave rise to that unique aspect of his elevation theory, the undulating fluid, and it was within the discussion of earthquakes rather than elevation that he may have first come upon the idea of an undulating fluid. Of particular interest to geologists in the 1830s when considering earthquakes was the nature of the motion of the land. It was believed that the correct interpretation of this motion would provide the clue to the cause. The motion was considered by some to be vibratory while others thought it was undulatory. At least one supporter of the undulatory theory, John Phillips, suggested that the undulation of the crust was the result of an "agitated" liquid beneath the crust.[26] That this agitated liquid of Phillips could be interpreted as an undulating fluid is attested to by the fact that Charles Darwin interpreted it in just that fashion. Darwin, in fact, seriously considered a relation between an undulating fluid and an undulatory movement of the crust during an earthquake and further considered their relation to elevation. He decided, however, that the concept of an undulating fluid was unjustified.[27] Rogers did not, and the result was his

[23] Rogers probably never thought of himself as either a uniformitarian or a catastrophist and chose from among the available ideas according to his own interpretations. His biographer J. W. Gregory believed him to be a uniformitarian (*Henry Darwin Rogers,* p. 23).

[24] In a letter to his brother William written from London on Feb. 14, 1833, Henry discussed his various meetings with English geologists in the home of Greenough, commenting that "none so awakened my admiration as Babbage." In 1834 in a paper read to the Geological Society Babbage gave his view on central heat. The paper was published in 1847: "Observations on the Temple of Serapis, at Pozzuoli, Near Naples, with Remarks on Certain Causes Which May Produce Geological Cycles of Great Extent," *Quarterly Journal of the Geological Society of London,* 1847, *3*:186–217.

[25] *Ibid.*

[26] John Phillips, *A Treatise on Geology,* Vol. II (London: Longman, Orme, Brown, Green and Longmans, 1839), p. 209.

[27] Charles Darwin, "On the Connexion of Certain Volcanic Phenomena in South America and on the Formation of Mountain Chains and Volcanos, as the Effect of the Same Power by Which Continents Are Elevated," *Transactions of the Geological Society of London,* 1840, *5*:620–621. In

theory of elevation. The influence of Babbage and Phillips is by no means a certain thing, but their thoughts and discussions on elevation and on earthquakes provide a reasonable basis for the development of his theory.

The reactions to Rogers' theory were varied, but for the most part it met with little acceptance. In America its delivery was hailed as one of the great American oratorical achievements,[28] but there is little evidence to suggest that it stirred the interest of other geologists. Certainly the theory was discussed by those at the 1842 meeting of the American Association of Geologists and Naturalists. However, the general textbook accounts of geology in the 1840s, including Edward Hitchcock's popular *Elementary Geology*, do not mention Rogers, and of the contemporary geological surveyors William Mather mentioned Rogers' theory in passing only.[29] The limited reaction characteristic of the United States geologists is in contrast with the reaction in Europe.

The European response was spirited and reflected the many different views of mountain elevation in vogue. The Manchester meeting of the British Association for the Advancement of Science at which the paper was read was well attended by leading European geologists. As might be expected, de la Beche, who had remained an adherent to Élie de Beaumont's theory, immediately argued that horizontal pressure alone was sufficient.[30] Adam Sedgwick thought the greater inclination of strata on the side farthest from the disturbing force, as outlined by Rogers, was not in keeping with the origin of the folds as suggested by Rogers. He argued instead that the effects of the disturbing force were dependent on the kinds of rocks affected, some being more easily contorted than others. Sedgwick also believed that the motion of the earth during an earthquake was vibratory and that the elevation of mountains had been gradual rather than paroxysmal in nature.

The views of de la Beche and Sedgwick were apparently typical of the reaction at Manchester, but support for the theory was not altogether lacking. John Phillips wrote to Rogers:

> I received your paper just before the Manchester meeting, at which my official occupations were excessive, and then commended it to the President of the Section, Mr. Murchison, Professor Sedgwick not having arrived. Against my expectation the paper was appointed for reading on the first day, when I was so entirely engrossed with my duties as not to be able to appear. Sedgwick, however, had arrived, and there was a lively and continued discussion. The opinions of the geologists present were apparently at variance with you, but I am very much of the opinion that, had your views been presented with large diagrams to show

---

discussing Phillips' ideas Darwin pointed out that the idea that an earthquake resulted from an undulating fluid had occurred to him in 1835.

[28] See *Life and Letters*, pp. 209–212, for J. L. Hayes' description of the impressive nature of the presentation.

[29] William W. Mather, *Geology of New York, Part I, Comprising the First Geological District* (Albany: Carroll and Cook, 1843), p. 631. Mather stressed a theory of his own that was quite different from Rogers'.

[30] A report on the meeting of the BAAS in which the discussion of Rogers' paper was summarized appeared in *The London Athenaeum*, July 1842, p. 592, and *Am. J. Sci.*, 1843, *44*:359–365.

> the mechanical reasoning, the result might have been different. There were several points in your argument which I approved and wished to advocate.[31]

Roderick Murchison also offered some support to Rogers' theories. Murchison, who was then president of the Geological Society of London, did not give an opinion of Rogers' theory at the Manchester meeting, but in his anniversary address to the Geological Society given at the beginning of 1843 and covering the society's activities of 1842 he spoke with some favor toward the theory. Noting that the American country was mostly unfamiliar to Europeans and that they had no illustrations or maps by which to follow Rogers' theories, he felt that those gathered at Manchester had been in a very poor position to judge the theory. Having himself become better acquainted with it he now found it "a praiseworthy effort to establish laws of phenomena regarding subterranean movements." Further, he found the paper was perfectly in keeping with "the spirit of inductive philosophy," that it was backed by "a wide and successful field survey," and that it was "generously confided to the friendly feelings of the geologists of England."[32] In his *Geology of Russia* (1845) his remarks on the South Urals conclude with the observation that no geologist has studied the sweeping ridges and valleys

> . . . without being led to think, certain wave-like undulations to which the whole has been subjected, were necessarily most rapid towards the disturbing centre, and gradually died away as they receded from it. In short, the views which American geologists have so admirably worked out in the Appalachian chain may, we think, be considered equally striking in the Southwestern Ural.[33]

Murchison continued to support Roger's theory for several years, and by the time his *Siluria* appeared in 1854 he was also leaning toward the acceptance of Rogers' coal theory, which he spoke of as an "ingenious" application of the paroxysmal theory.[34] There is some indication, however, that by 1848 Murchison felt some hesitation concerning the application of the theory. In his concern to explain inversion, especially enormous overthrows, and lateral folding in general in the Alps which could not be accounted for by large central masses of eruptive matter, cited by other geologists as causes for the phenomena, he introduced Rogers' views.[35] His comments concerned structural similarities between the Alps and the Appalachians, but he was decidedly not prepared, he said, to subscribe to the earthquake theory.

Although Phillips and Murchison found Rogers of some interest, it is clear

[31] *Life and Letters*, p. 218; letter dated Nov. 15, 1842.

[32] R. I. Murchison, [Anniversary Address of the President], *Proc. Geol. Soc. Lond.*, 1842–1845, 4:119.

[33] R. I. Murchison, *The Geology of Russia and the Ural Mountains, I. Geology* (London: John Murray, 1845), pp. 461–462.

[34] R. I. Murchison, *Siluria. The History of the Oldest Fossiliferous Rocks and Their Formation* (3rd ed., London: John Murray, 1859), p. 314.

[35] R. I. Murchison, "On the Geological Structure of the Alps, Apennines and Carpathians, More Especially to Prove a Transition from Secondary to Tertiary Rocks, and the Development of Eocene Deposits in Southern Europe," *Quart. J. Geol. Soc. Lond.*, 1849, 5:248–253.

that the majority felt differently, and a number of reasons may be singled out for this. Those like de la Beche and Sedgwick, both supporters of Élie de Beaumont, reacted on the basis of their own theoretical leanings. The contraction theory and the emphasis on lateral pressure alone as sufficient to explain folding and faulting was re-emphasized in 1846 and 1847 by the work of the American James Dwight Dana, who applied the theory specifically and in detail to the Appalachians. Theories like Rogers' which rested on a fluid interior supposed that this was covered by a relatively thin crust, but the work of the mathematician William Hopkins in the 1830s and 1840s suggested that the earth was far more solid than had been suspected. The tremendous interest in the process of mountain elevation that characterized geology during this period had culminated in a wealth of information on the structure of mountains. Most mountain systems of Europe are not structurally like the Appalachians, and therefore Rogers' theory could not be supported by reference to European mountains. Perhaps more important than any of these factors was the prominence of Charles Lyell. By the 1840s his uniformitarian argument had displaced much of the argument favoring any sudden or paroxysmal upheaval of the land; and of particular consequence for Rogers' theory, since it was based in part on an undulatory theory of earthquakes, was Lyell's interpretation of earthquakes as vibratory rather than undulatory.[36] Further, with regard to mountain elevation specifically Lyell could not "imagine any real connection between the great parallel undulations of the rocks and the real waves of a subjacent ocean of liquid matter, on which the bent and broken crust may once have rested."[37] Although Rogers had put together an interesting theory, there was little in the way of a receptive climate of opinion for that theory, and it disappeared rather rapidly from the discussions of the dynamics of elevation.[38]

[36] Leonard G. Wilson, *Charles Lyell. The Years to 1841: The Revolution in Geology* (New Haven/London: Yale University Press, 1972), p. 389.

[37] Charles Lyell, *Travels in North America, in the Years 1841–2; With Observations on the United States, Canada, and Nova Scotia*, Vol. I (New York: Wiley and Putnam, 1845), p. 78.

[38] Rogers upheld the theory for the rest of his life, much of which was spent as Regius Professor of Geology and Natural History at the University of Glasgow (1857–1866). He gave a detailed account of the theory in the final report of the Geological Survey of Pennsylvania that appeared in 1858: *Geology of Pennsylvania. A Government Survey*, Vol. II (Philadelphia: Lippincott, 1858), pp. 885–916.

Geology and Religion before Darwin: The Case of Edward Hitchcock, Theologian and Geologist (1793-1864)

*By Stanley M. Guralnick**

THE ROAD TO THE ACCEPTANCE of Darwinian evolution was paved with much agonizing over the relationship of nineteenth-century geological science to more timeless religious faith. On both sides of the Atlantic discussions about the role of the supernatural in the explanation of natural phenomena attracted the attention of many scientists, the most eminent as well as the most inconsequential, the theologically sophisticated as well as the naïve. Somewhere in the middle of both intellectual spectra stood the American Edward Hitchcock. An early leader in the development of the American Association for the Advancement of Science and one of the fifty founding members of the National Academy of Sciences, he exhibited an interest in the relationship of science and religion which extended over the whole of his professional career from the early 1820s to his death in 1864.[1] As one who was both a Congregational minister and a highly respected geologist, Hitchcock felt a special obligation to articulate his twin faiths in science and religion. While his efforts demonstrate little philosophical originality, the change in his attitudes, particularly toward miracles, is valuable as an aid to understanding the adjustments which scientists were making in those crucial decades before the Darwinian controversy broke full force. Although unique in the amount of energy expended in the quest for a rationale that would cover both historical Christianity and historical geology, Hitchcock was nevertheless thoroughly representative of the English-speaking scientific community in the values he held and the conclusions he reached.

Hitchcock's earliest opinions on the connection between geology and theology date from 1823–1824, while he was still a minister in Conway, Massachusetts, and just prior to his scientific apprenticeship in the laboratory of Benjamin Silliman at Yale. At that time he was decidedly uncritical in his acceptance of the thesis that science and

* Department of History, University of Bridgeport, Bridgeport, Connecticut 06602. The author gratefully acknowledges postdoctoral research support from the Smithsonian Institution for the academic year 1969–1970, when this study was prepared.

[1] There is no complete biography of Hitchcock. Shorter accounts of his life are listed in the *Dictionary of American Biography*, to which may be added the obituary notice in the *American Journal of Science*, 1864, *37*:302–304. An interesting aspect of his influence on American literature is discussed in Dennis R. Cohen, "Hitchcock's Dinosaur Tracks," *American Quarterly*, 1969, *21*:639–644.

theology were one. Not even then, however, did Hitchcock insist that a day of creation is literally equivalent to a modern twenty-four-hour period. But he confidently and unwaveringly expected that the history of Christian revelation and the history of the globe might be exactly related, that all scientific truth was ultimately revealed in the Bible, and, most significantly, that the sacred text might even be a starting point for scientific investigation. If the Bible and science could both agree (as they then did) that there was a universal Deluge, then the Bible ought to suggest to the paleontologist that he search for human remains only in the Biblical cradle of civilization.[2] If revelation concerned itself with history since man, then so too must geology. And such a scientific problem as the age of the earth prior to man was a "mere speculative enquiry of little practical importance," not very necessary "to explain existing phenomena."[3]

But while Hitchcock spent the next decade studying geology, teaching science at Amherst College, and heading the Massachusetts Geological Survey, explanations of both existing phenomena and the problem of dating in the whole realm of geological history, before and after the appearance of man, were brought into sharper focus. James Hutton's earlier theory that existing phenomena could be explained by reference to processes that occurred over an extensive period of time was elevated to a canon of geology by Charles Lyell.[4] The closer scrutiny of the location of fossils in relation to other geological formations, initially a jolt to both science and theology, provided a yardstick for the measurement of geological time.

Such a radical change in perspective among geologists made Hitchcock aware of the anachronism of his earlier position. It was clear to him now that the researches of science did not necessarily arise from Biblical concerns. The new strategy he adopted was to begin with nature or scientific findings and only then to proceed to the Good Book. He saw as his task to assure the world, especially any skeptics of science, that every finding of geology could then be corroborated in revelation, so that there was never any contradiction between the two accounts of nature. Unfortunately, this procedure too would eventually leave him intellectually dissatisfied.

Hitchcock's new detailed and mature discussion of the theological implications of geology was not to lie hidden in a book review or a lyceum address as were his earliest views; rather, it found voice in the *Biblical Repository*, the organ of Congregationalist Andover Theological Seminary. Spread over three articles and more than a hundred pages, his remarks predictably focused upon his belief that the "alleged disagreement [between geology and theology] is chiefly chronological."[5] He boldly asserted that the two records were equivalent and proceeded to give examples of how geological thinking paralleled interpretations of phenomena described in the Bible. He found that the revelation which depicted man as a recently created special being was supported by the prevailing scientific view that man was not very old. He found that the Bible spoke

[2] Edward Hitchcock, "Notice and Review of the Reliquiae Diluviannae," *Am. J. Sci.*, 1824, *8*: 338.

[3] Edward Hitchcock, *Utility of Natural History* (Pittsfield:Phineas Allen, 1823), p. 26.

[4] Lyell, incidentally, travelled out of his way to see Hitchcock's ichnology collection on his first American tour in 1841. See Charles Lyell, *Travels in North America, Canada, and Nova Scotia with Geological Observations* [1841–1842], 2 vols. (2nd ed., London: John Murray, 1855), Vol. I, pp. 251–256.

[5] Edward Hitchcock, "The Connection between Geology and Natural Religion," *Biblical Repository*, 1835, *5*:113–138; "The Connection between Geology and the Mosaic History of the Creation," *Bibl. Repos.*, 1835, *5*:439–451; "The Connection between Geology and the Mosaic History of the Creation," *Bibl. Repos.*, 1835, *6*: 261–332. Quotation is from the last article, p. 261.

of both fire and water as agents of change in the pre-Adamite earth, as did also the geologist. He suggested that the old doctrine of design with its heavenly order and divine purpose was not destroyed by the fact that geological change was so seemingly irregular, since that irregularity represented not "blind chance" independent of God's direction but revealed a larger "order by which the universe is sustained."[6] No less perfect were these geological operations over time with their deceptive discontinuities than the perturbations necessary to keep the astronomical bodies in their orbits. No less vital was God's immanent guidance of His system. In short, Hitchcock found pointless the suggestion that second causes or the operations of nature alone accounted for everything without the intrusion of the Great First Cause.

Expressed here in an unoriginal manner were the same confessions of faith that had sustained believing Protestant scientists from Newton onward—faith in God through acceptance of His word and faith in the efficacy of God's physical laws to regulate the works of His plan. Yet, since the time that Hume had been branded an atheist for suggesting that once it be granted that law controls, it is no longer necessary to postulate the existence of a Lawgiver, there had existed an incompatibility in these seemingly congruent beliefs. By insisting at once upon God as Lawgiver and Law-sustainer and also upon the sufficiency of physical laws to explain natural phenomena, Hitchcock found himself in the thick of this very dilemma, from which he would later try to free himself.

Hitchcock's position as self-appointed priest of American geology was quickly challenged by Moses Stuart, the most influential Congregationalist of his generation as Professor of Sacred Literature at Andover.[7] Stuart did not feel that science promoted atheism or that scientific pursuits tended to call the Bible into disrepute. Thus he was not the kind of divine for which Hitchcock's remarks had most obviously been intended. He was, moreover, eager to attack Hitchcock only upon the one point on which he had been most insistent—that science can and must be related to Scripture in detail.

Stuart presented his reply to Hitchcock in the same journal, beginning with a profession of personal ignorance of the details of geology but insisting, nevertheless, upon the pursuit of geology in all its minutiae, since "more light and knowledge on this subject, as well as on every other connected with the knowledge of the works of God, cannot fail in the end to be promotive of good."[8] His extensive sixty-page rebuttal formulated a single and consistent if not always self-evident viewpoint. He hid no subtle denunciation of Hitchcock or of any other men of science. He recognized that geologists were entitled to entertain their various views on scientific matters, just as theologians were allowed to hold their divergent views in their domain. He did not even charge scientists attempting to extend their concerns with either atheism or infidelity, granting to the geologists the same right as to other men of questioning the interpretations of the Bible.[9]

Stuart's thesis, rather, was that the Bible did not attempt to teach science, to tell a story translatable into more elaborate physical description as the details of science

[6] "The Connection between Geology and Natural Religion," p. 114.

[7] Moses Stuart, "Critical Examination of Some Passages in Gen. 1: with some Remarks on Difficulties that Attend some of the Present Modes of Geological Reasoning," *Bibl. Repos.*, 1836, 7: 46–106.

[8] *Ibid.*, p. 46.

[9] *Ibid.*, p. 55.

became clearer and clearer. He was a literalist, not in demanding that a Mosaic day as expressed in Genesis be considered a twenty-four hour period,[10] but in insisting that philology, not geology, was the only science to give better approximations to the intention of Moses' words. Unlike Hitchcock, Stuart did not believe it "impossible that [the Bible] should contradict . . . geology," in discussions of matters material, since Moses' words were never meant to be a scientific description or to contain within them the laws of any science.[11]

> Inspiration [as that received by Moses] does not make men omniscient. It does not teach them the scientific truths of astronomy, or chemistry, or botany, nor any science as such. Inspiration is concerned with teaching religious truths, and such facts or occurrences as are connected immediately with understanding them, or with impressing them on the mind. This is the object and extent of it; and to assume or suppose that it goes beyond this, is assigning a place to it which it was never designed to fill . . . modern science not having been respected in the words of Moses, it cannot be the arbiter of what the words mean which are employed by him.[12]

Hitchcock replied directly to Stuart's position in the next number of the *Biblical Repository* before shying away from an extended argument.[13] He was too active and too fanatically dedicated to his cause to see that some day he would have to seriously confront Stuart's contention. For the moment he concentrated on Stuart's statement that recent science could not help give the meaning of Moses' words. Hitchcock suggested that this contradicted Stuart's belief that modern philology could help interpret the meaning of the Bible. And even though Stuart had made it clear that "meaning" concerned moral judgments alone, Hitchcock continued to demonstrate how modern science had given a more precise meaning to the material references in the Bible. Was not the "rising of the sun," for example, better understood as a result of Copernican astronomy? Of course, if Stuart had chosen to prolong this debate he would have insisted that such precision is totally irrelevant to an understanding of the inspiration that was provided by the Biblical stories. Their lasting effect on the moral faculty was in no way altered by scientific findings.

It was obvious that there were to be three ways of interpreting the words of the Bible for believers of this generation. One, by literal interpretation of each word and phrase—a view associated with the fundamentalist minority. Another, by a symbolical interpretation with an insistence upon the equivalence of revelation and scientific facts—the position now held by Hitchcock. And a third, by a figurative connection between Biblical story and human moral condition—the position advocated by Stuart.[14] For example, by the first, or literal, a Genetical "day" of creation lasts exactly twenty-four hours; by the second, or symbolical, each "day" symbolizes some exact but longer period; and by the third, or figurative, a "day" is untranslatable, incommensurable with any literal or symbolical interpretation. Hitchcock's view, the one associated in the religion of geology with the postulate of geological epochs corresponding to

[10] Conrad Wright, "The Religion of Geology," *New England Quarterly*, 1941, *14*:344, describes Stuart as a "conservative literalist."

[11] Stuart, "Critical Examination," p. 54.

[12] *Ibid.*, pp. 80, 81.

[13] Edward Hitchcock, "Remarks on Professor Stuart's Examination of Gen. 1 in Reference to Geology," *Bibl. Repos.*, 1836, *7*:448–486.

[14] The three terms *literal*, *symbolical*, and *figurative*, which I have adopted for the sake of consistency, do not necessarily correspond to the same usage for these terms scattered through the works of the contemporary authors.

Biblical days, is the only one of the three which includes a belief that the findings of physical science can in any way actually improve our understanding of revelation. Indeed, it is the unifying theme of Hitchcock's labors—that science may help us to understand the Bible.

In the next decade, following the *Biblical Repository* series, Hitchcock published only a few articles on religion and science, all of which merely applied his thesis to other situations. He attempted, for instance, to demonstrate how the meaning of the Flood in Genesis could be properly circumscribed and made symbolic in order to correspond with the latest geological demonstrations that there could not have been an earth-covering deluge. But as academic theology had no further argument with geology after only some initial skepticism, Hitchcock, always eager to detect a new threat to his faith in either religion or science, happily reported in his address as first president of the Association of American Geologists at Philadelphia in 1841 that "such apprehensions are rapidly passing by."[15]

Hitchcock's tranquillity, his expectation that the future would consist in little more than fitting the chronology of the earth into that of Genesis, was rudely shattered in early 1845. The appearance of the American edition of the anonymous (Robert Chambers) *Vestiges of the Natural History of Creation* (1844) fixed his attention upon the problems of causation and miracle. And while not immediately abandoning the attempt to match chronologies, he nevertheless began to recognize that the connection between science and religion was not so simply stated.

The *Vestiges* supported the terrifying notion of development, by which natural processes alone could account for a progression to more advanced animal forms. It was not a new thesis. Those scientists who were not aware of this trend in biology directly from Lamarck, Geoffroy St. Hilaire, Bory St. Vincent, and Oken, were certainly aware of its exposition in Lyell's work. Yet Lyell and his successors, as well as Hitchcock, all had perfectly valid scientific objections to a belief in development; and so uneven and substandard was its presentation in the *Vestiges* that scientists, by rejecting it on their own grounds, no doubt unwittingly contributed to some religious apprehension that it was atheistic. In any case, the doctrine was shocking to most; for it combined the belief that nature was originally set into motion by God with the assertion that He subsequently did not interfere in its operation. And it provided no plausible scientific law to replace God in keeping the organic system working.

Yet with all the defects of the *Vestiges*, Hitchcock still called its thesis "ingeniously defended" and predicted that "a long drawn contest is yet before naturalists on this subject."[16] There were, of course, many issues involved, and among them were the following questions. How much of the incomplete spectrum of organic remains could one legitimately infer from the severely limited fossil evidence yet present in order to satisfy even the outward appearance of an evolutionary theory? Could the early postulates on the "inner drives" of organisms as framed by the developmentalists be the cause of uniform organic progression?

Retrospectively it is easy to disregard these problems and extract as the essence of the *Vestiges* the deduction that organic laws or mechanisms governed the entire realm

[15] Edward Hitchcock, *First Anniversary Address before the Association of American Geologists at their Second Annual Meetings* (New Haven: B. L. Hamlen, 1841), p. 45.

[16] Edward Hitchcock, *The Highest Use of Learning* (Amherst: J. S. & C. Adams, 1845), p. 31.

of life as inorganic laws governed inanimate substance. That idea, however, gained slow acceptance, for scientists were unwilling to allow laws different in form from those they customarily used in their other scientific work. Therefore, for the organic realm they held tenaciously to a belief in a Creator as essential for the production of new organic forms. As Gillispie has pointed out, in 1820 not a scientist in England would have denied that mechanism.[17] Even Lyell, whose work ironically prompted others toward a nontheistic conception of species formation, absolutely insisted that God was directly responsible for the creation of man. In 1845 there were few who disagreed.

The *Vestiges*, then, intensified the issue of how far one could venture into the organic world without calling on divine fiat. For some, such distinctions threatened the centrality of the human soul, but even without such moral arguments there still remained the vexing problem of whether life should be accounted for by scientific laws yet unknown or by divine fiat—fiat being nothing more than miracle or the suspension of ordinary law. Everyone, therefore, who sought to sustain any view of the origin and progress of organic life was compelled to reveal some attitude toward miracles.

Catastrophism, the regnant scientific position, often articulated a belief in miracles. Although usually associated with the catastrophe of the Biblical Flood, catastrophism generally de-emphasized this association and was more concerned with the "catastrophe of Creation," than of destruction, by which it served intellectually as a doctrine of progression. According to this school of thought, nothing but the personal operation of God, a procedure above and beyond known laws, could explain the absence of intermediate organic forms. Thus one could be scrupulously scientific and a devout catastrophist, as was Hitchcock. Indeed, biological catastrophism was the dominant mode of scientific thought among many who did not make the connection between miracle and creation explicit. Moreover, even noncatastrophist views were used to support a belief in some kind of miracle. As the Dutch historian of science Reijer Hooykaas has recently pointed out:

> . . . there is the generally received opinion that catastrophists are motivated by religious orthodoxy and, scientifically speaking, are on the wrong side, were it only on account of their mixing metaphysical arguments with scientific ones [but] allegiance to catastrophism may sometimes be a consequence of the desire to propound explanations conformable to facts only.[18]

With the customary unpredictability governing the deliverance of his views, Hitchcock took up the problems of law, miracle, and organic development in his inaugural address as president of Amherst College in 1845. Here he maintained that conclusions reached by science could not possibly promote atheism, and that any scientists who had joined the ranks of the infidels had done so "in spite of science, rather than through its influence."[19] From Hitchcock's observations it would seem that the *Vestiges*, now directing so much attention to organic substance, had provoked more interest among theologians than had the earlier shock of geological time. He

[17] Charles C. Gillispie, *Genesis and Geology* (New York: Harper Torchbooks, 1959), p. 96. See also Walter F. Cannon, "The Uniformitarian-Catastrophist Debate," *Isis*, 1960, *51*: 38–55, and "The Problem of Miracles in the 1830's," *Victorian Studies*,1960, *4*:5–32.

[18] Reijer Hooykaas, *Natural Law and Divine Miracle* (Leiden:E. J. Brill, 1959), p. 169.

[19] *The Highest Use of Learning*, p. 22.

assured them, however, that the long battle which scientists would wage among themselves

> . . . will not have so important a bearing on the cause of religion. . . . For even though these hypotheses of development should be established, an intelligent, spiritual, infinite Deity is quite as necessary to account for existing nature, as on the more common theory [catastrophism]. . . . I reject them, more because they have no solid evidence in their favor, than because I fear they will ultimately be of much injury to religion.[20]

He then recounted three generally accepted scientific conclusions supporting both his own objections to development and his continuing belief in God's governance as distinct from His mere existence.

> 1. There was a period, when no animals or plants existed on the globe, and, therefore an epoch when they were created.
> 2. There have been on the globe several nearly entire extinctions and renewals of organic life, each of which demands the agency of such a Being.
> 3. Man was only recently created,—almost the last of the animals; and since he is at the head of creation, nothing in nature has demanded a higher exercise of wisdom and power than his production; and, therefore, it must have required a Deity.[21]

In a number of occasional lectures delivered in the late 1840s Hitchcock reiterated these objections. Finally he incorporated them into his *Religion of Geology*, 1851.[22]

It is unfortunate that the *Religion of Geology* was the most widely circulated of Hitchcock's religious works and subsequently the easiest to find on library shelves. Even had Hitchcock not informed us in his autobiography that many of his journal articles on religion "cost [him] much more labor than some of his distinct volumes,"[23] we should certainly suspect it from the lack of organization which runs through the whole of the *Religion of Geology*. The work was obviously not intended to be Hitchcock's grand production on this theme, as his apology for it in the 1859 revision was to prove. But it does, nevertheless, at least focus our attention on the single theme of organic creation, putting an end to the bewildering array of evidence Hitchcock had combined with his poetical elegance to enjoin us from ascribing "operations of the natural world to nature's laws, instead of nature's God." Now we would be treated to subtle distinctions between miracle and scientific law.

Hitchcock's understanding of miracle forms part of his belief in God's total government—God as the arbiter at Judgment Day, the legislator who makes laws, and the executive who carries them out. Yet for all His power, God was not to be considered capricious in His use of it and could thus always be expected to perform the same miracle under the same circumstances; that is, Hitchcock believed in a law governing miracles. At first it may seem that an obvious contradiction exists between law and law of miracles, which is no more than a forced natural explanation for something admittedly supernatural. Hitchcock, nevertheless, discussed the difference in great detail and produced, for himself at least, some distinction between the two laws of common events and of miracles.

His reasoning, significantly, always begins with the statement that "every event in the universe takes place according to fixed laws," a thesis to which no scientist would

[20] *Ibid.*, p. 31.

[21] *Ibid.*, p. 32.

[22] Edward Hitchcock, *The Religion of Geology and Its Connected Sciences* (Glasgow: William Collins, 1851).

[23] Edward Hitchcock, *Reminiscences of Amherst College* (Northampton, Mass.: Bridgman & Childs, 1863), p. 391.

have objected. His next contention holds that natural law is "nothing more nor less than the uniform mode in which divine power acts," a somewhat vague and evasive statement, but one which the overwhelming majority of scientists endorsed. The last statement of belief, that "God has certain fixed rules by which he is regulated in the performance of miracles," probably did not concern many scientists unless they were also specifically interested in the Bible.[24]

Since the scientific observer is presented with events and not causes, Hitchcock rests his distinction between various methods or "causes" of divine governance on the appearance of the events. They are of three kinds: ordinary providence, special providence, and miraculous providence. Ordinary providence is another name for an event ultimately explicable according to mathematical formulas and physical theories, or the laws of nature. His use of the terms special providence and miraculous providence, not always clearly distinguished in the *Religion of Geology*, represents the battleground where his tensions were expressed.

Since miraculous providence is used as a synonym for special divine interposition, the two terms to be defined are actually "special providence" and "special divine interposition" (miracle). The basic distinction (not always adhered to) is that the special divine interpositions or miracles are events whose object is moral instruction but which definitely cannot under any circumstance be explained by natural laws.[25] Mere special providences, on the other hand, also have as their object moral instruction. But since they are produced by the conjunction of ordinary laws, they have the appearance of ordinary providences and hence we cannot tell when they have occurred. Of necessity, we must be aware that a miracle has occurred because no ordinary law or conceivable conjunction of them can account for the result. God produces these special divine interpositions according to His law, which cannot ever be known, in order to bring about results which cannot fail to be recognized.

If Hitchcock had not been truly sincere in this quest for faith and law, he could have rested with a definition of special providences alone. These would encompass phenomena like rainbows, which, though they have natural explanations, nevertheless would never have come about if God had not had a particular object in mind. Yet Hitchcock faced the difficulty of separating mere specials (natural) from miracles (supernatural) only that he might clearly insist upon a miraculous explanation for the organic creation, where no conceivable conjunction of natural occurrences could produce such a result as life. In all other circumstances except that of creation he did not find reliance upon law itself to be atheistic,[26] as did those few ready to condemn science for any of its revelations. He was, in fact, especially fond of the examples provided by Charles Babbage in his *Ninth Bridgewater Treatise* to demonstrate how calculating machines could be constructed to give results seemingly irregular, but actually according to the definite instructions of its manufacturer, an obviously possible analogy to the world situation.[27]

There are, of course, many questions that Hitchcock could not adequately answer. When, for instance, he realizes that "if miracles are performed according to law, as

[24] *Religion of Geology* (1851), p. 238.

[25] *Ibid.*, p. 269.

[26] *Ibid.*, p. 246.

[27] *Ibid.*, pp. 292–293; Charles Babbage, *Ninth Bridgewater Treatise. A Fragment* (1st American ed. from 2nd London ed., Philadelphia: Lea & Blanchard, 1841). This was not an official member of the Bridgewater series, and although Babbage intended it expressly to support religion, most thought it did otherwise.

much as common events . . . why is a present Deity any more necessary in the one case than in the other," he replies: "No event is any less God's work than if all were miraculous," thereby expressing his tension in holding views of both law and miracle.[28] Similarly a believer in the efficacy of prayer, he nevertheless recoils from suggesting that prayer's reward comes about by "suspension of law" and so relies upon "laws yet beyond our vision," which do not really fit into any of his classes of events. Thus he is forced to admit:

> True when we look at the subject philosophically, we must acknowledge that an event is just as really the work of God, when brought about by laws which he ordains and energizes, as by miraculous interposition. Still the practical influence of these two views of Providence is quite different.[29]

As a result of his concern to sustain the moral influence of Christianity by demonstrating how nature supported the same arguments, he was always dependent upon organic miracles. Yet, with the single exception of the revelation of the Word to man, he was totally convinced that everything was ultimately amenable to scientific law and description. As Hooykaas has suggested for many of the catastrophists, they were really acknowledging "the reign of Law . . . not only as a scientific but also as a metaphysical supposition."[30]

This unifying belief in law was what propelled Hitchcock into the description of Biblical tales, which, it must be emphasized, he never used as the arbiter of scientific theories. Committed to a belief in law, he was very reserved in his interpretation of miracles, especially during his careful analytical moments. For example, when speaking of the perplexing geological problem of the formation of rocks, he reasoned:

> If the rocks are an exception to the rest of nature (that is, if they are the effect of miraculous agency) there is no proof of it; and to admit it without proof is to destroy all grounds of analogical reasoning in natural operations; in other words, it is to remove the entire basis of reasoning in physical science.[31]

Similarly he cautioned against rushing to call the Deluge a miracle, in view of all the evidence against the universality of that purge.[32]

Most scientists, with all due piety characteristic of the times, did not have such detailed interest in the Bible and so probably found Hitchcock's examples neither provocative nor offensive, while the serious interpreters of the Bible were not working toward such scientific description. Thus only a middle ground, composed of those who for one reason or another feared that science did have irreligious tendencies, would be interested in the *Religion of Geology* as the blessing of a scientist upon religious belief. Still this middle ground grew in size, as a printing of the book appeared almost every year at either Glasgow, London, or Boston.

One of the most interesting reactions to Hitchcock's book came in a series of fourteen unsigned letters by an atheist, pseudonymously called Mr. Vindex, published in the *Boston Investigator* (1851).[33] This terribly vituperative and also anti-religious harangue accused Hitchcock of being "completely antagonistic to science, especially

[28] *Religion of Geology* (1851), pp. 240–242.

[29] *Ibid.*, p. 243.

[30] Hooykaas, *Natural Law*, p. 192.

[31] *Religion of Geology* (1851), pp. 31–32.

[32] *Ibid.*, p. 114.

[33] These letters are identified from the clippings preserved by Hitchcock in his papers, permanently house in the Robert Frost Library, Amherst College.

the science of Geology" and of trying to bring "science into the *subjection of the priesthood.*" In unsympathetic manner Mr. Vindex emphasized how absurdly contradictory the Bible was, maintaining that there was no single story to be reconciled with geology. Speaking specifically of Hitchcock's progression from a dismissal of the earth's history prior to creation, to a demonstration of the equivalence of Genesis and Geology by symbolical interpretation, to an emphasis on the creation of organic life, he engaged in such bombast as the following:

> . . . fifteen years ago this [emphasis on the historical age] was the language of the ablest geologists in and out of the church. In the church you denied entirely, or were making the days "periods of vast duration," but when you found that absolute fact forbid and demonstrated the impossibility, the fossil and skeleton remains of animals being found in other than the fifth and sixth periods and that there was an uninterrupted line of progressive development in all things, you hit upon the idea . . . [of] cycles of vast duration existing before the present species were created which now exist. You have now your six literal days for a last creative act, and your literal six thousand years. . . . Science will not permit . . . [it]. It is egregiously, impiously, infinitely absurd.[34]

Another anonymous reviewer in the *Bibliotheca Sacra*, who was, of course, no atheist, criticized Hitchcock for reliance upon outmoded Scriptural authorities and also for not recognizing the contradictions of the Bible.[35]

Hitchcock's argument had now formally been charged with being unnecessary for science and irrelevant to religion. Since Moses Stuart had pointed out the latter difficulty fifteen years earlier, it was perhaps fitting that in the same issue of the *Bibliotheca Sacra* which eulogized Stuart (who had died the year earlier) Hitchcock finally gave evidence of capitulation.

In an article entitled "The Relation and Consequent Mutual Duties between the Philosopher and the Theologian" he delineated the aims of the two branches of knowledge:

> The object of philosophy is to explain the phenomena of nature, mental, moral, and material; that of theology is exclusively to defend and enforce the moral relations of the universe. Hence the two subjects are almost entirely distinct in their aim.[36]

This was certainly not an original observation but only the first time that Hitchcock had made it. In strict orthodox form he defined the only overlap in subject matter for these two disciplines as the demonstration of the "character of man as a fallen being."[37] Only "incidentally" did they both treat of the subjects of creation, the Deluge, and organic races. Yet Hitchcock faltered in following his new doctrine to the logical conclusion, and in the end he acceded only equivocally to Stuart's position of almost twenty years earlier.

> But since revelation does not pretend to teach science, nor even to use language, in its strictly scientific sense, we ought to expect in such cases, only that there shall be no real, although there may be an apparent discrepancy between the two records.[38]

In this last effort to erase "apparent discrepancies" Hitchcock showed himself incapable of abandoning altogether his quest for consistency between the natural and the Biblical.

[34] *Ibid.*, Letter X.
[35] *Bibliotheca Sacra and Biblical Repository*, 1851, *8*:662–664.
[36] *Bibl. Sacra*, 1853, *10*:166–194.
[37] *Ibid.*, p. 167.
[38] *Ibid.*

In the next issue of the same journal Hitchcock attempted to straighten out some of his earlier views about miracles, demonstrating even more dramatically how thoroughly committed he was to law.[39] Here miracle is understood to be the "special divine interposition" of his earlier terminology, that is, an event which has a particular object in mind and which is not produced by some configuration of the ordinary laws of nature. He treaded dangerously close to the materialistic heresy that asserted the primacy of matter in describing physical reality when he decided to emphasize that God did not have to interfere directly in order to bring about the special result. He explained that "no new plan or motive of action can ever enter the Divine mind, and consequently whatever plans we find developed in God's government, must have been perfectly formed in the counsels of eternity."[40] God himself "never acts except under the guidance of those fixed principles which we call law." But the law of miracles, by which God is bound, does indeed "differ from all others, and this constitutes a miracle."

Would he then allow for a law of miracles juxtaposed to the laws themselves to produce only special providences, with miraculous providence never occurring at all? Not so quickly:

> Admit, if you choose, that all other events on the globe, even the creation of all other organic beings, might have been accomplished by ordinary laws; yet, so long as the great fact of man's creation stands out so conspicuously on our world's history, we need nothing more to establish, beyond cavil, the reality of Divine interposition in nature. . . . And this grandest miracle of nature is also the greatest of revelation.[41]

Naturally, since no man shakes off old habits of thought overnight, we can still find discrepancies in Hitchcock's pronouncements as late as 1863. But the view that he had enunciated in 1835—that the problem of relating geology and religion is "chiefly chronological" and that the six days of creation in Scripture correspond to geological epochs—has entirely disappeared. No longer would Hitchcock conceive of every geological fact as fitting into a neat Biblico-scientific framework; if the fossil record proved that there had been death in the world before sin (Adam), then Hitchcock would concede the truth of that discovery.[42]

The relationship between religion and science was discussed quite vigorously in the years between the first publication of the *Religion of Geology* in 1851 and its revision in 1859. Much of the attention which the subject received was inspired by the writings of the Scot Hugh Miller, whose works first appeared in American editions in 1850. In his *Footprints of the Creator* (1847) Miller relied upon detailed analysis of one of the known geological strata to present some cogent evidence against the slow progressive organic development somewhat dogmatically proclaimed in the *Vestiges*. Miller, who, like Hitchcock, was both demonstrably devout and scientifically creditable, gave new support to the doctrines of geological epochs corresponding to Pentateuchal days and to God's interference subsequent to creation. Finding many supporters among the

[39] Edward Hitchcock, "Special Divine Interpositions in Nature," *Bibl. Sacra*, 1854, *11*:776–800.

[40] *Ibid.*, p. 780.

[41] *Ibid.*, pp. 793–794.

[42] First announced in *Religion of Geology* (1851), p. 35.

religiously conservative, Miller was awarded an honorary degree by Brown University in 1854, two years before his death.[43]

Remarkably diverse personalities were to make Miller their rallying point in the ensuing wave of attention showered upon science and religion. The Academy of Natural Sciences in Philadelphia used the occasion of the renewed controversy to decry the fact that religious concerns diverted too much of the scientist's valuable attention away from science.[44] The American Association for the Advancement of Science seemed to contribute to the same voice; however regularly its early proceedings had been punctuated by the religious debates, the 1856 meeting at Albany heard the Reverend Mark Hopkins, president of Williams College, urge the abandonment of the tortured task of relating science to Scripture in minute detail. While fully endorsing scientific investigations, Hopkins insisted that

> . . . all knowledge is not scientific, or rather science is not all knowledge, nor can scientific knowledge in any case reach the essence of things. The inference from any particular science that there is, or is not, a God, is not a part of that science.[45]

But the religious community, now more than ever, showed itself not of one mind. Benjamin Silliman reported in 1853 that after a lecture on modern geology in St. Louis a local minister came up to him and "let down half the geologists as infidels—and confined the whole works to six common days, insisting that the word would bear no other significance."[46] On the other hand, Tayler Lewis, Professor of Greek at Union College, published an entire book to demonstrate that the Bible called for six epochs in the work of creation, expecting all to believe his contention, in utter disregard of geological discovery. As a scientist on the Union faculty predicted, this attitude caused a "stir among the theologians and Naturalists [both]."[47] In the ensuing debate a minister in New Haven gave a sermon on "Creation and [versus] Christianity," exhibiting Hugh Miller as the great Christian who slew the infidel *Vestiges*,[48] while a fundamentalist produced a book condemning all geologists to the flames, including Miller.[49] One reviewer of this last work, decrying the fact that anyone could still so unceremoniously laugh at scientific truth, claimed that such a view made even the "infidels scoff, the scientific laugh, and the judicious grieve."[50]

It was obvious that the debates were more widespread than they had been in previous decades. Yet, quite surprisingly, Hitchcock did not attempt to restate his particular position on the obvious issue of the times. The address which he delivered at the Albany meeting of scientists and elsewhere said little about Hugh Miller and nothing about the epochs; rather it concentrated on the miraculous origin of man.[51] When the *Religious Truth, Illustrated from Science* (1857) actually appeared[52]—a collection of

[43] *Historical Catalogue of Brown University, 1764–1914* (Providence:Brown University, 1914), p. 649.

[44] *Notice of Some Remarks by the late Mr. Hugh Miller* (Philadelphia, 1857), p. 4.

[45] Mark Hopkins, *Science and Religion* (Albany:Van Benthuysen, 1856), p. 24.

[46] MS, Benjamin Silliman to Edward Hitchcock, Oct. 13, 1853, President Hitchcock Collection, Amherst College.

[47] Tayler Lewis, *The Six Days of Creation or the Scriptural Cosmology* (Schenectady:Riggs, 1855). MS, Jonathan Pearson: Diary, July 15, 1855, Union College Archives.

[48] MS, Chauncey Allen Goodrich [1857], Goodrich Family Collection, Yale University Library.

[49] Thomas A. Davies, *Answer to Hugh Miller and Theoretic Geologists* (New York:Rudd & Carlton, 1860).

[50] *New Englander*, 1860, *18*:1103.

[51] Edward Hitchcock, *The Religious Bearings of Man's Creation* (Albany:Van Benthuysen, 1856).

[52] Edward Hitchcock, *Religious Truth, Illustrated from Science, in Addresses and Sermons on Special Occasions* (Boston:Phillips, Samson, 1857).

essays containing this address and other works dating back to the 1840s—one clerical reviewer seized upon the opportunity to insist that Hitchcock had still not realized that the Bible "expresses scientific error" and that the findings of geology are "irreconcilable with the [contradictory] narrative in its present state."[53] With respect to Hitchcock's restated belief that we ought to expect that "science rightly understood, should not contradict statements of revelation, correctly interpreted," Reverend Ellis said that this is indeed "*more* than we ought to expect." Now, from a Unitarian, Hitchcock had been told again that the interpreters of the Bible "make no stand for his [Moses'] literal accuracy, whether in scientific or historical matters."[54]

Not until 1859 did Hitchcock publish the second edition of his *Religion of Geology*.[55] If we compare this 1859 Boston edition with the Boston edition of 1852, we find that the first 511 pages, with one slight but significant difference, are identical in all details of text and footnotes, the later edition having probably been pressed from the same plates as the earlier.[56] Only page 67 in the second edition has been inconsequentially altered in the text so that a footnote might be changed without destroying the pagination. In the first edition the note refers the reader to the 1835 *Biblical Repository* articles for the essential statement of the author's view of creation. In the second edition that note is replaced with the statement: "I have been in the habit of thinking it safer to stop with the theory of a long interval between the 'beginning and the demiurgic days,' than to regard the days as figuratively long periods" (symbolical, in my terminology).[57] There then appears a reference to the end of the edition where an entirely new "synoptical" chapter of sixty-eight pages appears, in which Hitchcock makes his views upon certain matters explicit—views that we had seen developing as early as 1853. Here he admitted that the Bible was meant to be a popular account of truth and that

> . . . if we expect to find in it all the principles of modern science, or of systematic theology, we shall be disappointed. . . . [It is] sufficient to satisfy any reasonable mind that the Scriptures teach nothing contrary to science, although we may find nothing in it about our favorite theories. If we can only be satisfied with general principles . . . without attempting to find something in Scriptures corresponding to all the details of science, or something in nature corresponding to every particular in revelation, we shall find harmony and mutual corroboration where an unwise and unauthorized attempt to extend the parallelism to details might leave us in doubt and perplexity. . . .[58]

It must have been a welcome relief for Hitchcock to rid himself of the burden of exact correspondences. Now instead of remaining outside of the debates about Hugh Miller, Hitchcock could admit that he shared the same religious faith as Miller, while at the same time pointing out all the scientific errors inherent in that man's attempts to relate the two chronologies.[59] In that context Hitchcock again insisted that "if we attempt to descend from these generalities and to show how all the details of the six days' work are likewise placed in exact chronological order, I think that we must involve ourselves as well as the sacred text, in inexplicable difficulties."[60] The *Biblio-*

[53] Rufus Ellis, "Buchanan and Hitchcock on Religion and Science," *Christian Examiner and Religious Miscellany*, 1857, *62*:445–460; quotation from p. 460.

[54] *Ibid.*, pp. 458–459 (my italics).

[55] *Religion of Geology* (new ed., Boston: Phillips, Samson, 1859).

[56] *Religion of Geology* (Boston:Phillips, Samson, 1852).

[57] *Religion of Geology* (1859), p. 67.

[58] *Ibid.*, p. 526.

[59] *Ibid.*, pp. 539–545.

[60] *Ibid.*, p. 545.

theca Sacra, reviewing the second edition, could now give him the praise he so long sought: "Prof. Hitchcock resists the attempt to interpret the first chapters of Genesis as scientific statements of geological truth."[61]

Hitchcock's final literary effort on science and religion came in 1863 just before his death. At that time he confronted the issue of evolution as expressed by Darwin—a problem in science that was certainly not new to him. Unlike the body of his *Religion of Geology*, with its old and often imprecise if not contradictory statements, his last article was written "that there may be no mistake as to our [his] meaning."[62] This paper was concerned with the necessity of God to sustain His works. Insistent upon Calvinist doctrines of depravity and redemption, Hitchcock made few specific references to the Bible. He was at that time decidedly anti-Darwinian, although, it is important to emphasize, only for the lack of human paleontological evidence; his final opinion on evolution was thus that "the real question is, not whether these hypotheses accord with our religious views, but whether they are true."[63]

In this final judgment Hitchcock clarified the nature of his objections to the Darwinian theory of evolution, revealing that his skepticism grew from scientific rather than religious grounds. His integrity as a scientist had never permitted him to dismiss well-founded evidence out of religious scruples: even throughout the early discussions of organic evolution and the purely geological debates that had preceded them, Hitchcock had maintained, not that religious views should modify our explanations of nature, but that "the discoveries and inferences of geology [may] . . . modify our views of any religious truth, natural or revealed."[64] Thus, here again, in his specific response to the problem of Darwinism, he took a clear stand on the side of scientific method above religious leaning. Doubtless had Hitchcock lived to see, decades later, the physical confirmations of Darwin's hypotheses, his peculiar religious concerns would not have stood in the way of his accepting them as scientific fact. As it is, he stands as an interesting example of the highest level of intellectual development possible for the skeptical scientist of the pre-Darwinian age.

Indeed, Hitchcock is thoroughly representative of the scientific temper of his times, however bizarre the twentieth century may think his belief in the miraculous catastrophe of human creation or his passion for reconciling, so far as possible, Biblical and geological records. Scientists of his day were generally in perfect agreement with his major metaphysical presuppositions, even if they expressed them less frequently; for the beliefs that God exists, that man enjoys a relationship with God, and that an observable order exists in God's world were faiths common to the Christian, the scientist, and the Christian scientist together. Thus Hitchcock and his contemporaries gave wide acceptance even to the now long discredited notion of divine intervention in the regular order of nature to provide some special meaning for human history. As one American scientist put it, there was not "anything in the constitution of Nature so far revealed to us by the discoveries, at all at variance with it."[65] And in England the

[61] *Bibl. Sacra*, 1860, *17*:229.

[62] Edward Hitchcock, "The Law of Nature's Constancy Subordinate to the Higher Law of Change," *Bibl. Sacra*, 1863, *20*:489–561.

[63] *Ibid.*, p. 524.

[64] *Religion of Geology* (1859), p. 516.

[65] George I. Chace, *The Relation of Divine Providence to Physical Law* (Boston:Ticknor & Fields, 1854), p. 6.

mathematician Babbage pronounced his *a priori* acceptance of the notion that

> It is more probable that any law, at the knowledge of which we have arrived by observation, shall be subject to one of those violations which according to Hume's definition constitutes a miracle, than that it should not be so subjected.[66]

It may seem strange to this century that many eminent scientists of the last were so religious as to feel moved to express their persuasions in scientific discussions now considered inappropriate as vehicles for such views. Yet such protestations of faith were common to the nineteenth century, and we must guard against hastily concluding that a naïve relationship existed between intensity of religious belief and degree of scientific acumen.[67] It was as common in Hitchcock's day, for instance, for believing Christians to accept the Darwinian theory of evolution, in part or in full, as it had earlier been unextraordinary for religious skeptics to reject the findings of Darwin's predecessors. A brief survey of nineteenth-century scientific opinion, in fact, reveals that those who wrestled with the problems of science *vis-à-vis* religion reached scientific conclusions as varied as those who did not. Hitchcock, himself a believer, was able to find scientific merit even in the *Vestiges* where an ardent Darwinian and skeptic like Thomas Henry Huxley, as Lovejoy pointed out, had been unable to penetrate scientific error to look for scientific truth.[68] Asa Gray, who as the first Darwinian in America eagerly denied repeated acts of divine intervention for being inconsistent with the evolutionary concept of new species production, nevertheless relied upon the "divine will" to supply the energy for Darwin's "natural selection." And a recent historian has even provided us with an interesting example of a debate in which a theologian is cast in the role of defender of development and a scientist appears as the champion of older scientific orthodoxy.[69]

Just as religious opinions would provide a faulty guide for scientists who sought to employ them in their theoretical work, so the mere existence of extrascientific beliefs is of little help to the historian who seeks to understand how scientists reacted to particular situations. Anti-Darwinians in 1860, therefore, ought not be dismissed as simple-minded fundamentalists or religious apologists, for those scientists—Hitchcock among them—had left no room in their debates over the previous decades for the arbitration of scientific theory by faith alone.

[66] Babbage, *Ninth Bridgewater Treatise*, p. 145.

[67] Joseph Leon Blau, *Men and Movements in American Philosophy* (New York:Prentice-Hall, 1952), p. 79, writes in this vein. Other hasty conclusions about Hitchcock specifically are found in Howard Mumford Jones, *Ideas in America* (New York:Russel & Russel, 1965), pp. 133 ff.; John Dillenberger, *Protestant Thought and Natural Science* (New York:Doubleday, 1960), p. 214; and most recently in Paul F. Boller, Jr., *American Thought in Transition: The Impact of Evolutionary Naturalism, 1865–1900* (Chicago: Rand McNally, 1969), p. 12. It is therefore refreshing to acknowledge that Walter P. Metzger, *Academic Freedom in the Age of the University* (New York:Columbia Univ. Press, 1955), pp. 18 ff., does advocate caution in studying pre-Darwinian scientists in the context of religion *versus* science.

[68] Arthur C. Lovejoy, "The Argument for Organic Evolution before the 'Origin of Species,'" *Popular Science Monthly*, 1909, *75*:499–514 and continued pp. 537–548. See pp. 502 ff.

[69] Morgan B. Sherwood, "Genesis, Evolution and Geology in America before Darwin: The Dana-Lewis Controversy, 1856–1857" in *Toward a History of Geology*, ed. Cecil J. Schneer (Cambridge, Mass.:M.I.T. Press, 1969), pp. 305–316. See pp. 312 ff.

Nineteenth-Century State Geological Surveys: Early Government Support of Science

*By Walter B. Hendrickson**

AMONG the earliest efforts of American government to give financial support to science in the nineteenth century was the authorization by state legislatures for geological surveys.[1] Both the idea and the method of conducting a geological survey were conceived in Europe, especially by James Hutton and William Smith in Great Britain at about the turn of the eighteenth century.[2] Pioneer attempts at comprehensive surveys in America were made by Johann Schöpf, Constantin F. Volney, and William Maclure;[3] and surveys of limited areas were carried out by Samuel L. Mitchell, Benjamin Silliman, Samuel Akerly, Horace H. Hayden, Samuel L. and James F. Dana, Edward Hitchcock and Amos Eaton.[4] They were privately financed, often by the doctors, ministers and college professors who did the work and who had, in the eighteenth century tradition, a real, but dilettante interest in natural science.

Another group of rudimentary surveys were those made by United States government exploring parties sent into the Mississippi Valley and into the Far West; and the reports of Lewis and Clark, Henry Schoolcraft, Stephen H. Long and others contain a mass of information, not always accurate, on the rocks and ores that were located by more or less casual inspection of the country through which they traveled. Perhaps the most ambitious of these surveys was that made by Schoolcraft in 1819, and written up as *A View of the Lead Mines of Missouri; Including Some Observations on the Mineralogy, Geology, Geography, Antiquities, Soil, Climate, Population, and Productions of Missouri and Arkansas and other Sections of the Western Country.*

* MacMurray College, Jacksonville, Illinois.

1 A. Hunter Dupree, in *Science and the Federal Government* (Cambridge, 1957), has discussed federal aid to science. There is no comparable study of state aid, and it is hoped that this paper may show the way toward such a project by examining some of the factors involved in the establishment of geological surveys.

2 Archibald Geikie, *Founders of Geology,* 2d ed. (London and New York, 1905), pp. 381-396; Frank D. Adams, *The Birth and Development of the Geological Sciences* (Baltimore, 1938), pp. 269-275; Carroll L. and Mildred A. Fenton, *The Story of the Great Geologists* (Garden City, 1945), pp. 78-83; "The Museum of Practical Geology," reprinted from *Frazier's Magazine,* n. d., *Eclectic Magazine,* 1851, *23:* 480.

3 Josiah D. Whitney, "Geographical and Geological Surveys," Part II, *North American Review,* 1875, *121:* 291. See also George P. Merrill, *The First One Hundred Years of American Geology* (New Haven, 1924), pp. 1-74, *passim.* Cited hereafter as Merrill, *One Hundred Years.*

4 Merrill, *One Hundred Years,* pp. 1-74, *passim.*

Schoolcraft also explored through the Great Lakes and the upper waters of the Mississippi in 1821 and 1822.[5] In his reports there was much emphasis on the economic value of the minerals and other natural resources to prospective settlers in the West, not so much from the viewpoint that individual settlers might get rich, but rather that as these abundant resources were developed, the wealth and greatness of the United States as a nation would be enhanced.

But the direct forerunner of the state survey was the detailed systematic geological survey of Rensselaer County, New York, made by Amos Eaton in 1821 and paid for by the land-wealthy Stephen Van Rensselaer. This survey had a definite economic purpose, and in addition to making a geological section and giving a lithological description, Eaton noted the kinds of soil, the most suitable crops and the best methods of cultivating them—all matters of concern to a landowner. In the next year, again under Van Rensselaer's patronage, Eaton made a survey of the entire route of the Erie Canal, and his findings were published in 1824.[6]

By the 1820's there was thus the beginning of the idea that the geologist, in addition to being concerned with the description of rocks and minerals, their origin, and their history, could also locate and evaluate mineral and soil resources so that they might be exploited as a source of wealth to the individual and to the state.

The first state geological surveys were those of North and South Carolina in 1823 and 1824. The prime mover in North Carolina was Denison Olmstead, a native of Connecticut and a graduate of Yale, where he had been a student of Benjamin Silliman. In 1817 he had accepted the post of professor of chemistry at the University of North Carolina.[7] Here, as in other Eastern states, there was much concern with the problem of improving transportation routes, particularly by dredging rivers and digging canals. In 1819 the legislature created a Board of Internal Improvements to prepare plans and surveys. At the same time, the Board was to make a recommendation on whether or not there would be enough economic advantage to justify the cost. It was on this point that Olmstead hoped to shed some light when he wrote to the Board asking for an appropriation of not more than a hundred dollars to pay his expenses while he made a geological survey of the state during his summer vacation. He was actually offering to supply the Board of Internal Improvements with information as a sort of by-product of his summer's work, but he thought that for this he was entitled to some financial reimbursement for his expenses. He was not seeking to make any profit himself, he said, and he wrote to the Board:

> The acquisition of knowledge, by which I might be better able to fulfill the duties of my profession and the opportunity of furnishing a geological descrip-

[5] *Travels in the Central Portions of the Misssissippi Valley: Comprising Observations on its Mineral Geography, Internal Resources, and Aboriginal Population* (New York, 1825).

[6] Ethel M. McAlister, *Amos Eaton* (Philadelphia, 1941), pp. 298-310; James Hall, "The New York Geological Survey," *Popular Science Monthly*, 1883, *22:* 816.

[7] Merrill, *One Hundred Years,* p. 94; Whitney, "Geographical and Geological Surveys, Part II," *North American Review,* 1875, *121:* 298.

tion of this hitherto undescribed country to the American Geological Society of which I have the honor to be a member, would be all the recompense I require; and the collection of specimens to illustrate my lectures, as well as increased ability to impart information to my pupils respecting their native state, would be the means of securing some advantage to the university.[8]

Thus, rather vaguely, the idea was taking form in the minds of scientists that the state should support research because it would be economically advantageous to the state, and at the same time the scientist would be able to carry on his private research at the expense of the state. This view was expressed publicly and more concretely by Olmstead's friend and former teacher, Benjamin Silliman, the influential editor of the *American Journal of Science,* who wrote:

From the intelligence, zeal and scientific attainments of Professor Olmstead, we cannot doubt, that (*if adequately encouraged by the local government, or by patriotic individuals*) the enterprise will produce important advantages to science, agriculture, and other useful arts, and will prove honorable to the very respectable state of North Carolina [Italics in the original].[9]

Although Olmstead did not succeed in getting any pecuniary aid from the state on his first try, he persisted in his efforts and in 1823, the legislature authorized the State Board of Agriculture to have a geological survey made, and appropriated $250 a year for four years.[10] In the next year, Lardner Vanuxem, professor of chemistry and geology at the South Carolina College, was the recipient of a $500 appropriation for "making a geological and mineralogical tour during the recess of the college and furnishing the specimens of the same." While these surveys provided useful information for agencies of the state governments, they gave nothing of permanent public value because the legislatures did not provide money for publishing the geologists' reports.[11]

The first full-blown state supported geological survey was that of Massachusetts, authorized by the legislature in 1830. Massachusetts, like other states, was tackling the problem of internal improvements, and as a guide to a solution, the legislature established a trigonometrical geographical survey. Just as in North Carolina, it was felt that roads and canals should be located where there would be the most commodities to transport. Further, in order to justify large expenditures for roads it was desirable that new needs for them should be developed. Governor Levi Lincoln said to the legislature:

[8] Reprinted from the minutes of the Board of Internal Improvements in George P. Merrill, ed. and comp., *Contributions to a History of American State Geological and Natural History Surveys,* Smithsonian Institution, United States National Museum, *Bulletin 109* (Washington, 1920), 365. Cited hereafter as Merrill, *State Surveys.* This is a collection of source materials: state laws, governors' messages, reports and historical accounts by state geologists.

[9] "Notice of a Geological Survey of North Carolina," *American Journal of Science,* 1822, *5:* 202.

[10] Merrill, *State Surveys,* p. 365.

[11] Merrill, *One Hundred Years,* p. 122; Merrill, *State Surveys,* p. 459; Whitney, "Geographical and Geological Surveys, Part II," *North American Review,* 1875, *121:* 299; Charles W. Hayes, comp., *The State Geological Surveys of the United States,* United States Geological Survey, *Bulletin 465* (Washington, 1911), 136. This is another collection of state laws and historical accounts by state geologists.

> Much knowledge of the natural history of the country would be thus gained [by a geological survey] especially the presence of valuable ores, the extent of quarries, and of coal and limestone, objects of inquiry so essential to internal improvements, and the advancement of domestic prosperity would be discovered, and the possession and advantage of them given to the public.[12]

At first, following the pattern laid down in North Carolina, the legislature appropriated only a hundred dollars to pay the geologist's expenses and to provide for printing the "geological features of the state" on the map the topographical survey was to make. In succeeding years, however, $2500 to $8200 was appropriated annually for salary and expenses, and for publication.[13]

The man appointed by Governor Lincoln to make the geological survey was Rev. Edward Hitchcock, professor of natural science and chemistry, and later president of Amherst College. Hitchcock was vigorous physically and mentally. He encompassed the whole range of natural science in his interests, writing fourteen volumes, five tracts and seventy-five papers on botany, mineralogy, geology, physics and chemistry, to say nothing of extensive publications on religion and current social and political problems. A careful and thoughtful scientist, he became an authority on fossil footprints, and wrote an elementary geology textbook that went through thirty editions.[14]

In many ways Hitchcock was the ideal state geologist. Well-trained in his field, he had a social and intellectual standing that was respected by the legislature, a sense of responsibility to the public, and a smooth, politic manner that readily won him friends and enabled him to influence people. Hitchcock said that his "commission and instructions" meant that he "was to have principally in view, in [his] examinations practical utility; not neglecting, however, interesting geological facts, which have an important bearing on science."[15]

Hitchcock made his first report on his examination of the geology of Massachusetts in 1832. It was subtitled *The Economic Geology of the State,* and it located and described the classes of rocks that lay under the soil, and rocks and minerals that were "useful in the arts": building stones, slates, clays, marls, peat, coal, graphite, and the ores of lead, iron, zinc, copper, manganese, silver, and gold. It must have been a disappointment to the legislators that of these, so Hitchcock reported, only building stones were present in valuable quantities, although he left the matter open by saying that there were many other natural products whose value could be determined only after further prospecting.[16]

In fulfilling his instructions quite exactly, Hitchcock gained the confidence of the legislature, the survey was continued, and its scope was enlarged to include a list and description of plants and animals, and a report on soils. When Hitchcock finished his work, the results were published in two quarto

[12] Edward Hitchcock, *Report on the Geology, Mineralogy, Botany, and Zoology of Massachusetts* (Amherst, 1833), iii; Merrill, *State Surveys,* p. 149.

[13] Merrill, *State Surveys,* pp. 149-155; Edward Hitchcock, *Final Report on the Geology of Massachusetts,* 2 vols. (Northampton, 1841), I, v-vi.

[14] Merrill, *One Hundred Years,* p. 152.

[15] "Report on the Geology of Massachusetts," *American Journal of Science,* 1832, *22:* 2.

[16] Hitchcock, *Report on the Geology, etc.,* (1833), *passim.*

volumes with about half the space given to economic geology, and half to "scientific" geology, including what was later issued as his textbook on elementary geology.[17]

Following the lead of Massachusetts, twenty other states established surveys in the decades of the 1830's and '40's.[18] In some states — Maryland, Connecticut, Virginia, Georgia, Indiana, North Carolina, South Carolina and Massachusetts — there was a close tie between the demand for systems of internal improvements and the authorization of geological surveys. In all states a major purpose was to locate, describe, and publicize such natural resources as salt and mineral springs, building stones, shales, clays, slates, coal, and ores.[19] With this information in hand, any person would have a basis for judging how successfully they might be exploited.

Usually the geologist in charge of the survey was required to collect mineral, rock, and ore specimens and display them in some central place, and he was frequently instructed to distribute suites of specimens to academies and colleges within the state. Finally, the geologist had to make a formal report, which, in some cases, was printed and put into the hands of state legislators and other officers, as well as being sent to educational and scientific organizations.

Back of the concern of state legislatures that geological surveys should produce useful information lay the doctrine of mercantilism which held that "the state has a broad sphere of action in the economic field."[20] In a period when the frontier was rapidly moving westward, when the industrial revolution was in full swing, and when material success was held to depend on individual initiative and enterprise, laissez-faire philosophy would seem to explain the economic system. It was true that the number of men who made direct profit from the findings of the geologist was exceedingly small, since only the capitalist and business entrepreneur could invest the money necessary to develop mines and quarries and so get rich. It was, however, believed that any aid given by the state to a particular group would help all other groups since "all were bound together in an indissoluble community of interest."[21] As Professor and Mrs. Oscar Handlin have shown, the laissez-faire idea "that the people individually, and not the government are the judges of their interests . . . was

[17] See "Final Report on the Geology of Massachusetts," *North American Review*, 1843, *56:* 435-51.

[18] These are the states, with date of establishment of the first survey: North Carolina (1823), South Carolina (1824), Massachusetts (1830), Tennessee (1831), Maryland (1833), New Jersey (1833), Connecticut (1835), Maine (1836), New York (1836), Ohio (1836), Pennsylvania (1836), Virginia (1836), Georgia (1836), Delaware (1837), Indiana (1837), Michigan (1837), Kentucky (1838), New Hampshire (1839), Rhode Island (1839), Vermont (1845), Alabama (1848), Mississippi (1850).

[19] For some examples see Merrill, *State Surveys:* Massachusetts, p. 149; Tennessee, p. 464; Maryland, pp. 138-139; New Jersey, p. 308; Connecticut, p. 46; Virginia, p. 509; Georgia, p. 56; Delaware, p. 51; Indiana, pp. 72-73; Michigan, p. 158; Kentucky, p. 102; New Hampshire, p. 299; Rhode Island, pp. 456-57; Vermont, p. 497.

[20] Louis Hartz, *Economic Policy and Democratic Thought: Pennsylvania, 1776-1860* (Cambridge, 1948), p. 4. Missouri, which felt that it could not afford a geological survey, appealed to Congress to authorize one for the state. See James W. Primm, *Economic Policy in the Development of a Western State: Missouri, 1820-1860* (Cambridge, 1954), p. 97.

[21] Oscar and Mary Flug Handlin, *Commonwealth: A Study of the Role of Government in the American Economy: Massachusetts, 1774-1861* (New York, 1947), pp. 55-56.

subversive to the end and aim of all government and so was . . . utterly impracticable."[22]

There are many mercantilist statements that refer specifically to the exploitation of natural resources. For example, a note in *Niles Register,* quoting the Bucks County (Pennsylvania) *Intelligencer,* said:

> The coal regions of our state possess much interest in a domestic, a political, and a scientific point of view, and claim the fostering attention of those who delight in the development of our natural resources of wealth, and the prosperity of our enterprising citizens.

The writer goes on to say that, because a few "public spirited citizens" have opened up producing mines,

> the coal trade has assumed an aspect of greatest importance in regard to individual comfort, state commerce and state economy. Under these circumstances it is of some consequence that the people should be acquainted with the coal interest, as it is intimately connected with the state prosperity.[23]

The *American Daily Advertiser* was quoted by the writer, because it pointed out that there must be many unknown rich coal beds — the "hidden treasure" of the state — "which might reward the labor of a full topographical and geographical survey of the state, which has been so frequently urged upon the attention of our legislatures by individuals and scientific associations."[24]

"A Farmer," also writing in *Niles Register,* urged that there should be a more general knowledge of science, and advocated county "Museums" for public education. He said that scientists had made such important discoveries of natural resources that it was clear "that the prosperity of individuals and of the nation is in proportion to the industry, the skills, and the general intelligence, which is applied in unfolding and appropriating these gifts of nature. . . ."[25]

That these ideas motivated public policy is evidenced by this resolution of the Maine legislature:

> That . . . a geological survey of this state, upon a basis commensurate with the magnitude and variety of its territory, . . . is an enterprise that may rightfully claim the encouragement of every class of industry, as involving more or less of probable utility to each other and is intimately connected with the advancement of the arts and sciences, of agriculture, manufacturers, and commerce.[26]

Governor Noble of Indiana was even more positive in declaring that the many as well as the few, would benefit from a geological survey. He told the legislature that "in this State, we have external indication of large beds of coal, and other natural deposits; but for want of the proper test of science, their extent and value are unknown." He said that he was "satisfied that our mineral resources, properly developed, will give employment to thousands, subserve the

[22] *Ibid.,* p. 55. See also Oscar Handlin, "Laissez-faire Thought in Massachusetts, 1790-1880," *Tasks of Economic History* (New York, 1943), pp. 55ff.

[23] "The Coal Regions of Pennsylvania," *Niles Weekly Register,* 1834, *46:* 386.

[24] *Ibid.*

[25] *Niles Weekly Register,* 1834, *46:* 418-419.

[26] Merrill, *State Surveys,* p. 129. The date was 23 March 1836.

purposes of commerce, contribute to the support of our public works [internal improvements] and add greatly to the wealth of our citizens and the State."[27]

As noted above, state legislatures were also moved to establish geological surveys by the argument that they would advance education, the members believing, as did most "liberal" men of the day, in the Jacksonian idea that education should be available to all men. Since the seventeenth century, natural sciences had been a part of the learning of the educated man. It was the same in the nineteenth century, except that now that every man was educated, or could be, scientific knowledge should be widely available. At the same time it was believed that education was not solely for the intellectual uplift of the common man; but was also a means of maintaining and enlarging the economic prosperity and well-being of the people. In Massachusetts, for example, the state legislature provided funds for a library, an experiment station, and a college to develop better farming methods and to disseminate information about them. The geological survey was a natural extension of this kind of education.[28] In this period, too, there was a strong and successful movement to broaden the curriculum of the secondary schools and colleges to include the teaching of science. As this movement succeeded, the study of natural science moved out of the hands of the amateur scholar into the schools and colleges, and men found lifetime careers as teachers of geology, chemistry, mineralogy, botany, etc. These teachers, and scientists in general, found enthusiastic audiences, not only among college students, but in the public at large. Amos Eaton, Benjamin Silliman, and David Dale Owen all found that it was easy to attract audiences when they took to the lyceum platform to explain the principles of science.[29] In such an intellectual atmosphere, the argument that the state would be supporting education if public funds were spent for a geological survey carried great weight.

Behind both the idea that geological surveys would be economically significant, and that they would be of educational value, was the belief of the nineteenth century that science could do great things if applied practically; that is, that science should be useful to society. To acknowledge this and to promote education in science was a manifestation of "liberalness." For example, Charles T. Jackson, one of the leading geological surveyors of the 1830's and 1840's said:

> The Geological Surveys which have been made, or are now in progress, and the numerous calls that are made on our state governments for similar investigations of their respective territories, demonstrates that the communities are fully aware of the advantages, which must necessarily accrue to them, from a scientific examination of their mineral resources. It is certainly a source of congratulation that we find the American people so liberal and enlightened respecting the application of science to the arts, than the people of any European state. This is no doubt to be attributed to the generous diffusion of knowledge in our country.[30]

[27] *Indiana House Journal,* 1836-1837, pp. 26-27. See also Walter B. Hendrickson, *David Dale Owen, Pioneer Geologist of the Middle West.* Indiana Historical Bureau, *Collections,* XXVII (Indianapolis, 1943), p. 27.

[28] Handlin and Handlin, *Commonwealth,* p. 256.

[29] McAlister, *Amos Eaton,* pp. 180-211; Hendrickson, *David Dale Owen,* pp. 63-65.

[30] Charles T. Jackson to Benjamin Silli-

It was Jackson's suggestion that each of the states establish museums of natural history and exchange specimens with each other. Such a museum would be a "practical school" at which children would "learn the book of nature"; it would impress foreign visitors, and be of practical value to miners; and, said Jackson, "such an institution would favor the growth of science, and I may safely add, that the morality of our country would be very much improved by the diversion of young and ardent minds, from idle or vicious amusements to solid and useful learning."[31]

But no matter how "liberal" state legislatures might be in favoring science education, or how much they might want to promote the economic interests of citizens, they were always faced with the problem of justifying the spending of public funds. When times were good and money was easy, geological surveys were authorized; when there was a panic, state expenses were curtailed and geological surveys — as will be pointed out below — always slightly suspect as frills by a few hard-headed legislators, were suspended.

In the decade 1830-1840, for example, only one state had a survey in 1830, but by 1836, ten states had surveys in progress, and in 1837, fourteen states; in 1840 twelve states still conducted surveys. In the next decade, however, when hard times caused by the panic of 1837 still held on, the number of surveys dropped to three in 1849. This was also in part due to the bankruptcy, or near bankruptcy, of those states that had engaged in extensive internal improvement programs. With the bustling expansion and optimism of the 1850's, the number of state surveys in progress climbed to a high point of fourteen in 1855, and reached this figure again in 1859 and 1860. During the 1860's the number declined, but in 1873, eighteen surveys were going on. The depression of that year and of succeeding years reduced the number to eleven by 1885.[32]

It is clear, therefore, that the final determination of whether or not a state would have a geological survey was economic. Surveys, when states could afford them at all, were authorized because they would serve the practical purpose of improving the material well-being of the farmer and the manufacturer, as well as that of the capitalist and laborer; at the same time they would advance education and promote scientific knowledge, and the end result of both would be a better and fuller life for all men. These themes were played over and over again. Petitions of "liberal-minded" citizens pointed them out to governors and legislatures; supporting letters from scientists reiterated them; committees repeated them in their reports to legislatures; and, finally, state geologists summarized them in their published reports.[33]

---

man, Boston, 12 March 1836, "On the Collection of Geological specimens and on Geological Surveys; by Charles T. Jackson," *American Journal of Science,* 1842, *43:* 203. Note also that Great Britain established a geological survey in 1835, but it was not until 1851 that a museum was open to the public. See "Museum of Practical Geology," *Eclectic Magazine,* 1851, *22:* 478.

31 Charles T. Jackson to Benjamin Silliman, Boston, 12 March 1836, *American Journal of Science,* 1842, *43:* 203.

32 This analysis was derived from the histories of the state surveys in Merrill, *State Surveys.* A graph was made of the number of surveys in existence at any one time, and this graph was superimposed on one showing the ups and downs of the business cycle. It should be noted that legislative action lagged behind the peaks of business activity, and a boom would be almost over before a survey was authorized, and, similarly, a panic would be almost over before a survey was suspended.

33 See the histories of the state surveys in

Whether or not a state legislature got the results it expected from a survey was the responsibility of the state geologist, and state geologists, as a group, were very active in urging that legislatures appropriate money to support surveys. They wrote letters in support of each other to governors and legislative committees, their editorials in scientific journals argued for surveys, and in their official reports they pointed with pride to their accomplishments, and announced how much more they could do if the legislature would continue the survey. There was considerable self-interest here, since, through his work on the survey, the geologist received money and fame — the more money a legislature appropriated, the longer the survey would last, or the more extensive it would be, and consequently the more fame to be won. In the mind of every state geologist there was a conflict between his ambition, or his devotion to science, and his responsibility to give the public concrete and practical information about natural resources. The way the geologist resolved this dilemma determined whether or not the economic and educational objectives of the survey were attained.

But exactly what was economically useful information was never determined by either geologist or state legislature. An objective answer could be based only on "before and after" statistics of the wealth and productivity of a state, and even if anyone then had thought of doing this, it would have been as impossible as it is now to draw valid conclusions, because so many other factors were operating — internal improvements, population changes, technological advances, and alternations of the business cycle. What the legislature accepted as being "useful" was the result of subjective interpretation of the geologist's report. If legislators saw that the geologist had located seams of coal exactly, not according to the strata of rock with which it was associated, but in the geographical terms of the rectangular survey, or with relationship to local features such as streams and hills, and if he had determined the thickness, direction, and inclination of the vein, and stated if it were workable or not, and if he gave a chemical analysis and stated clearly whether the coal could be used for heating, smelting or something else, then the geologist's work was "useful." Much the same argument applies to the estimate of educational value. If the legislator could see an orderly display of rocks, minerals and fossils, each specimen plainly labelled, it had value, because it answered the questions people asked about the physical world in which they lived.

There were some geologists like Edward Hitchcock in Massachusetts and David Dale Owen in Indiana who were very conscientious about giving the legislature assurance that the taxpayer was getting his money's worth. Both men understood that the first reason for legislative appropriation was that some practical purpose would be served, but they also believed that a state-

Merrill, *State Surveys*. More detailed accounts are available for some surveys, as follows: James A. Clarke, *James Hall of Albany* (Albany, 1921), pp. 48-51; James Hall, "The New York Geological Survey," *Popular Science Monthly*, 1883, *22:* 815-818; Willard R. Jillson, "A History of the Kentucky Geological Survey, 1838-1921," Kentucky State Historical Society, *Register*, 1921, *19:* 90-95; James P. Lesley, *Historical Sketch of Geological Explorations in Pennsylvania and Other States*, Second Geological Survey of Pennsylvania, *Volume A*, 1878 edition (Harrisburg, 1878), pp. 110-23; Harold S. Cave, *Historical Sketch of the Geological Survey of Georgia, Bulletin 39* (Atlanta, 1922), pp. 3-6; Hendrickson, *David Dale Owen*, pp. 25-28.

supported survey could secondarily provide additions to the fund of "pure" scientific knowledge. Owen stated his views thus:

> I have considered it my duty while surveying a country as new as ours [Indiana], to remember, that a state just settling is like a young man starting in life, whom it behooves to secure to himself a competency before he indulges in unproductive fancies. I have considered it the most important object, to search out the hidden natural resources of the State, and to open new fields of enterprise to her citizens. That object affected, time enough will remain to institute inquiries (which a liberal policy forbids us not to overlook) of a less productive and more abstract character; inquiries which are interesting in a scientfic rather than a commercial point of view.[34]

Owen was always very sensitive to legislative feelings, and sought to forestall objections by giving lectures on geology and geological surveying to the members of the Indiana legislature, and by answering questions as fully and as promptly as possible. He prefaced his first report with a brief account of basic geology so that the reader would be able to understand what followed.[35]

Edward Hitchcock, even before Owen, fully understood the legislature's demand for utilitarian results. The first part of his report, published in 1832, was "The Economic Geology of the State," and he said of it: "If I have not misunderstood my commission and instructions, I was to have principally in view, in my examinations, practical utility; not neglecting, however, interesting geological facts, which have an important bearing upon science."[36] Following this policy, Hitchcock devoted more than half of his final report to paleontology and "scientific" geology. Both Hitchcock and Owen, with their understanding of the legislative mind first gave something practical and then went on into the higher atmosphere of "pure" science.[37] Both also saw that there was no hard

[34] David Dale Owen, *Report of a Geological Reconnoisance* [*sic*] *of the State of Indiana; Made in the year 1837*... (Indianapolis, 1838), p. 4. See also Hendrickson, *David Dale Owen,* pp. 34-35, p. 133-34. These matters were remarked upon in my biography of Owen, but at the time they did not seem especially unusual. Further study of state-supported geological and natural history surveys shows Owen to have been among the few who sincerely followed the policy here outlined. See also the chapter on Owen—"Immigrant Innovator"—in Fenton and Fenton, *The Story of the Great Geologists,* pp. 165-178, in which much is made of Owen's concern with economic geology. This chapter, however, contains no material not found in my biography.

[35] Hendrickson, *David Dale Owen,* pp. 28, 33-34.

[36] *Report on Geology,* 1833, i.

[37] For .other examples of harmony between geologist and the legislature that existed because the geologist kept a balance between practical economic utility and science for science' sake, see "Dr. Jackson's Report on the State of Maine, and on Public Lands Belonging to Maine and Massachusetts," *American Journal of Science,* 1839, *36:* 143-156; "Notice of a Report on the Geological Survey of Connecticut, by Prof. Charles Upham Shepard, M. D. Etc., with Extracts and Remarks by the Editor," *American Journal of Science,* 1838, *33:* 151-175; Newton H. Winchell, "The History of the Geological and Natural History Survey of Minnesota," *Bulletin 1* (St. Paul, 1889), p. 25; Henry M. Chance, "A Biographical Notice of Peter J. Lesley," Mary Leslie [*sic*] Ames, ed., *Life and Letters of Peter and Susan Lesley,* 2 vols. (New York and London, 1909), II, 491-92; "Report on a Projected Geological and Topographical Survey of the State of Maryland; by Julius T. Ducatel, M. D., Professor of Chemistry, University of Maryland, and John H. Alexander, Esq., Topo, Eng.," *American Journal of Science,* 1835, *27:* 1-38; see also a review of a later report by Ducatel in *Niles National Register,* 1840, *58:* 230; Ernest F. Bean, "State Geological Surveys of Wisconsin," Wisconsin Academy of Sciences, Arts, Letters, *Transactions,* 1937, *30:* 214-215; Eugene W. Hilgard, "An Historical Outline of the Geological and Agricultural Survey of the State of Mississippi," *American Geologist,* 1901, *27:* 292-298.

line between "pure" and "applied" science. Owen, for example, told how a knowledge of paleontology would enable the miner to determine the lower limits of coal-bearing strata: no coal existing below strata containing the fossil Archimedes.[38]

On the other hand, there were geologists who looked upon the state-supported survey as a heaven-sent opportunity to make personal collections of rocks, fossils and minerals for themselves, and as a means of accumulating scientific data which could be published at state expense and so enhance their professional standing. There were some who paid lip service to the legislative demands for practical results, but who were really concerned with "pure" science. It was not that they were dishonest but rather that they saw the publicly financed survey as a means to the more important end of the enrichment of knowledge.

The classic example of the man who subordinated practical geology to scientific geology was James Hall, the perennial state geologist of New York, who also headed for brief periods surveys in Iowa and Wisconsin. Hall may have been encouraged to follow his interest in paleontology because the economic survey of the part of New York to which he was first assigned was soon completed; he found no coal and few minerals there. (Except for coal, this situation did not hold true for the rest of the state.)[39] Hall began his work as a young man just out of Rensselaer Polytechnical School in 1836, when the New York survey was instituted. He was one of the four geologists who were each given a district of the state to survey. In 1840 he was made State Paleontologist, and from then on he devoted most of his time to the study, description and naming of fossil invertebrates, rather than to economic geology. Hall succeeded in getting the legislature to appropriate large sums for engraving plates and publishing elaborate reports. He acknowledged his paramount interest in paleontology when he said:

> In *nearly all* respects [my italics, because the New York survey was authorized on economic grounds] the survey has been carried out according to the original conception and plan. It has resulted in far larger and more interesting collections; and in far more interesting and valuable publications than could have been anticipated by the original promoters.[40]

Hall's objectives were not shared by all of his associates on the New York survey. William Mather, who was in charge of the First District, wrote:

> Details and facts, belonging strictly to pure scientific geology, will not be made public until the final report. The object of the annual reports is to give publicity to such facts and localities as may be of practical utility, so that the benefit may be derived from a knowledge of them during the progress of the survey.[41]

[38] Owen, *Report of a Geological Reconnoisance [sic] of the State of Indiana; Made in the Year 1837 . . .*, 13.

[39] "Citations from an Abstract of the Geological Reports of the State of New York. . . ," *American Journal of Science,* 1839, *36:* 1-49.

[40] James Hall, "The New York Geological Survey," *Popular Science Monthly,* 1883, *22:* 824.

[41] Quoted in "Citations from an Abstract of the Geological Reports of the State of New York. . . ," *American Journal of Science,* 1839, *36:* 15.

Hall was always more interested in describing fossil shells and in putting out handsome and expensive volumes with many steel-engraved plates than in doing the humdrum work of finding and testing building stones and other natural resources. He was forever asking the legislature for more time to prepare a volume for printing, or for more money to pay the engraver. James A. Gould, a fellow geologist, sympathized with Hall in his efforts to pry money out of the legislature for these purposes:

> A scientific book cannot be written like a sermon or a box of travels. Ask any of the politicians how much longer it would take him to write a quarto volume of Fourth of July toasts, which he expected to be clapped for their pith and brevity, or a volume of congressional speeches, or newspaper leaders, and he will have some idea of the difference between theirs and yours [Hall's] geology. Surely it would be just cause for the scientific world and all lovers of enlightened legislation the world over to denounce the Empire State if they should throw obstacles in the way of work of so much importance, so anxiously anticipated, so satisfactorily executed thus far, and so advanced toward completion.[42]

In spite of difficulties Hall did succeed in getting appropriations as long as he lived. He had a most persuasive way about him, and he earned such a high place in paleontology that legislators hesitated to refuse his requests for fear that they would be considered stubborn and ignorant. Hall also continually pointed out that if additional funds were not appropriated, what had already been spent would be wasted.

Hall was not so fortunate in Iowa and Wisconsin where, because of his great reputation as a scientist, he was named state geologist. He did not give much attention to the work of either state, being more interested in the honor — and the salary — of heading several state surveys at the same time, as well as having first choice of all fossils found. Partly because the professors at the State University of Iowa protested that the state wasn't getting much in return for the sums spent by Hall, the legislature discontinued the survey after two years. Hall did succeed, however, in publishing a magnificent volume on the fossils of Iowa.[43] In Wisconsin, Hall received a similar rebuff, only here the legislature did not reimburse him for money spent out of his own pocket for engravings to illustrate his report. In both states the legislators objected to spending state funds for expensive books which were of value only to a small group of paleontologists and conchologists.[44]

Other state geologists also were at odds with their legislatures. John S. Newberry, director of the second survey of Ohio, was unsuccessful in getting a $60,000 appropriation for a third volume of reports on paleontology, and some of the scientists whom he employed worked without compensation and had to have their work published in scientific journals.[45] The first survey of

[42] Justin A. Gould to James Hall, July [?], 1849, Clarke, *James Hall of Albany,* pp. 231-232. For Hall's troubles with the legislature see Clarke, *James Hall of Albany,* pp. 466-467.

[43] *Ibid.,* p. 279; Merrill, *State Surveys,* p. 91.

[44] Clarke, *James Hall of Albany,* pp. 336-62. Hall was but one of several state geologists who were paleontologists first, and economic geologists second. See Thomas C. Mendenhall, to the Editor, "Beginnings of American Geology," *Science,* 1922, *56:* 661-663; George H. Williams, "Some Modern Aspects of Geology," *Popular Science Monthly,* 1889, *35:* 640-648.

[45] Edward Orton, "Preface," *Geological Survey of Ohio,* 7 vols. (Columbus, 1883),

Michigan was curtailed in 1840 "on account of straitened finances and the hostility of the legislature to labors which promised no early practical benefit to the material progress of the state,"[46] and the second survey was suspended in part because the legislature disapproved of the spending of public funds for botanical and zoological activities.[47] Willis S. Blatchley, state geologist of Indiana, speaking to the Indiana Academy of Science in 1916, remarked upon the small sum the legislature had appropriated for the first state geological survey in 1838, and said:

> This, then, was the beginning of that shortsighted, parsimonious policy which has continued toward not only the geological department, but every scientific bureau of the State of Indiana, from that day to this. . . . The average politician who is chosen as a "representative" (mark the word) of the dear "peepul" of the State of Indiana knows nothing about science — sees no connection between science and future development of the State and is afraid to vote an extra dollar in a worthy cause for fear that he will be snowed under at the next election. The geologist or other scientist who is dependent solely upon political appropriations to do good work in the State of Indiana has indeed a rocky road to travel.[48]

As a last example of the misunderstanding between a legislature and a scientist over the purposes of a survey, take the case of William B. Rogers, who had, in 1835, succeeded in persuading the legislature to institute a geological survey of which he was appointed head.[49] When it came time to appropriate money to publish his report in 1838, a few legislators objected and there was a heated debate on the floor of the House. Rogers wrote his brother, Henry D. Rogers, then state geologist of both New Jersey and Pennsylvania:

> This proceeding strikes me as being very indelicate, at the same time that it is obviously absurd. How can these gentlemen pretend to judge my reports? I take it for granted that some sneers have been cast upon my labors, and the thought of a legislative body employing itself in venting spleen or exercising wit upon a paper of which but few of them have any adequate comprehension really fills me with indignation.[50]

William's troubles were echoed by Henry, who was trying to get the Pennsylvania legislature to make an appropriation for the survey, a matter in which he eventually succeeded:

VII, v; Thomas C. Mendenhall to the Editor, "The Beginnings of American Geology," *Science*, 1922, *56:* 661-663.

[46] R. C. Allen and Helen M. Martin, "A Brief History of the Geological and Biological Survey of Michigan, 1837-1920," *Michigan Historical Magazine*, 1922, *6:* 676.

[47] Alexander H. Winchell, the head of the third Michigan survey, planned such a broad program that the Geological Board of the state considered it impractical, and said that it would not give "direct and immediate benefits." Because the Board would not back him up, Winchell resigned. *Ibid.*, pp. 689-690.

[48] Willis S. Blatchley, "A Century of Geology in Indiana," Indiana Academy of Science, *Proceedings, 1916* (Ft. Wayne, 1917), pp. 97-98.

[49] William B. Rogers to Henry D. Rogers, William and Mary College, 11 February 1835, *Life and Letters of William B. Rogers*, edited by his wife, with the assistance of William T. Sedgwick, 2 vols. (Boston and New York, 1896), I, 116.

[50] William B. Rogers to Henry D. Rogers, University of Virginia, 8 March 1838, *Life and Letters of William B. Rogers*, I, 118.

> To give you some idea of the tribunal to which I have to bow, one senator, who a few days ago told me the survey ought by all means to be finished, uttered himself thus, "Mr. Speaker, I shall vote against the appropriation, on the ground of its unfairness to other sciences of like nature with this geology. The bill, sir, makes no provision for phrenology, animal magnetism, and the highly important science of *water-smelling;* it is partial and I will vote against it."[51]

There are many more instances of difficulties between geologists and state legislatures involving the purpose of geological surveys.[52] Legislatures and geologists were often at cross purposes; as has been shown, the final determination for the appropriation of funds for surveys rested on what the legislators considered to be economic grounds, and even the appeal to aid education and advance knowledge was denied or granted almost wholly on the basis of whether or not a public purpose would be served. It is little wonder, then, that the geologist, generally dedicated to the idea that knowledge must be sought for the sake of knowledge, often ignored the legislature's responsibility to the public when it came to spending money raised by taxation, and that the legislatures failed to appreciate the virtues of "pure" science. It is doubtful whether Sir Charles Lyell, the eminent English geologist who visited the United States in 1841, fully understood the situation when he said:

> The American state surveys were not exclusively confined to the practical bearing of geology on economical improvements, although this was *professedly their chief end;* but the surveys, *unfettered by narrow utilitarian restrictions,* were liberally encouraged to collect facts and publish speculations which they deemed likely to illustrate the general principles of the science and lead to sound theoretical conclusions [My italics].[53]

It is true, as Sir Charles says, that in hundreds of volumes of state geological surveys produced in the nineteenth century, one finds much to "illustrate the general principles of the science," and often "sound theoretical conclusions" were drawn. But such good scientific work was done in spite of the purposes for which the legislatures appropriated money. The reports were very uneven. Some geologists maintained a good balance between locating and describing the qualities of natural resources and the identification of fossil plants and animals. Other geologists clearly directed their reports to fellow geologists, and paid little attention to the legislative desire that the reports should be useful to the general public.

In the last two or three decades of the nineteenth century, these conflicts declined as the idea of the continuing survey, with its permanent staff and its alliance with state-supported educational institutions, prevailed. The establishment of the United States Geological Survey and its close cooperation with state surveys, whereby much of the theoretical work was taken over by the

[51] Henry D. Rogers to William B. Rogers, Harrisburg, Pennsylvania, 1 May 1841, *Life and Letters of William B. Rogers,* I, 190.

[52] For some general accounts see Josiah D. Whitney, "Geographical and Geological Surveys, Part II," *North American Review,* 1875, *121:* 271-275; Albert R. Leeds, "State Geological Surveys," *Popular Science Monthly,* 1873, *3:* 226-229; Nathaniel S. Shaler, "The Value of Geological Science to Man," *The Chautauquan,* 1894, n. s. *11:* 170-174.

[53] "Account of Meeting of the British Geological Society on 17 February 1843," *Niles National Register,* 1843, *64:* 149.

national agency, left the state survey free to undertake local practical work. And finally, the increasing number of outlets for publishing the results of research — the journals of the professional organizations, and the monograph series of universities, for example — made it less necessary for the geologist to depend on legislatures.

Louis Agassiz and the Races of Man[‡]

By Edward Lurie *

WHEN Louis Agassiz came to the United States from Switzerland in 1846, he soon became identified with a scientific controversy which raged for nearly two decades and affected religion, politics and social theory in its varied ramifications. The point of contention concerned the question whether mankind had originated in a single pair in a particular place, or whether human beings were of different species, having originated in many different centers of creation.[1]

The two schools of thought were divided into believers in the basic "unity" of mankind, or in the "plurality of origin" of human species.[2] One of the most significant aspects of the dispute was its relationship to the problem of Negro slavery, as the pluralist school insisted men were separately created and therefore endowed with innate physical and mental differences. Conversely, the advocates of unity urged that men were of one species, and usually they insisted upon the equality of mankind.[3] Another important product of the controversy was that significant discussions concerning hybridity, variation, inheritance, the nature of species and the meaning of geographical distribution took place for a long while prior to the publication of the *Origin of Species*. Agassiz's role in the dispute was intimately related to his views concerning the separate and specific creation of species. Moreover, Agassiz's position demonstrated the manner in which a scientific theory was made to serve a social doctrine, thus illustrating the important relationship between science and society in nineteenth-century America.

In the eighteenth century, the problem of the origin of mankind had been debated in America, England and on the continent, but without the acrimony that characterized the discussion in the United States during the nineteenth century.[4] In this earlier period, most important European naturalists held that mankind was composed of one species, although they offered different explana-

‡ This is a revised version of a paper read at the 28 December 1953 meetings of the History of Science Society.

* Massachusetts Institute of Technology.

[1] For a good discussion illustrating modern and historic evidence for the basic unity of mankind, see Ashley Montagu, *Statement on Race* (New York, 1951), pp. 22–33.

[2] These groups might be more properly termed *monogenists* and *polygenists*. However, since the disputants continually referred to themselves as "pluralists" or advocates of "unity," these terms will be employed throughout this discussion.

[3] J. L. Cabell, Professor of Comparative Anatomy and Physiology at the University of Virginia, was an exception to this rule. In *The Testimony of Modern Science to the Unity of Mankind*, (New York, 1859), he upheld the unity of man, but defended slavery as a temporary expedient designed by God to preserve the Negro. Cabell claimed that such inferiority was only a temporary state, and that colored peoples would soon achieve equality with the whites. I am indebted to Professor Conway Zirkle for calling my attention to Cabell's volume.

[4] See Conway Zirkle, "Father Adam and the Races of Man," *Journal of Heredity*, *45*, 29–34, 1954 for various traditional interpretations of the origin of human diversity.

tions for the physical differences between races.[5] As white relations with subservient colored populations became a pressing social and political problem in the nineteenth century, the unitarian-pluralist debate increased proportionately, particularly in England and America.[6] One result of American scientific interest in the problem was the incorporation into European writings of a body of observation from the New World relating to Indians,[7] Negroes and animal life.[8]

After Blumenbach's classification of mankind into five races,[9] Sir William Lawrence [10] and James Cowles Prichard [11] became the outstanding European spokesmen for the unity school. Prichard's classic, *Researches Into the Physical History of Mankind*, was published in four editions between 1813 and 1847.[12] It is significant that the 1836 edition of this work was written with the express purpose of exposing those pluralists such as Virey, Desmoulins and Bory de St. Vincent, who maintained that since mankind was composed of different species, the Negro was naturally inferior to the white. Prichard labelled such views as a distinct apologia for white mis-treatment of native populations.[13] In so doing, he characterized the emerging social issue which was to provide a unique background for the American debate on the origin of man a decade later.

It is also of interest that such advocates of the "unity" of man as Prichard, Lawrence and William Charles Wells [14] defended their basic interpretation in

[5] John C. Greene analyzes eighteenth- and early nineteenth-century opinions on this question in "Some Early Speculations on the Origin of Human Races," *American Anthropologist*, *54*, 31–41, 1954.

[6] An early American advocate of the unity of man was the Reverend Samuel Stanhope Smith of the College of New Jersey, who attacked the pluralist position in *An Essay on the Causes of the Variety of Complexion and Figure in the Human Species . . . and Strictures on Lord Kaims's Discourse, on the Original Diversity of Mankind* (Philadelphia, 1787).

[7] It is of interest to note that the American Indian, when compared to the Negro by pluralist writers, was usually endowed with moral virtues such as courage and bravery, which were held to be absent in the Negro. Similarly, it was claimed that Indians died or their spirit was broken in captivity, while Negroes were happier in a slave status than as free individuals. These assertions were made so as to "prove" that the "natural" constitution of Negroes fitted them for slavery. For one interpretation of white intellectual attitudes toward Indians in the eighteenth and nineteenth centuries, see Roy H. Pearce, *The Savages of America: A Study of the Indian and the Idea of Civilization* (Baltimore, 1953). This analysis should be compared with that of David Bidney, "The Idea of the Savage in North American Ethno-history," *Journal of the History of Ideas*, *15*, 322–327, 1954.

[8] Darwin, for instance, devoted most of Chapter VII, Part I, of *The Descent of Man*, New York, I, 1871, pp. 206–228, to proving that "all the races of man are descended from a single primitive stock." (p. 220.) He presented both sides of the argument, and in so doing noted the work of the Americans Nott, Gliddon, Agassiz and Bachman.

[9] See Thomas Bendyshe (ed. and transl.), *The Anthropological Treatises of Johann Friederich Blumenbach* (London, 1865), especially pp. 305ff.

[10] William Lawrence, *Lectures on Physiology, Zoology, and the Natural History of Man* (Salem, 1828), especially pp. 212–235. These lectures were delivered in 1819 to the Royal College of Surgeons.

[11] Prichard's first work on this problem took the form of a dissertation at Edinburgh, *De Humani generis varietate*, 1808. It is noteworthy that this university was an important center for the study of racial diversity, since Prichard and Samuel George Morton had studied there. For Prichard's career, see *Dictionary of National Biography*, *16*, 344–346. Lucile E. Hoyme, in "Physical Anthropology and its Instruments: An Historical Study," *Southwestern Journal of Anthropology*, *9*, 412–415, 1953, discusses medical education and phrenological interests at Edinburgh.

[12] London: 3 vols., 1813; 2 vols., 1826; 5 vols. (enlarged), 1836–1847; 4 vols., 1841–1847.

[13] Prichard, *Researches Into the Physical History of Mankind* (London, 1836), *1*, pp. v–viii, 6–7; see also Prichard, *The Natural History of Man* (London, 1843), p. viii.

[14] For William Charles Wells, of Charleston, S. C. and London, who appended a statement on natural selection to his *Two Essays . . .* (London, 1818), pp. 425–439, see Richard Harrison Shryock, *The Strange Case of Wells' Theory of Natural Selection (1813): Some Comments on the Dissemination of Scientific Ideas* (n. p., 1946), reprinted from *Studies and Essays in the History of Science and Learning*

a manner that was in some respects markedly anticipatory of Darwinian evolution.[15] The pluralist argument, as best typified by Agassiz, simply rested on the assumption that the Creator had ordained the existence of separately and specifically created human species.[16]

Although Charles Caldwell, a Philadelphia physician, had written *Thoughts on the Original Unity of the Human Race* in 1830 as a defense of pluralism, the argument did not receive its main impetus in the United States of this period until Samuel George Morton had published *Crania Americana* in 1839.[17] Morton was typical of medical men like Joseph Leidy, Jeffries Wyman, Oliver Wendell Holmes and John C. Warren, whose pioneer efforts in the field of comparative anatomy did much to elevate that science during the nineteenth century. Born and educated in Philadelphia, Morton continued his medical training in Edinburgh and Paris. Returning to his native city in 1826, he began an active career of teaching, medical practice and research in natural history that won him a high rank in the cultural and scientific community of the city and the nation.

As a leading member of the Academy of Natural Sciences of Philadelphia, Morton used this institution as headquarters for a collection of human crania which he had begun in 1830. This interest in osteology was stimulated by the opportunity to examine the specimens of the Lewis and Clark expedition. By 1839, Morton had amassed nearly 900 human skulls — the outstanding collec-

---

in Honor of George Sarton, 1946; Conway Zirkle, "Natural Selection Before the 'Origin of Species'," *Proceedings of the American Philosophical Society*, *84*, 106–109, 1941; Greene, *op. cit.*, 36. Shryock (p. 10) speculates that Wells wrote of human rather than animal variation because this was "the concern of the moment." Such an interpretation would seem quite justified, considering the congruence of writings on the origin of mankind (Prichard, 1808, 1813; Lawrence, 1819, 1828; Wells, 1813). To go one step further, it seems likely that Wells wrote about the "varieties of man" from an "original stock" with the same motivation as Prichard (*cf.* note 13), namely to disprove pluralist, separate-creationist theories advanced to explain human differences. Clearly, there was a marked concern in this period with such problems (cf. Shryock, *op. cit.*, p. 10, note 22). While there is no direct evidence that Wells knew of Prichard's work (Shryock, p. 8, assumes some awareness of it, but Zirkle, *op. cit.*, pp. 106–107, merely places Wells as Prichard's contemporary), it is difficult to imagine Wells not having read Prichard or not having been influenced by the same cultural climate.

[15] Prichard, *Researches* . . . (1826), *2*, pp. 573, 590–591: ". . . variation . . . is not merely an accidental phaenomenon, but a part of the provision of nature for furnishing to each region an appropriate stock of inhabitants, or for modifying the structure and constitution of species, in such a way as to produce races fitted for each mode and condition of existence. . . . it may have happened, that a new stock has sprung from a few individuals who happened themselves to be characterized by some peculiarities; such peculiarities may have been transmitted by the parents to their offspring, and by the subsequent increase and multiplication of a family, may have become the prevalent character of a whole tribe or nation." For Lawrence, see *op. cit.*, pp. 179, 309, 313, 348, 375.

[16] "We must acknowledge that the diversity among animals is a fact determined by the will of the Creator, and their geographical distribution part of the general plan which unites all organized beings into one great organic conception." Agassiz, "Sketch of the Natural Provinces of the Animal World and their Relation to the Different Types of Man," in J. C. Nott and Geo. R. Gliddon, *Types of Mankind* (Philadelphia, 1854), p. lxxvi.

[17] For other early pluralist arguments, see Richard Colfax, *Evidences against the Views of the Abolitionists* (New York, 1833), and J. J. Flournoy, *An Essay on the Origin, Habits . . . of the African Race, incidental to the propriety of having nothing to do with Negroes* (New York, 1835). For Morton's career, see Daniel Moore Fisk, "Samuel George Morton," *Dictionary of American Biography*, *13*, 265–266; Charles D. Meigs, *A Memoir of Samuel George Morton, M. D.* (Philadelphia, 1851); Henry S. Patterson, "Memoir of the Life and Scientific Labors of Samuel George Morton," in J. C. Nott and Geo. R. Gliddon, *Types of Mankind* (Philadelphia, 1854), pp. xvii-lvii.

tion of its type in the nation.[18] The *Crania Americana* was based upon a critical analysis of such materials.[19]

The most significant aspect of this volume was Morton's assertion that there were innate differences in the varied human types inhabiting the Americas. These distinctions, particularly the alleged divergent cranial capacities of Negro and Caucasian skulls, were "aboriginal" in that they were not to be explained as due to varied environmental conditions. From this beginning, Morton attempted to prove during the next decade that such evidence as the physical characteristics of the American Indians,[20] the history and archeology of ancient Egypt [21] and the true nature of hybrid variation and animal species [22] demonstrated without question the plural origin of mankind. Morton was aware that in so doing he controverted the accepted theological interpretations of the Scriptures.[23] Like Agassiz, however, Morton did not seek an open break with revealed religion, but felt instead that science should not draw back in the face of tradition and accepted belief.

The unity — plurality controversy took on its real significance when Morton's initial efforts were supported by other scientists and popularized by publicists and statesmen. The net result of such labors was to supply a "scientific" basis for a theory of racial inequality. Once this had been accomplished, it was possible for apologists for slavery to assert that since history, physiology, anatomy and ethnology proved that the Negro was not genetically related to the white, then his status as a slave was a "natural" one, and not subject to criticism in the light of conventional moral codes. Such a rationalization, supported by Agassiz's authority, resulted in a vigorous religious, social and scientific reaction. This reaction was in large measure conditioned by the ideology and motivation of pluralist social philosophers.

The most typical of these was Josiah Clarke Nott, a student of Morton's, who succeeded in synthesizing science and social doctrine to such a degree that he transformed his teacher's researches into a theory justifying racial inequality.[24]

[18] Meigs, *op. cit.*, p. 23; *Catalogue of Skulls of Man and the Inferior Animals in the Collection of Samuel George Morton* (Philadelphia, 1849); W. S. W. Ruschenberger, *A Notice of the Origin, Progress, and Present Condition of the Academy of Natural Sciences of Philadelphia* (Philadelphia, 1852), pp. 31–33. In 1846, Agassiz estimated that Morton's collection was composed of 600 skulls, which might have been a more accurate figure. Agassiz [to Rose Mayor Agassiz], [2] December 1846, Elizabeth Cary Agassiz (ed.), *Louis Agassiz His Life and Correspondence* (Boston, 1885), *2*, p. 417.

[19] *Crania Americana; or a Comparative view of the Skulls of various Aboriginal Nations of North and South America; to which is prefixed an Essay on the Varieties of the Human Species* (Philadelphia, 1839). Hoyme, *op. cit.*, pp. 415–417, notes Morton's contributions to the development of physical anthropology and his allied interest in phrenology. Agassiz had the highest respect for Morton as a naturalist, and stated that the collection of crania alone was worth his visit to America. Jules Marcou, *Life, Letters, and Works of Louis Agassiz* (New York, 1896), *2*, 28–29; Elizabeth Cary Agassiz, *op. cit.*, pp. 417–418.

[20] See, for example, his article "Some Observations on the Ethnography and Archaeology of the American Aborigines," *American Journal of Science and Arts*, *2*, 2nd Series, 1–17, 1846.

[21] *Crania Ægyptiaca; or Observations on Egyptian Ethnography, derived from History and the Monuments* (Philadelphia, 1844).

[22] A typical paper of this type was his "Value of the Word Species in Zoology," *American Journal of Science and Arts*, *11*, 2nd Series, 275–276, 1851. As evidence of the controversial nature of Morton's views, the editors of the *American Journal of Science* took the unusual step of prefacing one of his articles with the statement: "In receiving this paper, we commit ourselves . . . to none of the opinions of the author. . . ." *Ibid.* *3*, 2nd Series, 39, 1847.

[23] Meigs, *op. cit.*, pp. 36, 38–39.

[24] For Nott's biography, see George H. Ramsey, "Josiah Clarke Nott," *Dictionary of American Biography*, *13*, 582–583; William Leland Holt, "Josiah Clarke Nott of Mobile, An

After receiving his degree in medicine from the Pennsylvania Medical College in 1827, Nott spent two years as an intern and demonstrator of anatomy in Philadelphia, and then practiced medicine for six years in Columbia, South Carolina, the city of his birth. After a sojourn in Europe, he returned to America in 1836 and made his home in Mobile, Alabama, where he soon became one of the leading physicians of the South, doing important work on the cause and incidence of yellow fever.[25] As an ardent champion of the South and her institutions, Nott was motivated by a sort of missionary zeal to convince the general public and the scientific world of the plural origin and consequent physical and mental differences of human types. Nott's conception of the value of his efforts was buttressed by the conviction, which he shared with Morton and Agassiz, that the labors of the pluralist school were but another battle in the age-old conflict between the progressive forces of science at war against the encrusted tradition of religion and conservatism.[26]

In 1844, Nott published *Two Lectures on the Natural History of the Caucasian and Negro Races*.[27] This was the first of a long series of Nott's writings which appeared during the next two decades in book form and in journal articles, and which relied on the authority of Morton and Agassiz to support a belief in the plural origin of mankind.[28] The 1844 work appeared just in the period when the southern pro-slavery argument was becoming most militant and aggressive; in the same degree as Nott's pronouncements on this subject from the lecture platform, the book provided a scientific standard and rationalization for the ideology of inequality, by citing many alleged "scientific facts" in order to prove the essential difference between the Negro and white races, based on the assumption of plural origins.[29] It is of interest to note that while

American Prophet of Scientific Medicine," *Medical Life*, *35*, 487–504, 1928; Emmett B. Carmichael, "Josiah Clark Nott," *Bulletin of the History of Medicine*, *22*, 249–262, 1948. For Morton's appreciation of Nott's worth as a scientist and the importance of his contributions to the theory of plural origins, see Morton to Nott, 29 January 1850, Patterson, *op. cit.*, p. 1.

[25] Nott, "On the Pathology of Yellow Fever," *American Journal of the Medical Sciences*, *9*, new series, 277–293, 1845; "An Examination into the Health and Longevity of the Southern Sea Ports of the United States, with Reference to the Subject of Life Insurance," *Southern Journal of Medicine and Pharmacy*, *2*, 1–19, 121–145, 1847; "Yellow Fever Contrasted with Bilious Fever — Reason for Believing it a Disease of Sui Generis . . . Probably Insect or Animalcular Origin," *New Orleans Medical and Surgical Journal*, *4*, 563–601, 1848; Carmichael, *op. cit.*, 250–251.

[26] Nott to Lewis R. Gibbes, 21 November 1850: "I have little patience with prostitutors of science . . . the parsons have gored me into controversies . . . I have fought back at them . . . because I thought it best to agitate and keep the subject before the world." The Papers of Lewis R. Gibbes, Manuscripts Division, Library of Congress. Cf. also, Nott to Ephraim George Squier, 14 February 1849, The Papers of Ephraim George Squier, Manuscripts Division, Library of Congress; and Nott to Joseph Leidy, 4 October 1854, The Papers of Joseph Leidy, Academy of Natural Sciences of Philadelphia (on loan to the Museum of Comparative Zoology, Harvard University).

[27] Mobile, Alabama, 1844.

[28] For example, *Southern Quarterly Review*, *8*, 148–190, 1845; *9*, 1–57, 1846; "The Slave Question," *The Commercial Review of the South and Southwest*, *4*, 287–289, 1847; *Two Lectures on the Connection Between the Biblical and Physical History of Man* (New York, 1849); *An Essay on the Natural History of Mankind, Viewed in Connection with Negro Slavery* (Mobile, 1851).

[29] The "peculiarities" of the Negro which Nott usually cited as proof for the concept of innate and permanent inequality were: the ability to withstand warm climates, enlarged genitalia, bent knees, woolly hair, and such "moral" deficiencies as promiscuity, laziness and lack of intelligence. The influence of Nott's "science" is best represented by such works as John H. Van Evrie, M. D., *Negroes and Negro "Slavery"; the First, an Inferior Race — the Latter, its Normal Condition* . . . (Baltimore, 1853); and Josiah Priest, *Slavery, as it relates to the Negro . . . and Causes of his state of*

Morton and Agassiz wrote of racial differences between caucasoids and mongoloids in the same way that they compared whites with Negroes,[30] Nott and other southern propagandists extracted from such writings only that material which dealt with Negro-white divergencies, and unfairly identified the work of these naturalists as directed solely toward this end. In addition to interpretations of Morton's work, such "facts" as Nott offered were predicated on his experiences with Negro patients as a physician and surgeon. Coming from the pen of a medical man, such statements attracted considerable notice in the southern press.[31]

The believers in plurality could not fully develop their argument, however, until it had achieved increased stature as a scientific theory. It remained for Morton and Agassiz to supply this respectability during the years 1846 to 1851. Once this had been done, pluralists could plead that their argument exemplified the latest results of scientific research. Morton's first contribution to this cause took the form of two essays on hybridity, published in 1847 in the *American Journal of Science and Arts*.[32] He argued that the sterility of hybrid crosses between species was not a proper mark of species, since many fertile hybrids had been produced through crossing. Such reasoning served to supply an important deficiency in the pluralist argument. If the sterility of hybrid crosses was accepted as the standard of specific identity, then mankind, whose different "races" produced fertile "crosses" could not be classified into different species. But once the assertion was made, with a degree of scientific impartiality, that crosses between different species had produced fertile hybrid offspring, then pluralists could argue that successful race mixture did not contradict the fact that man was composed of more than one species.

As Morton's assertion struck at the core of the unitarian argument, it elicited a storm of protest from naturalists and laymen of the opposite conviction. The most outspoken defender of the unity concept in general and the nature of hybridity in particular was John Bachman, the co-author, with Audubon, of *Quadrupeds of North America*.[33] Besides being a highly competent researcher

---

Servitude . . . with strictures on Abolitionism (New York, 1845). According to Priest, for instance, "the baleful fire of unchaste amour rages through the negro's blood;" he was therefore sub-human, and given to acts of perversion and licentiousness (p. 152).

[30] See Agassiz in *The Christian Examiner*, *49*, 111–113, 135, 144, 1850.

[31] Thomas Smyth, *The Unity of the Human Races . . . with a Review of the Present Position and Theory of Professor Agassiz* (New York, 1850), 353–354; Arthur Young Lloyd, *The Slavery Controversy 1831–1860* (Chapel Hill, 1939), p. 232 and note 41; Nott, "An Examination into the Health and Longevity of the Southern Sea Ports . . . ," *Southern Journal of Medicine and Pharmacy*, *2*, 1–2, 1847; William Sumner Jenkins, *Pro-Slavery Thought in the Old South* (Chapel Hill, 1935), pp. 257–258. Southern fundamentalists attacked Nott for contradicting the Bible. Cf., *Southern Quarterly Review*, *9*, 372–391, 1846; *Southern Literary Messenger*, *20*, 660–668; 1854; *21*, 30, 1855. In 1851, the American Medical Association appointed a special committee headed by Dr. R. D. Arnold, of Savannah, Georgia, to report on "The Physiological Peculiarities and Diseases of the Negroes." See *Charleston Medical Journal and Review*, *6*, 586–587, 1851; and Samuel A. Cartwright, M. D., "The Diseases and Physical Peculiarities of the Negro Race," *ibid.* 643–652.

[32] "Hybridity in Animals, Considered in Reference to the Question of the Unity of the Human Species," *American Journal of Science and Art*, *3*, 2nd Series, 39–50, 203–211, 1847. Nott may have advanced the pluralist concept of hybridity before Morton (cf. Jenkins, *op. cit.*, pp. 256, 265, note 67); but he clearly relied on Morton's authority as a naturalist in this regard. See Nott in *Charleston Medical Journal and Review*, *3*, 102, 1848, and Morton in *ibid.*, *5*, 339, 755, 1850.

[33] For Bachman's career see C. L. Bachman,

and writer, Bachman was also Professor of Natural History at the College of Charleston, and a Lutheran clergyman in that city. He had decided fundamentalist views on the interpretation of the Bible, and pointed out that the concept of the plural origin of man contradicted the accepted meaning of the Book of Genesis. For the most part, however, Bachman argued with Morton and other pluralists on purely scientific grounds, maintaining that all evidence of the fertility of hybrid crosses was actually proof of *varieties* produced in the same species.[34] Such emphases inspired interesting discussions on both sides relating to definitions of species, effects of domestication, and the nature of variation. Morton and Bachman carried on their controversy in nearly every issue of the *Charleston Medical Journal and Review*, from early in 1850 until Morton's death in the spring of 1851. In fact, from 1846 onward this publication served as the primary forum for the pluralist-unitarian debate.[35]

In 1845, the year before Agassiz came to the United States, he had written an article attempting to prove that there existed specific "zoological provinces" in the natural world which were notable for distinct flora and fauna and for particular human types. There were, he reasoned, Negroes peculiar to the torrid zone, Caucasians living in the temperate zone, and Eskimos in the arctic zone. Yet he affirmed that despite this particular geographic distribution of species, mankind, in distinction to animals and plants, was composed of not many, but one original species.[36] Bachman was so impressed with Agassiz's affirmation of belief in the unity of mankind that he included the Swiss naturalist's name as an authority on the question in a volume written in 1850,[37] in refutation of the writings of Nott and Morton. But, as Bachman soon discovered, Agassiz had experienced a decided conversion to the pluralist position during the period 1846 to 1850.

John Bachman, The Pastor of St. John's Lutheran Church . . . (Charleston, 1888). John James Audubon and The Reverend John Bachman, *The Viviparous Quadrupeds of North America*, 3 vols. (New York, 1846–1854).

[34] See, for example, Bachman, "A Reply to the Letter of Samuel George Morton . . . ," *Charleston Medical Journal and Review*, *5*, 466–508, 1850; "Second Letter to Samuel G. Morton . . . ," *ibid.*, 621–660; "Additional Observations on Hybridity in Animals . . . being a reply to the essays of Samuel George Morton," *ibid.*, *6*, 383–396, 1851.

[35] This journal was begun by two Charleston doctors, J. Lawrence Smith and Seaman D. Sinkler, in 1846, as *The Southern Journal of Medicine and Pharmacy*. The title was changed with volume 3, 1848. Although the editors gave equal space to Bachman in his debates with Morton, their sympathies were clearly with the pluralist school. See volume 6, 98–100, 1851. On Morton's death in May, 1851, D. J. Cain and F. P. Porcher, then editors of the *Journal* wrote: "We can only say that we of the South should consider him as our benefactor, for aiding most materially in giving to the negro his true position as an inferior race." —*6*, 597, 1851. Articles representative of the Morton-Bachman controversy appeared in *5*, 168–197, 328–344, 466–508, 621–660, 755–805, 1850; *6*, 145–152, 301–308, 373–383, 383–396, 1851.

[36] Louis Agassiz, *Notice sur la Géographie des Animaux* (Neuchatel, 1845), reprinted from the *Revue Suisse* for August, 1845. Nine years later, when Agassiz was a convinced pluralist, he made the strange assertion that he had written of plurality as early as 1845, when in fact he had expressed quite the opposite views. Cf., Agassiz, "Sketch of the Natural Provinces of the Animal World . . . ," in Nott and Gliddon, *Types of Mankind* (Philadelphia, 1854), lviii. Bachman was quick to point out this contradiction in *Continuation of the Review of Nott and Gliddon's "Types of Mankind"* (Charleston, 1855), pp. 10–11 (reprinted from *Charleston Medical Journal and Review*, *10*, 1855). See Thomas Smyth, *op cit.*, pp. 348–354, for an analysis of Agassiz's changing position on plurality during the years 1845–1850.

[37] John Bachman, *The Doctrine of the Unity of the Human Race Examined on the Principles of Science* (Charleston, 1850), especially pp. 35–36. Bachman identified Agassiz as a believer in unity from statements made in Agassiz and Gould, *Principles of Zoology* (Boston, 1848), pp. 154–181. These carried the same meaning as Agassiz's 1845 article,

Two basic reasons impelled Agassiz to change his opinion on the origin of man. The first, and most important, was that his belief in the successive, separate and independent creations that had characterized the natural history of animals and plants forced the same interpretation with regard to man, in the interests of consistency. Had Agassiz admitted a common parentage for human beings, he would have approached a position totally alien to his entire philosophy of nature, a system of belief which continually emphasized that the Wisdom of the Creator had been responsible for the permanence of independently ordained species. Indeed, it was Agassiz's assertion of his belief in specific zoological zones of creation that precipitated his first debates with Asa Gray on the evolution question.[38] The unity-plurality argument, therefore, provided a significant background for Agassiz's role in the controversy over the *Origin of Species*.

Another important factor which motivated Agassiz's reasoning on the nature of mankind was his first experience with Negroes in America. So amazed was he at the observed differences between Negro and white which he saw during his first contact with colored peoples in Philadelphia and Charleston, South Carolina, during 1846 and after, that he soon began to change his opinion on the unity of mankind.[39] Thus, in Agassiz's first American lectures delivered in Boston during the winter of 1846–1847, he held that Negroes quite probably were not descended from the same stock as whites, and that their ancestry could not be traced to the sons of Noah. These unorthodox scientific and religious views caused John Torrey to inquire of his fellow botanist Asa Gray if Agassiz was going to offend conservative opinion in his forthcoming New York lectures.[40] Gray replied with characteristic perspicacity that Agassiz's respect for religion was of the highest, but that the Swiss naturalist did not understand the peculiar social climate in the United States with regard to the question of the nature of the Negro.[41]

Late in 1847, Agassiz visited the South for the first time, in order to give a series of lectures in Charleston, South Carolina. His coming had been eagerly awaited by such local naturalists as Lewis R. Gibbes, Josiah E. Holbrook, John Tuomey, Robert W. Gibbes and Francis S. Holmes. Nott's articles and lectures on the differences between Negro and white and the plural origin of man were causing much discussion in the intellectual community of Charleston.[42] The naturalists of the area thus looked forward to learning Agassiz's opinion on the

---

with phrases like "man . . . a cosmopolite . . . is everywhere one identical species." Such words indicated Agassiz's reluctance to announce in print the views expressed in his Boston lectures of 1846–1847 and his Charleston lectures of 1847. Although Bachman heard Agassiz's public profession of belief in plurality at the AAAS meetings in March, 1850, and knew of his printed affirmation (*Christian Examiner*, *49*, 110–145, 1850), his book on unity had already gone to press. See Bachman, *Charleston Medical Journal and Review*, *6*, 385, 1851.

[38] *Proceedings of the American Academy of Arts and Sciences* (Boston, 1857–1860), *4*, pp. 130–135, 171–179, 195–196 (11 January, 22 February, 22 March, 1859).

[39] Elizabeth Cary Agassiz, *op. cit.*, *2*, 497–498; Robert W. Gibbes to S. G. Morton, 28 October 1847, 31 March 1850, Manuscript Collections, Library Company of Philadelphia.

[40] John Torrey to Asa Gray, 11 January 1847, Historic Letter File, Gray Herbarium, Harvard University. An article by "E. H. S." entitled "The African Race," *Christian Examiner*, *41*, 33–48, 1846, illustrates that Agassiz's views were not too exceptional or unorthodox.

[41] Asa Gray to John Torrey, 24 January 1847, Asa Gray Papers, Gray Herbarium, Harvard University.

[42] R. W. Gibbes to S. G. Morton, 30 November 1847, Manuscript Collections, Library Company of Philadelphia.

subject, since they cherished his views as the latest expression of truth in the realm of natural history. In December, Agassiz lectured on the problem to the Literary Club of Charleston, whose membership included the leading theologians and scientists of the city. To the great delight of the pluralist school, Agassiz emphasized in this lecture that the Negro and white were physiologically and anatomically distinct species.[43]

By early 1850, Agassiz was even further prepared to announce publicly his views on the origin of man, and to incorporate his opinions into a larger theoretical framework. From 1847 on, he had spent the winter months lecturing at the College of Charleston, and observing Negroes in the fields and cities of the region. Moreover, he had been very impressed with Morton's collection of human cranial types, which he had inspected with great care during various visits to Philadelphia. As if to set the stage for his most public and complete conversion to pluralism, Agassiz wrote an article for the March, 1850, issue of the *Christian Examiner*, a Boston Unitarian journal of liberal religious views.[44] The purpose of the article was to point out that an unorthodox interpretation of human origins did not necessarily contradict a truly religious attitude. As a naturalist, Agassiz maintained that a belief in a common center of origin for species was the greatest obstacle to the intelligent study of the geographical distribution of animals. A statement such as this was noteworthy for it demonstrated Agassiz's realization that pluralism was an essential basis for a view of nature which denied development and change. He stated further that it was impossible to believe, with those who interpreted the Bible in literal terms, that all present-day animals were created in one place. Affirming that Genesis only referred to those animals and plants placed near Adam and Eve by the Creator, he claimed that there had been as many as a dozen separate creations, and that all this proved the power of the Deity to be far *greater* than the believers in "unity" would allow. These assertions on Agassiz's part were equally significant. In order to defend his basic belief in the separate creation of species, it was necessary to go against accepted theological interpretation. To justify such an apparent heresy, he pleaded that such a view would in effect encourage greater faith and religious belief, since the Creator was thus endowed with a wider control over the origin of life.[45]

At the conclusion of this article, Agassiz noted that man was a special case, and promised to take up the question of human origins at another time. Such an opportunity soon presented itself in the same month of March, 1850, at the third meeting of the American Association for the Advancement of Science, held in Charleston. These meetings took place just at the time when the unity-plurality controversy was nearing its most explosive stage. Agassiz had offended

[43] R. W. Gibbes to Lewis R. Gibbes, 24 June 1846, The Papers of Lewis R. Gibbes, Manuscripts Division, Library of Congress; Michael Tuomey to S. G. Morton, 28 February 1847; R. W. Gibbes to S. G. Morton, 4, 10 March, 24 October 1847; George R. Gliddon to S. G. Morton, 9 January 1848, Manuscript Collections, Library Company of Philadelphia.

[44] "Geographical Distribution of Animals," *The Christian Examiner*, *48*, 181–204, 1850.

[45] Agassiz's success in this regard was demonstrated by the way N. L. Frothingham, an advocate of unity, defended the naturalist's right to interpret the Bible as he saw fit. Cf., "Men Before Adam," *The Christian Examiner*, *50*, 79–96, 1851. For a typical fundamentalist reaction, see Thomas Smyth, *op. cit.*, 375.

some theologians with his deviation from the standard interpretation of the Scriptures,[46] Morton and Bachman were exchanging heated arguments in the public prints,[47] while Nott's writings were a source of irritation to fundamentalist clergy in both sections of the country.[48] The scientists gathered at Charleston clearly looked forward to a definite pronouncement on the problem from Agassiz.[49] They were not disappointed.

The sessions of 15 March began with a paper by Nott, entitled "An Examination of the Physical History of the Jews, in its Bearings on the Question of the Unity of Races." [50] A condensation of views Nott had expressed many times before, the main point of the address was that a "race" such as Nott considered the Jews to be had remained "pure" because its vitality had not been marred by intermixture with other "species" of races. It is evident that this sort of reasoning became the basis of the southern argument against racial equality, which was advocated as sound social and scientific doctrine before and after the Civil War. One can easily detect in this sort of "science" the social and intellectual outlook that prompted Nott to supervise the publication of the American edition of Count Arthur de Gobineau's *Essay on the Inequality of Human Races*,[51] and which motivated race-conscious southerners to hail "science" as proving races were distinct and hence unequal. That Louis Agassiz should have allowed the weight of his international scientific standing to lend support to this type of ideology was indeed tragic. It is clear that he never fully realized the extent to which his views served the partisan purposes of racists, nor fully appreciated the sensitivity of the majority of Americans to the slavery question. Obviously, Agassiz's underlying motivation in the unity-plurality controversy was to strengthen his position in regard to the separate creation of species and to strike a blow for scientific freedom as opposed to an intellectual dogmatism which relied on traditional beliefs. It is significant

[46] For evidence of this reaction and the defense of Agassiz by liberal theologians, see "The Tendencies of Modern Science," *North American Review*, *72*, 114–115, 1851; *Christian Examiner*, *50*, 79–82, 96, 1851. Such statements should be compared with that of Smyth, *op. cit.*, p. 375.

[47] The Morton-Bachman debate was precipitated by the editors of the *Charleston Medical Journal*, through the publication of a highly favorable review of Morton's *Catalogue of Skulls*, *5*, 84–86, 1850. Bachman responded with an article in the next issue attacking Morton's hybridity theories of 1847, *ibid.*, 168–197. The editors commented that they were pleased to present Bachman's article as it bore "directly upon the great mooted question of the present day, the Unity or Plurality of the Human Race," *ibid.*, 274.

[48] George R. Gliddon to S. G. Morton, 9 January 1848; R. W. Gibbes to S. G. Morton, 21 January 1850, Manuscript Collections, Library Company of Philadelphia.

[49] R. W. Gibbes to S. G. Morton, 31 March 1850, *ibid.*

[50] *Proceedings of the American Association for the Advancement of Science* (Charleston, 1850), *3*, 98–106. Cf., Josiah C. Nott, *Two Lectures on the Connection Between the Biblical and Physical History of Man* (New York, 1849). These lectures had been delivered at the University of Louisiana in 1848. It is interesting to note that this sort of argument was one of the main emphases in the southern attack upon racial amalgamation. Thus Nott held that mulattoes (hybrids) were unhealthy and less prolific, and resembled the Negro more than the white. This was also the view of Agassiz (cf. note 67). This despite the fact that the fertility of hybrid crosses was a basic article of faith with pluralist thinkers.

[51] This appeared as: Joseph Arthur de Gobineau, *The Moral and Intellectual Diversity of Races . . . their respective influence in the Civil and Political history of mankind . . . to which is added an appendix containing a summary of the latest scientific facts bearing upon the question of the unity or plurality of races, by J. C. Nott.* (Philadelphia, 1856). Nott to Joseph Leidy, 24 August 1855: "I think the work of Gobineau will do more good in liberalizing the public than any yet written" (Leidy Papers).

that in later years Agassiz maintained that American Negroes had a right to political equality, but that such equality should be granted only with a full realization of the permanent physical, social and mental differences which made the Negro inferior to the white.

After Nott's paper had been read, Agassiz rose and stated that "the differences . . . between the races were . . . primitive . . . races did not originate from a common center, nor from a single pair." He then related this assertion to his previously expressed conception of distinct zoological provinces with characteristic animal inhabitants. These provinces were also marked by particular and distinct human races, which, like animals and plants, had been successively created in different parts of the world at different times.[52]

Such a public profession of belief in the plural origin of mankind stimulated a heated discussion from the audience of scientists and laymen that heard Agassiz. While Bachman refrained from joining the argument, the concept of unity was defended in a speech by Thomas Smyth, a Presbyterian clergyman who shortly thereafter published a volume attacking Agassiz for his conversion to pluralism.[53] The net result of Agassiz's Charleston pronouncement was to provide an ultimate scientific justification for a social theory. Consequently, his announced beliefs disturbed believers in unity as much as they delighted members of the pluralist camp.[54] Nott was particularly appreciative, and clearly aware of the importance of Agassiz's support. He wrote Morton: "With Agassiz in the war the battle is ours. This was an immense accession for we shall not only have his name, but the timid will come out of their hiding places. I have been agitating, agitating till I have got . . . Agassiz into the fight . . . The parsons now are certainly in the way of being licked." [55]

In order to reaffirm his Charleston statements in the public prints, and to establish his respect for religion, Agassiz published his views in the July, 1850, issue of the *Christian Examiner*.[56] This was a significant article, for Agassiz gained many adherents to the concept of pluralism by force of his authority as a naturalist and his contradiction of fundamentalists who had challenged his religious sincerity. Agassiz insisted that naturalists had a right to examine the question of human origins without reference to religious or political doctrines. At the same time he asserted that while there was a genetic difference between human types, mankind was all one brotherhood by virtue of its spiritual and moral *unity* with the Creator. Taking note of the fact that he had been accused of supporting the doctrine of slavery with his pluralist interpretation, Agassiz pointed out that he was concerned not only with Negroes, but also with Chinese, Malayans, and other human types. Pleading the role of the true scientist, he

[52] *Proceedings of the American Association for the Advancement of Science* (Charleston, 1850), *3*, pp. 106–107.

[53] See Smyth, *The Unity of the Human Races . . .* (New York, 1850), pp. 353–354, for an account of the Charleston meeting.

[54] *Christian Examiner, 49*, 110–116, 1850; Nott to Morton, 25 July 1850, Manuscript Collections, Library Company of Philadelphia; John Carey to Asa Gray, 26 August 1850, John Torrey to Asa Gray, 30 August 1850, Historic Letter File, Gray Herbarium, Harvard University.

[55] Nott to Morton, 4, 26 May 1850, Manuscript Collections, Library Company of Philadelphia; also Nott to Squier, 4 May 1850, Squier Papers.

[56] "The Diversity of Origin of Human Races," *Christian Examiner, 49*, 110–145, 1850.

claimed that his task was only to examine facts, and that he was not responsible for what "politicians" did with such facts. It was noteworthy that Agassiz's emphasis upon the spiritual brotherhood of mankind, and his reminder that he was not concerned solely with Negro-white differences, were conveniently ignored by pluralist propagandists. Such assertions, however, were evidence that Agassiz had been deeply hurt by the charges of heresy and bigotry which had been made against him. At this point in the controversy he is reputed to have exclaimed to a student, "Why, there is no freedom for a scientific man in America." Agassiz's desire to promote the freedom of inquiry necessary to science was very understandable. The fact of the matter was, however, that despite all of Agassiz's pleading to the contrary, there was no middle ground on the question of plurality in the America of the 1850's, and any views were bound to be claimed as proving the cause of one side or the other.[57]

Actually, Agassiz's desire to elevate the neutrality of science in the midst of an ideological debate was contradicted by his own statements in this article of July, 1850. There were original differences between men, he wrote, and it was not really very important whether human types were designated as "races," "species," or "varieties"; what was important was a recognition of the fundamental differences between them. Agassiz maintained, therefore, that in addition to physical peculiarities, Negroes were by nature "submissive," "obsequious," and "imitative"; it was thus mere "mock philanthropy" to consider them as equal to whites. Africans, for instance, had been in contact with whites for thousands of years, yet were still adverse to civilized influences. White relations with colored peoples would be conducted more intelligently if the fundamental differences between human types were realized and understood. Agassiz's real motivation for the statements made in this article was underscored by another piece which appeared in the January, 1851, number of the *Christian Examiner*. At this time, the logic of Agassiz's pluralism was made clear when he wrote that a belief in the separate origin and independent creation of both human and animal species was necessary in order to demonstrate the supreme creative power of the Deity and to disprove any evolutionary or "developmental" philosophy.[58]

The unity-plurality controversy died down during the years 1851–1854, mainly because Nott and his co-believers were hard at work producing the classic statement of pluralist philosophy. This was the volume, *Types of Mankind*, published in 1854 by Nott and George R. Gliddon, an archeologist who had furnished the pluralist school with much material. There is some evidence to indicate that the collaboration of Nott and Gliddon was not a

[57] Bachman claimed that Agassiz's *Christian Examiner* articles of 1850 had been responsible for a shift in popular opinion in favor of plurality. See Bachman, *A Notice of the "Types of Mankind"* (Charleston, 1854), p. 13 (reprinted from *Charleston Medical Journal*, *9*, 1854). See "Natural History of Man," *Democratic Review*, *26*, 27, 327–345, 41–48, 1850, for proof of Bachman's contention. Nott to Morton, 25 July 1850: "I have seen Agassiz's two articles . . . and hope he will go on — they are superb" (Manuscript Collections, Library Company of Philadelphia). For Agassiz's complaint about restrictions upon scientific freedom, see William Dallam Armes (ed.), *The Autobiography of Joseph Le Conte* (New York, 1930), p. 140.

[58] "Contemplations of God in the Cosmos," *Christian Examiner*, *50*, 1–17, 1851.

happy one; in later years Nott laid the blame for the volume's weaknesses upon Gliddon, and charged that the virulent attacks against revealed religion were owing to that author's prejudices.[59] The 738-page tome was dedicated to the memory of Morton, who had died in 1851. It contained his published and unpublished writings on comparative anatomy and hybridity in relation to plural origins, emotional attacks upon religious prejudice by Gliddon, pseudo-scientific analyses of racial types from the pen of Nott, and an essay by Agassiz. The volume was a conglomeration of all sorts of questionable science, offered as convincing proof of the inferiority of the Negro to the white. It had a wide popular sale, and was very favorably received by defenders of slavery. Much of the success of the work was due to the authority of Agassiz.[60]

In his essay Agassiz distinguished eight primary human types — the Caucasian, Arctic, Mongol, American Indian, Negro, Hottentot, Malay and Australian. He noted the extent to which these "types" inhabited specific zoological provinces, and pointed out that these zones were in turn characterized by particular flora and fauna. Agassiz repeated what he had urged before, that such types of humans were distinct, and that their differences were original and primitive. He argued in a very revealing fashion that unless one embraced a belief in Lamarckian "development" from one primary stock, the only alternative was acceptance of a belief in the original diversity of mankind, predetermined by the Creator and exemplified in the laws of geographical distribution. Since any form of developmental philosophy was at complete variance with Agassiz's philosophy of nature, it followed that the doctrine of the unity of man was "contrary to all the modern results of science." As the most decisive Agassizian statement on plurality, this essay is unmistakeable evidence of the relationship of such a belief to the naturalist's opposition to any concept of evolution.

In 1857 Nott and Gliddon published another volume, *Indigenous Races of the Earth.* This work, intended as a supplement to *Types of Mankind*, was more moderate in tone and superior in organization to the 1854 volume. As such, the book was the culmination of Nott's efforts dating back to 1844, and directed towards gaining respectability for his views by identifying them with

[59] Nott to Joseph Leidy, 4 October 1854, 6 April 1856, Leidy Papers. The tone of Nott's letters and his own attacks on the clergy contradict his assertions in these letters.

[60] Henry R. Schoolcraft to John Bachman, 23 September 1854: "The types are . . . the fruits of the mountain that was in labor. From one end of the land to the other, subscribers have been drummed up for this work, and when it came forth it is a patch-work of infidel papers . . . if this be all that America is to send back to Europe . . . it were better that the Aborigines had maintained their dark empire undisturbed" (quoted in C. L. Bachman, *op. cit.*, p. 317). There were nearly 1000 original subscribers to "Types," and it appeared in ten editions between 1854 and 1871. Agassiz to Dana, 18 July 1856: "As to your allusion to my paper in . . . Types of Mankind. . . . I do not regret contributing it. Nott is a man after my heart, for whose private character I have the highest regard. He is a true man, and if you knew what he has had to suffer in Mobile from the criminations of bigots . . . you would not wonder at his enmity to such men. I know him to be a man of truth and faith. Gliddon is coarse. . . . But I would rather meet a man like him . . . than any . . . who shut their eyes against evidence" (Manuscript Collections, Rare Book Room, Yale University). Agassiz, "Sketch of the Natural Provinces of the Animal World and their Relation to the Different Types of Man," *Types of Mankind* . . . (Philadelphia, 1854), pp. lvii-lxxviii.

the most advanced scientific doctrine. This attempt once more received the sanction of Agassiz's name and intellect.[61]

In his remarks it seemed as if Agassiz were writing in clear anticipation of his forthcoming debates with Asa Gray at the American Academy of Arts and Sciences, debates which signalled the actual beginning of the argument over evolution in America.[62] For Agassiz emphasized what by this time was quite plain, namely, that problems concerning the origin of mankind embraced the entire question of origins in all of natural history. In noting that this was an area of vital concern to all zoologists, Agassiz demonstrated that the controversy which was to divide naturalists into opposing factions but two years later, was in the real sense a culmination of a traditional speculative interest in such matters. Agassiz did not offer any new or startling evidence in support of his contention concerning the plural origin of man, but did insist on the truth of his position in stronger terms than he had ever used before. As proof for the pluralist position, Agassiz cited Richard Owen's researches upon monkeys, in the course of which the English anatomist had distinguished three separate species.[63] Agassiz maintained, with questionable logic, that since anthropoids had been classified in this manner, and since such species differed from each other in the same degree that marked human racial divergencies, therefore mankind should also be characterized into different species.[64] In an equally positive manner Agassiz discounted one of the basic arguments of the unity school by stating that linguistic affiliations were no proof of common human origins since obviously distinct animal species shared communicative resemblances.

Thus, in the space of seven years Nott and his followers had succeeded in establishing pluralism as an hypothesis based upon a general scientific framework. Such professional respectability would not have been possible without the important contributions of Agassiz. Although Bachman was still one of the few public advocates of unity,[65] it was clear that the end result of the work

[61] Agassiz, "Prefatory Remarks," *Indigenous Races of the Earth* (Philadelphia, 1857), pp. xiii-xv. Nott and Gliddon made every effort to reinforce the authority of Agassiz and Morton with that of Joseph Leidy. Nott to Leidy, 4 October 1854, 6 April 1855, 6 April 1856, Leidy Papers. Leidy, however, would not commit himself in favor of plurality, but did allow a letter he had written to Nott (Leidy to Nott, 10 February 1857, Leidy Papers) to be included in the volume. *Indigenous Races*, pp. xxvi-xxix.

[62] See note 38.

[63] Agassiz had advanced such views before in *Proceedings of the American Academy of Arts and Sciences* (Boston, 1852–1857), *3*, pp. 7–8 (22 June 1852).

[64] Owen actually took an equivocal position on the problem. In his Presidential Address to the BAAS (*Report of the Twenty-Eighth Meeting of the British Association for the Advancement of Science . . . 1858*, London, 1859, pp. xciii-xcv) he stated that "the human species is represented by a few well marked varieties," but also praised Agassiz's concept of zoological provinces.

[65] Bachman renewed his attack on the pluralist school, which he had ceased after Morton's death, when the "Types" appeared. This took the form of three articles, appearing in volumes 9 and 10, 1854 and 1855, of the *Charleston Medical Journal*, and totaling 105 pages. Cf. Bachman, *A Notice of the "Types of Mankind"* (Charleston, 1854); *An Examination of Professor Agassiz's Sketch of the Natural Provinces of the Animal World . . .* (Charleston, 1855); *Continuation of the Review of Nott and Gliddon's "Types of Mankind"* (Charleston, 1855). Nott to Leidy, 25 July 1855: "I have just sent on to (the) . . . Med. Journal a . . . reply to Bachman's blackguard review of Agassiz . . . If his articles be . . . read only by Naturalists they would do no harm; but they are read and intended for the people, and bigots — I think it the duty of some one occasionally to roll back these masses of filth from the road of science" (Leidy Papers).

of Morton, Gliddon, Nott and Agassiz was to lend the sanctity of science to a social and political doctrine based upon the inequality of colored and white races. As much as Agassiz protested that science was neutral with reference to politics and society, southern statesmen and publicists had taken full advantage of pluralist writings to reinforce their defense of slavery as a justifiable institution.[66]

After 1857, Agassiz centered his attention on the affirmation of special creationism with reference to the animal kingdom. Nevertheless, his views on the separate origin and plurality of human species and on the physical, social and mental inferiority of the Negro were apparent in the advice he gave regarding governmental reconstruction policy in the South.[67] Agassiz's racial interpretations were also apparent in his report of a journey to Brazil, in which he cited the physical and mental differences of Indians, Negroes and mulattoes as proof for pluralism.[68]

It is obvious that Agassiz's writings on plurality and his announcements on the subject from the public platform were an integral part of his total opposition to any concept of evolution in human or animal life. Charles Lyell, a firm believer in the unity of mankind, best explained Agassiz's position on this question in his volume dealing with human development: "Were we to admit," he paraphrased Agassiz's reasoning, " a unity of origin of such strongly marked varieties as the Negro and European . . . how shall we resist the arguments of the transmutationist, who contends that all closely allied species of animals and plants have . . . sprung from a common parentage. . . . Where are we to stop, unless we take our stand . . . on the independent creation of distinct human races . . . ?"[69]

In so defending his philosophy of nature, Agassiz had provided a scientific rationalization for the existence of slavery. While this was not his intention, such a result might have been avoided or postponed had he confined his statements to professional journals. This was not the case, as Agassiz's most significant pronouncements were made from the platform or appeared in popular magazines or volumes. Yet it was one of the most noteworthy aspects of Agassiz's personality that he exhibited a great desire for popular support and

[66] Nott claimed that Calhoun, when Secretary of State in 1844, had consulted Gliddon and Morton regarding scientific evidence for the inequality of races. Nott stated further that Calhoun, a convinced pluralist, incorporated such sentiments into his diplomatic negotiations with England and France, in which he asserted the scientific justification for the institution of slavery in the United States. See Nott, *Types of Mankind,* pp. 50–52. Alexander H. Stephens, Robert Toombs and Robert B. Rhett were three leading southern politicians who adopted the argument from science in their defense of the slave system. Cf. Jesse T. Carpenter, *The South as a Conscious Minority 1789–1861* (New York, 1930), pp. 11–12 and note 14; Merle Curti, *The Growth of American Thought* (New York, 1943), pp. 445–446. Nott to Squier, 4 May 1850: ". . . all the . . . articles I have written on *niggerology* have been eagerly sought for at the South, and in the present excited state of the political world I think the thing will go well . . ." (Squier Papers).

[67] Agassiz to Samuel Gridley Howe (of the American Freedmen's Inquiry Commission) 6, 9, 10, 11 August 1863, Autograph File, Houghton Library, Harvard University.

[68] Agassiz, "Permanence of Characteristics in Different Human Species," in Professor and Mrs. Louis Agassiz, *A Journey in Brazil* (Boston, 1868), pp. 529–532.

[69] Charles Lyell, *The Geological Evidences of the Antiquity of Man, With Remarks on Theories of the Origin of Species by Variation* (London, 1863), pp. 387–388.

acclaim in matters of scientific dispute.[70] Asa Gray and James Dwight Dana, however, although avowed believers in the unity of mankind, never broached their opinions on this subject in a popular fashion.[71]

Apart from its social, political and religious significance, the real importance of the unity-plurality debate for the history of American natural history is to be found in the fact that for fifteen years prior to the publication of Darwin's volume, the American public was exposed to many of the same arguments that were to be employed in the evolution debates. With the one school of thought insisting on the origin of man from one common stock, and the other group emphasizing the separate and specific creation of human types, such significant questions as the nature of variation, the importance of hybridity and the meaning of geographical distribution were all raised by Agassiz and the other disputants. It was unfortunate that such discussions were conducted with reference to the equality or inferiority of the American Negro, but the fact that they were, demonstrates the great degree to which scientific problems were of social and cultural import in nineteenth-century America.

[70] Agassiz followed the same procedure in the evolution controversy, wherein he published but one article in a scientific journal (*American Journal of Science and Arts*, *30*, 2nd series, 142–154, 1860), with the rest of his statements appearing in popular form.

[71] Dana challenged Morton's concept of fertile hybrid mixtures in "Thoughts on Species," *American Journal of Science and Arts*, *24*, 2nd series, 305–316, 1857. He stated: "We have . . . reason to believe from man's fertile intermixture, that he is one in species . . ." (p. 311). Dana wrote Gray his belief in the unity of man in: Dana to Gray, 11 December 1856, Historic Letter File, Gray Herbarium, Harvard University. Gray made clear his firm agreement with Dana in: Gray to Dana, 13, December 1856, Asa Gray Papers, Gray Herbarium, Harvard University.

Science in the Civil War

The Permanent Commission of the Navy Department

By Nathan Reingold *

WITHIN a period of less than a month during the course of the Civil War two organizations concerned with the provision of scientific aid for the Government were created. On 11 February 1863, Gideon Welles, the Secretary of the Navy, approved the organization of a Permanent Commission to advise the Department on "questions of science and art." The members of the Commission were Rear Admiral Charles Henry Davis, Chief of the Bureau of Navigation; Alexander Dallas Bache, Superintendent of the Coast Survey; and Joseph Henry, Secretary of the Smithsonian Institution.[1] On 3 March 1863, President Lincoln signed the act incorporating the National Academy of Sciences and permitting it to "investigate, examine, experiment, and to report upon any subject of science or art" referred to it by the Government.[2] Bache was the first President of the National Academy.

The appearance of these two organizations indicates a lively interest in the role of science in the Civil War, at least among the small group involved in the two agencies. Most of what follows is an attempt to explain why the Permanent Commission and the National Academy of Sciences accomplished so little during the Civil War in terms of scientific aid for the Government, that is to say, why the Civil War was not a "scientific" war.

The crucial characteristic of a "scientific" war is that a deliberate attempt is made to have the best scientific talents, usually basic scientists, improve existing or devise new weapons, equipment and processes, and that, as a consequence, new or drastically improved weapons appear in battle in sufficient quantity to alter the tactics and strategy of the armed forces. Three antecedent conditions are necessary for a "scientific" war. First, there must be an opponent capable of waging a "scientific" war. The reason why the Navy was relatively more progressive than the Army during the Civil War was that only in naval warfare did the Confederacy threaten the Union forces with novel devices. Because the United States encountered opponents capable of waging a "scientific" war only in the present century, its armed services generally lagged behind European powers in research and development until recently. Certainly, Indian wars or scuffles with the Mexicans did not call for super-weapons. Second, there must be sciences at stages amenable to significant applications and scientists versed in those fields. Although these considerations are outside the scope of this paper, the relative scarcity of physical scientists in Civil War America, and the fact that physics and chemistry in the mid-nineteenth century were not obviously pregnant with warlike possibilities, are significant bench marks. The third condition is the existence of industries requiring and performing

* The National Archives.

[1] Gideon Welles to Charles Henry Davis, 11 Feb. 1863, Letters Received, Permanent Commission, Naval Records Collection of the Office of Naval Records and Library (RG 45), National Archives. Unless otherwise cited, all records are in the National Archives. "RG" is the symbol for record group.

[2] 12 Stat. 808.

research and capable of translating research data into military hardware. In the absence of such industries the Federal Government was forced to rely on the chance, unreliable labors of inventors and amateurs of science. The position of the private amateur inventor in relation to the professional scientist was the crucial issue in the role of science in the Civil War. This paper will attempt to demonstrate how the role of the inventor largely determined official actions and policies in utilizing science in warfare.

Evaluating innovations in military technology was a familiar peacetime problem to the Army and Navy. Stimulated by a mixture of patriotic fervor and avarice too complex for historical analysis, inventors literally besieged official Washington after the outbreak of war. By the end of 1861 Secretary of the Navy Welles established a Naval Examining Board to appraise inventions. His instructions stressed that recommendations to adopt an invention should state "the advantages and the economy that will result from its use and the total expenditure that it will occasion." For, as the Secretary had previously admonished, "the money appropriated by Congress for the Navy cannot be applied to any experimental purpose but only for objects of undoubted utility." The Board functioned from 2 January to 10 July 1862 and accomplished very little.[3]

In 1863 the Navy Department's need for a means of reviewing inventions coincided with more ambitious stirrings in the American scientific community — notably Davis, Bache, and Henry. Rear Admiral Davis was one of the few naval officers of that period with any substantial scientific competence. His Bureau of Navigation, embracing the Naval Observatory, the Hydrographic Office, and the Nautical Almanac Office, was intended to be the Navy's scientific bureau.[4] Bache, a West Pointer who became a geophysicist, was a first-rate scientist with pronounced abilities as an organizer and administrator. He was the leader of an influential group of scientists and had favored the organization of an American society of savants at least since 1851.[5] Joseph Henry was a very cautious protector of the Smithsonian endowment from proposals conflicting with his organization's primary purpose, "the increase and diffusion of knowledge among men." During the Civil War he readily gave his services and the facilities of the Smithsonian Institution to the war effort. Henry was simply seeking a way of accomplishing a disagreeable but necessary task.[6]

The bare outline of the events leading to the formation of the Commission and the Academy is as follows: Some time after Davis returned to Washington in November 1862, the scientific coterie around Bache began to discuss the problem of utilizing science in the war effort. By January two specific proposals had progressed from talk to action. Because of Joseph Henry's initial

[3] See Minutes of the Naval Examining Board and Inventions Referred to the Naval Examining Board (RG 45). Welles' instructions of 27 Dec. 1861 are attached to the front of the Minutes.

[4] For Davis, see C. H. Davis, Jr., *Life of Charles Henry Davis, Rear Admiral, 1807–1877*, Boston, 1899.

[5] Frederick W. True, ed., *A History of the First Half-Century of the National Academy of Sciences*, Washington, 1913, 7–8. The possibility of splitting the American scientific community, many scientists being antagonistic to Bache's group the "Lazzaroni," was recognized by the founders of the Academy. Louis Agassiz wrote Bache on 6 March 1863 (Rhees Collection, Huntington Library), "How shall the first meeting be called. [sic] I wish it were not done by you, that no one can say this is going to be a branch or of the Coast Survey and the like." See also Merle M. Odgers, *Alexander Dallas Bache, Scientist and Educator, 1806–1867*, Philadelphia, 1947.

[6] An example of this attitude: on 15 July 1863 he wrote Assistant Secretary of the Navy Fox, advising tolerance of an inventor whose letter "is more that of a visionary enthusiast than that of a modest benefactor of his race." He went on to state, "I never seek to be employed on public commissions since they are generally attended with more disagreeable than pleasant consequences. On the other hand I never refuse to lend my services gratuitously to the Government in any way which they may think of importance." In Fox Papers, New York Historical Society.

opposition to the proposed Academy on the grounds that Congress would not pass such an act and that a National Academy would arouse jealousies among scientists, the National Academy was ostensibly dropped in favor of a Permanent Commission, a select standing body to advise the Navy. On 26 January, 1863, Bache, Henry, and Davis conferred with Fox. Further details were discussed by the three on 5 February. Two days later Henry transmitted to Fox a "programme," which was almost faithfully copied in the 11 February letters sent over Welles' signature to the three original members of the Commission. But without Henry's knowledge, Bache, Davis, Agassiz, and the astronomer Benjamin Apthorp Gould were actively promoting the National Academy. Henry heard of these efforts only after the bill for the incorporation of the Academy was drawn up, and learned of the bill's passage only on 5 March.[7]

While the legislation was pending, the Commission was already in operation. Its first meeting was held on 20 February 1863.[8] Two days later it received 41 proposals from the Navy Department; the earliest was dated 3 July 1862, and the latest 5 February 1863.[9] Some of the early proposals, as well as others subsequently transferred to the Permanent Commission, had originally been referred to the Naval Examining Board. The last meeting of the Commission of which any record survives occurred in April 1865, but the last report, bearing only the signature of Rear Admiral Davis, is dated 21 September 1865. Between 20 February 1863, the date of the first meeting, and 21 September 1865, the date of the last report, the Commission held approximately 109 meetings and submitted 257 formal reports on proposals evaluated in those sessions.[10] Eighty-two of these meetings occurred in the period up to 24 February 1864, the date of the last meeting for which minutes survive. During that year and four days, 187 of the reports were prepared. After this period of its greatest activity, the Commission began to flag, and by the summer of 1864, it entered a period of marked decline. On the basis of the surviving records a reasonable estimate would be that some 300 inventions were transmitted to the Commission.

[7] Based on the following: Henry to Bache, 26 Jan. and 5 Feb. 1863, Bache to Henry, 14 Feb. 1863, and "Locked Book" extract of 28 Oct. 1863, Henry Papers, Smithsonian Institution; No. 103, Miscellaneous Letters Received of the Secretary of the Navy, Feb. 1863, vol. I, (RG 45) is Henry's letter transmitting the "programme," followed by the enclosure and Fox's penciled draft of the 11 Feb. 1863 reply of Welles'; Davis, *op. cit.*, 289–292; Andrew Denny Rogers III, *John Torrey*, p. 274, Princeton, 1942.

For information concerning the founding of the Permanent Commission and the National Academy, I am indebted to Dr. A. Hunter Dupree (University of California, Berkeley), who has examined the Asa Gray and Benjamin Peirce Papers at Harvard University. Dr. Dupree has specifically called my attention to a letter of Henry's to Louis Agassiz, 13 August 1864, Peirce Papers, which outlines Henry's connection with the founding of the Academy.

[8] Minutes, Permanent Commission, 2 (RG 45). The records of the Permanent Commission in the National Archives upon which this paper is largely based consist of the following: 3 vols., Jan. 1861–Dec. 1865, of Proposals Referred to the Commission; Minutes of the Permanent Commission, Feb. 1863–Feb. 1864, 1 vol.; Letters Received, Feb.–June 1863, 1 vol.; Letters Sent, March 1863–Sept. 1865, 2 vols. of press copies. The originals of the Reports of the Permanent Commission to the Secretary are also in RG 45 among the records of the Office of the Secretary of the Navy. The three volumes of proposals do not contain all the proposals considered by the Commission as many were returned to their authors or referred to other agencies. On the other hand, many in the volumes were never considered or were still pending when the Commission's activities ceased. One minor mystery connected with these records is the absence of any Minutes after Feb. 1864, especially since the surviving volume of Minutes contains several hundred blank pages.

[9] Minutes, Permanent Commission, 6–7 (RG 45).

[10] See the vol. of Reports of the Permanent Commission (RG 45). The last record of a meeting is in Letters Sent, vol. 2, which has a letter, 12 April, 1865, calling for a meeting the following Friday. As Report No. 256, 10 July 1865, is signed by the full Commission, it is probable that a meeting or meetings occurred after April. The number of reports is from the Reports volume. The number of meetings is based on a count from correspondence plus the 82 meetings in the Minutes.

The operations of the Permanent Commission were quite informal, as befitted the deliberations of a group of old friends. Secretary Welles' instructions, based on Henry's "programme," simply authorized the use of associates and prohibited compensation for members as well as the associates.[11] The Commission itself adopted only two procedural resolutions during its lifetime. At the 13 April 1863 meeting it accepted a resolution of Bache's that "it is not expedient" for members to receive personal communications from persons having proposals before the Commission. It then approved Davis' resolution that the Commission "confine itself to plans and descriptions presented to it," i.e., those coming through the Office of the Secretary of the Navy.[12]

During its greatest period of activity the Commission held meetings as often as three times a week, usually rotating among the Bureau of Navigation, the Smithsonian Institution, and the Coast Survey offices. On occasions when demonstrations were called for, the Commission met at the Washington Navy Yard or elsewhere in Washington. Inventors or their representatives were encouraged to appear personally and explain their proposals. Those meriting special attention were assigned to one or more members for study before preparation of a report. Others were summarily rejected. In spite of all the serious business transacted at these sessions, they were also rather pleasant social occasions. (On 10 February 1864 the minutes gravely noted, "Mrs. Bache being unwell, the Professor regretfully announced that there were no oysters." [13])

The membership of the Commission was soon expanded to five by the addition of Joseph Saxton and Brig. Gen. John Gross Barnard. Saxton was Bache's assistant for work on standards of weights and measures. Initially an associate of the Commission in the summer of 1863, he became a full member, at Henry's urging, by the end of the year.[14] Although Henry had written the War Department at the same time as the Navy,[15] Barnard's presence was simply due to the fact that many subjects considered by the Commission were also of interest to the Corps of Engineers (i.e., harbor defences). Prior to July 1863 when he appeared on the scene, the War Department and the Permanent Commission had exchanged inventions. Barnard played a minor role in the Commission's deliberations. According to Joseph Henry, after the General joined the group, "our music is not quite as entirely as harmonious as before." [16]

Apparently the Civil War public was somewhat confused as to the relation of the Permanent Commission to the National Academy of Sciences. Both were founded at about the same time and in part for the same purpose. It was certainly confusing to have Bache, as a member of the Commission, vote to have a question referred to himself as President of the National Academy. The confusion was further compounded by the appointment of Academy committees including the very members of the Permanent Commission who had voted to refer the question to the Academy in the first place! The relationship between the two organizations apparently troubled the Commission members

[11] See fn. 1, above. The absence of any profit motive in the legislation authorizing the National Academy was probably one of the reasons that the *Scientific American* opposed the Academy. See the issue of 23 May 1863, *8*: 329. The magazine favored private, rather than public, support, especially in "practical" matters.

[12] Minutes, 26 (RG 45).

[13] Minutes, 151 (RG 45).

[14] Henry to Bache, 13 August 1863, Henry Papers, Smithsonian Institution; Minutes, 141.

[15] Locked Book Extract, 28 Oct. 1863, Henry Papers, Smithsonian Institution. These extracts were apparently copied by Henry's daughter after his death from a volume containing a diary and/or copies of outgoing correspondence. The original volume was presumably destroyed afterwards.

[16] Minutes 81, 94; AGO Special Order 275, 22 June 1863, in Letters Received, Permanent Commission; Henry to Bache, 13 Aug. 1863, Henry Papers.

also, and they took pains to explain the situation in the only public announcement issued on the Commission's activities:

> The present members of the Commission are also members of the National Academy of Sciences; and the Commission itself would probably never have been created if the Academy had been in existence at that time, since they both have the same objects, and are designed to perform similar duties; it is not impossible that the former may at some time be resolved into a Committee of the latter.[17]

But for the awkward fact that it was founded earlier and within the Navy Department, the Permanent Commission might have served as the operating arm of the Academy. That is to say, the Permanent Commission was an abortive National Research Council.

The questions referred to the National Academy by the Permanent Commission stemmed from the changes which revolutionized naval warfare in the nineteenth century — the introduction of armor and steam. The Navy, for example, encountered a serious difficulty in compass deviations due to the large masses of iron in its new vessels. The problem fell within the jurisdiction of the Bureau of Navigation, which furnished compasses and other navigational devices to the fleet. The same day that Congress incorporated the National Academy, it authorized the Navy Department "to make experiments for the correction of local attraction in vessels built wholly or partly of iron." [18] The matter was referred to the Permanent Commission, which appointed Wolcott Gibbs, Bache and Henry to witness the correction of compass deviations in the steamer *Circassian* by the application of methods agreed on in conference with "practical" experts.[19] The three met in New York, where the *Circassian* lay, on 19 March 1863. A few days later Welles approved the appointment of the committee to conduct the experiments called for by Congress. This body consisted of the three from the *Circassian* group plus Benjamin Peirce and W. P. Trowbridge. The five met in New York on 21 April 1863, shortly before the first meeting of the National Academy. After the Academy was organized, the Permanent Commission referred the problem of magnetic deviations to it on 8 May 1863.[20] The Academy committee consisted of the five from the Navy group and Fairman Rogers and Charles Henry Davis. The problem was obviously of great interest to Bache and Henry. But although the Academy committee labored diligently, it did not make any signal contributions. The English astronomer Airy had previously proposed a means of coping with the problem, and several individuals, some of whom were hired by the Navy, were already applying Airy's method.[21] Nevertheless, the report of the Academy committee filled a definite need for expert appraisal of a significant practical problem which impinged on basic issues in physics. The Bureau of Navigation immediately established stations to correct compasses by Airy's method.

The experiments on the expansion of steam had origins similar to the work

[17] *Scientific American*, 12 Mar. 1864, *10*: 165. The Commission's statement was elicited by a previous item in the magazine (*ibid.*, 27 Feb. 1864, *10*: 135) which informed readers that inventions were to be sent directly to the Commission. The Commission was forced to return inventions so received and asked that they be addressed to the Secretary of the Navy. The 12 March announcement was mainly intended to correct this and other inaccuracies in the earlier item. See also Henry to Mrs. Alexander Dallas Bache, 16 July 1864, and Henry to Joseph B. Varnum, 8 April 1865, Henry Papers, Smithsonian Institution.

[18] See True, *op. cit.*, 215–217.

[19] Minutes, 9–10 (RG 45).

[20] Minutes, 19, 36. The Permanent Commission proposed the appointment of the five on 25 March. The Secretary's consent was received two days later (RG 45); True, *op. cit.*, 216.

[21] See B. F. Greene, ed., *The Magnetism of Ships and the Deviations of the Compass*, a collection of papers on the subject by Airy, Poisson, and others issued by the Bureau of Navigation in 1867.

on magnetic deviations. They also stemmed from an act of Congress authorizing experiments, passed on 3 March 1863,[22] whose origins can be traced to the labors of "practical" experts, in this case Horatio Allen, president of an iron works, and B. F. Isherwood, Chief of the Navy Department's Bureau of Steam Engineering. As in the case of the magnetic deviations, members of the Permanent Commission were also interested in the general subject area. (One of Bache's earliest triumphs was the investigation of the explosion of steam boilers at the Franklin Institute.)[23]

The experiments were originally referred to a Commission composed of Allen and Isherwood, but the instructions drafted by them were sent to the Permanent Commission. What happened between March 1863, when the Commission considered the proposed experiments, and February 1864, when the National Academy got into the act, is not clear. On hearing from Isherwood that the plans for the experiments were in the hands of the Permanent Commission, Allen wrote Bache that he would be most happy to cooperate with the Commission.[24] But no action was taken until February of the next year. At Fox's suggestion, the original two-man group was supplanted by a nine-man tripartite body composed of representatives of the Navy, the National Academy, and the Franklin Institute.[25] Davis, along with Allen and Isherwood, represented the Navy. The experiments were conducted at Allen's plant by Navy engineers and were never completed. The project was topheavy; there was little need for a tripartite commission to supervise a fairly routine testing operation.

The committee on protecting ironclads from corrosion and fouling in salt water offers by far the most interesting example of Navy–National Academy relations during the Civil War. There is a reasonable doubt that the Department even considered the Academy's committee as acting in its behalf. It involved the Commission in the first of its two exasperatingly delicate encounters with Professor Eben N. Horsford, a food chemist who resigned in 1863 from the faculty of the Lawrence Scientific School at Harvard to engage successfully in the manufacture of baking powder and who during the war devised methods of condensing milk and producing field rations. One of the first actions of the Commission at its initial meeting was to make Horsford an associate to report on means of protecting ironclads from corrosion. On 27 April 1863, Horsford dispatched a long, comprehensive report on possible methods of combatting corrosion and fouling. Horsford devoted most of his attention to the possibility of electroplating vessels with copper, but he also discussed possible use of paints. The experiments on electroplating would take four to eight months and would use ships already constructed. On 8 May 1863, the Commission referred Horsford's report to the National Academy of Sciences and so informed Horsford.[26]

After an interval, Professor Horsford penned a letter to Welles on 25 June, protesting against the Commission's action:

> I beg respectfully to suggest to the Department that my report contained

[22] 11 Stat. 751; $20,000 was appropriated.

[23] See, *Report of Franklin Institute of Philadelphia, in relation to the explosion of Steam Boilers*, Washington, 1836. House Document No. 162, 24 Cong., 1 Sess. (Serial 289).

[24] Minutes, 22–3 (RG 45). Allen to Bache, 3 April 1863, Rhees Collection, Huntington Library.

[25] Henry to Bache, 6 March 1864, Henry Papers, Smithsonian Institution. Davis to Bache, 24 Feb. 1864, Letters Sent, Permanent Commission (RG 45).

[26] Minutes, 2, 35; Horsford to Permanent Commission, 27 April, 1863, Letters Received; Permanent Commission to Horsford, 8 May 1863, Letters Sent (RG 45). See "Eben Norton Horsford," *Dictionary of American Biography*, IX, 236–237. For other examples of his war activities, see Davis to Fox, 30 April 1863, Fox Papers, New York Historical Society, and *Annual Report of the Smithsonian Institution*, 1862, 45.

> the results of much thought, considerable labor and some expenditures, which are embodied in proposed methods of protection to which I attach great value, and which I am unwilling should pass to the knowledge of many persons, particularly in their unverified state.

Horsford asked that his report be restricted to members of the Permanent Commission; he would carry on all research at his own expense and forward the results to the Department.[27] The Commission immediately ordered the withdrawal of Horsford's report from the Academy on 9 July, 1863.[28]

The next day Davis wrote Bache:

> Please send me the report of Mr. Horsford, "on protecting the bottoms of ironclads," for reasons which I will explain to you at some very distant future period — say a hundred years hence.
>
> Your very truly,
> C. H. Davis
>
> P. S. Lest this should appear too mysterious, I will mention the real reason, viz. that Mr. Horsford has written to the Secretary (under date of June 25th, & admitting the receipt of our letter of the 8th of May informing him of the reference of his Report to the National Academy), to say that he considers it confidential & does not wish it to go beyond the Commission.

Although Davis assured Horsford on the same date that his report had not gone beyond Bache, he subsequently learned that the efficient Bache had already sent out a copy to the committee of the National Academy. Bache returned his copy on 17 July and, after being reminded by Davis, returned the copy furnished the committee.[29]

One consequence of Horsford's action was that the committee of the Academy, formed to consider his report, had nothing to deliberate upon. As the members of the committee were not versed in the problem, their report was inconclusive. Henry, however, offered facilities in the Smithsonian Institution to carry on research if Congress appropriated funds.[30]

The Commission's second encounter with Horsford began on 3 August 1863, when Gideon Welles referred to it Horsford's proposal for a submarine. During the previous winter Horsford had conducted experiments on underwater transportation at the Washington Navy Yard — interestingly enough at the very time the *Alligator*, the first Navy submarine, was tested there[31] — and had discussed his vessel with Fox. On 8 June, 1863, he wrote Fox: "I have no doubt of my ability to open the way for the Monitor fleet to Charleston Harbour." The next day he dispatched a letter to Gideon Welles giving additional details: "I am prepared to contract to remove the obstructions to Charleston Harbour. I understand the offer of a million of dollars will be made for the accomplishment of this object." No such offer was ever made, but the rumor certainly acted as a stimulant to Horsford's imagination. He proposed to build two submarines (the extra one in case of an accident to the first) in six weeks. The Government was to transport them to Charleston and to provide an ironclad or a monitor to aid the submarine. In the event of an early peace

[27] Horsford to Welles, 25 June 1863, Letters Received, Permanent Commission (RG 45).

[28] Minutes, 93 (RG 45).

[29] Davis to Bache, 10 July 1863, Letters Sent, Permanent Commission (RG 45). Bache to Davis, 13 July and 31 July 1863, Letters Received, Permanent Commission; Davis to Bache, 28 July 1863, Letters Sent, Permanent Commission (RG 45).

[30] *Annual Report of the National Academy of Sciences, 1863*, 4–5, 21–23. House Miscellaneous Document No. 79, 38 Cong., 1 Sess. (Serial 1200).

[31] Louis H. Bolander, "The *Alligator*, First Federal Submarine of the Civil War," *U. S. Naval Institute Proceedings*, June 1938, *64*: 845–854.

Horsford asked only for reimbursement of costs.[32] In the following month the plans for the submarine, the *Soligo*, arrived in Washington. The cost was $54,000, and Horsford now estimated that construction would require from three to four months.[33] Welles referred the plans to Commodore Joseph Smith, Chief of the Bureau of Yards and Docks, who was unimpressed by their originality, utility, or clarity but thought the price reasonable if Horsford guaranteed success.

The day after his proposal was received by the Commission, Horsford appeared before them to explain his invention. The Commission actually wrote two reports on the Horsford submarine. The first, dated 7 August 1863, gingerly noted that, although ingenious solutions were offered for purely scientific difficulties, the Commission hesitated to rule on the practicability of the vessel for warfare. Ten days later the Commission raised several criticisms. The method of steering in a vertical plane was questioned, as was the proposed use of telescope and compass to aid the navigator. The Commission finally pointed out the ease with which the skin of the submarine could be ruptured or penetrated.[34]

Henry wrote Bache:

> He [Horsford] thought the contract would be entered into without submitting the matter to the Commission, but in this he was mistaken. The plans were referred to the Commission, which, though it wished as far as possible to deal kindly with the Professor, could not indorse the invention. The result was Mr. Fox could not accept the proposition; for said he, we have no special appropriation for this object. . . . I informed the Professor that, although I was sorry that he should be troubled in regard to the matter, I was not sorry that his proposition . . . was not accepted.

Henry felt that Horsford had "considerable suggestive power and abounds in kind feeling." Henry further stated that Horsford was making a fortune in baking power but that submarines were beyond his depth.[35]

Undaunted by the initial rejection, Horsford once more proposed the construction of his submarine in 1864. This time the Commission agreed to test it because Congress had voted an appropriation for the tests of submarine inventions.[36] The Commission was willing to pay part of the construction expenses. But Horsford, who was trying to interest the Governor of Massachusetts and the Mayor of Boston in subscribing $10,000 and $5,000, respectively, was apparently unable to raise funds in time. The *Soligo* remained high and dry in Horsford's imagination.[37]

Contacts with scientist inventors like Horsford illustrate some of the awkward situations confronting the Commission. But it is the less dramatic rela-

[32] Horsford to Fox, 8 June 1863 and Horsford to Welles, 9 June 1863, Miscellaneous Letters Received of the Secretary of the Navy, June 1863, vol. I (RG 45).

[33] Horsford to Welles, 11 July 1863, in Proposals Referred volume marked, "Unfinished Business, 1862–1864," Permanent Commission (RG 45).

[34] Reports, Nos. 156 and 157. Smith's report is enclosed in Report 157. Permanent Commission (RG 45).

[35] Henry to Bache, 19 Aug. 1863, Henry Papers, Smithsonian Institution. Bache to Henry, 25 Aug. 1863, Rhees Collection, Huntington Library.

[36] By this was meant any method or device for producing an explosion below the water line of ironclads. Civil War usage lumped mines, torpedoes, and submarines into one category.

[37] Horsford to Welles, 7 Mar. 1864, Miscellaneous Letters Received of the Secretary of the Navy, March 1864. Horsford was referred to Chief Engineer W. W. W. Wood of the Bureau of Construction and Repair. Davis to Horsford, 27 July 1864, Letters Sent, Permanent Commission, announced the decision to construct the submarine; Davis to Horsford, 21 Aug. 1864, Letters Sent, asked that the plans be sent in immediately. Horsford to Davis, 21 Aug. 1864, replied to this request with the details of Horsford's attempts to have the State of Massachusetts support his project and also enclosed Wood's letter of 1 Aug. 1864, praising Horsford's proposals — in Proposals Referred volume marked, "Unfinished Business, 1864–1865, "Permanent Commission. All in RG 45. 13 Stat. 329 (4 July 1864).

tions with the Navy Department that provide significant clues for an understanding of the use of science in the Civil War. July 24, 1863 was probably the most crucial day in the life of the Commission. On that date it approved a report on William Norris' design for an ironclad, recommending that it be examined by a "professional board" of navy officers. The report stated: "It would take the liberty, in this connection, also to suggest that it would, in general hardly venture to express its views upon purely technical questions, but rather to act upon such as involve principles or applications of science not forming a part of familiar knowledge." [38]

Bache yielded to the views of the majority while doubting the propriety of the Commission's reasons for not considering the plans: "True they involve much study & the aid of a professional board for the special occasion may be needed but there must be principles involved of which in my opinion the board should take cognizance." [39] The Commission had previously turned Norris down in an eleven-page refutation of his claims.[40] Referring Norris to a board of officers was a neat way of "passing the buck" to the Navy.

But the weakness in the position of the Permanent Commission majority was that most of their labors did not involve "principles or applications of science not forming a part of familiar knowledge." Except for the topics referred to the National Academy of Sciences and a few inventions such as submarines and balloons, the proposals evaluated largely fell within the area of "familiar knowledge" of the Navy and Army. Most could be rejected on the basis of experience; few were rejected solely on theoretical grounds. Bache, the West Pointer who became a geophysicist, was alone in sensing the desirability of having a board of officers test specific devices in cooperation with a high level scientific group.

On 24 July 1863, the Commission also asked the Navy Department's permission to refer plans for experiments on armored vessels to the National Academy.[41] To this proposal the Secretary replied with an unequivocal rejection:

> I think the subject somewhat professional, and should be investigated and reported upon by Officers connected with the Department, or a commission instituted by it, rather than the National Academy of Sciences.
>
> A transfer of duties from the Department to that institution does not strike me favorably. I know not what funds the Academy has for extensive experiments, the Department certainly has none which it would feel justified in placing at their disposal. . . .[42]

The Commission had received many proposals on methods of armor-plating vessels and the design of ironclads. "Practical" experts had proposed solutions that were in dispute. The members of the Permanent Commission had obvious interests in the problem — Davis, as a naval officer who had commanded armored vesels and had served on the board which recommended the construction of the *Monitor*,[43] Henry because of his service on the board evaluating the Stevens Battery, and Bache and Barnard as engineers designing the coastal fortifications against the dreaded depredations of Confederate raiders. The most significant difference between the proposed tests on armored vessels and the steam expansion and magnetic deviation tests was the lack of statutory

[38] Report No. 131, Permanent Commission (RG 45).

[39] Bache to Davis, 1 Aug. 1863, Letters Received, Permanent Commission (RG 45).

[40] Report No. 22, 16 April 1863, Permanent Commission (RG 45).

[41] See Minutes for those dates. The letter to Welles is Report No. 136, dated 25 July 1863 but actually signed and sent out on 29 July.

[42] Welles to Permanent Commission, 29 July 1863, Letters Received (RG 45).

[43] Baxter, *op. cit.*, 245–250.

authority buttressed by an appropriation. What Welles' reply said, in effect, was that the time for commissions and boards was past — Davis and Henry had already served on these — and now the Navy was concerned with the problems of production and combat utilization. The Department had a war to win and could no longer wait upon the deliberations of scientists.[44]

The difference of opinion among the members of the Commission persisted. The first overt sign was Report No. 174, 4 February 1864, on the use of corrugated iron armor, which laconically noted that the members "are not unanimous" in rejecting the proposal. Shortly afterwards a report rejecting John Ridgway's "Vertical revolving battery" was sent out at Bache's request without his signature. Ridgway's turret was presumably capable of firing in any direction or elevation, a most beguiling idea but fraught with practical difficulties. Ridgway had overcome the objections of the Naval Examining Board, which recommended a trial. When the Permanent Commission came into existence, it referred the device to Benjamin Peirce, the noted Harvard mathematician. Peirce submitted a report on 23 March 1863, certifying that the invention was theoretically sound, but stating, "I have made no allusion to the difficulties of its construction . . . because I regard the enquiries upon this point as belonging to a board of practical engineers and constructors." This report probably did not impress the Department, which had also received a report from one of its engineers, A. C. Stimmers, pronouncing the invention impracticable. It must have been galling to the Commission to have the opinion of Peirce set aside by the opinion of a man who was regarded as a partisan of Ericsson's and was subsequently the builder of a little fleet of unfloatable ironclads.

The Commission was handicapped by the uncertainty of the boundary between the labors of its members as "scientific gentlemen" and the labors of "practical" men. In Ridgway's case there was such a weight of authoritative opinion behind the invention that Bache was moved to open protest. Even the Commission's first report acknowledged the possible merits of the device. On 23 July 1864, the Commission reversed itself and reported that improvements in the design justified tests. The recommendation was disregarded.[45]

By raising his objection in the field of ironclads, Bache was applying pressure to an exposed nerve of the Navy Department. Navy policy had already committed public funds to an extensive construction program of monitors and other ironclads — in the words of Gideon Welles, all "objects of undoubted utility." Within the Department there was considerable doubt about these vessels, which were designed not for duty on the high seas but for harbor assault and defense.

[44] Bache to Davis, 1 Aug. 1863, Letters Received and Davis to Bache, 7 Aug. 1863, Letters Sent, Permanent Commission (RG 45).

The author has located only one other Navy proposal for using the National Academy during the Civil War. The Bureau of Ordnance asked for and received the Department's permission to enlist the Academy's aid in determining whether fulminate of mercury in contact with metal, as in cartridges, changes into the fulminate of copper; if so, the Bureau desired to know under what circumstance, and at what speeds, and what precautions should be taken. The records of the Bureau of Ordnance in the National Archives contain no indication that any formal approach to the National Academy was ever made, nor do they explain why the Bureau dropped the subject. If, and when, the National Academy of Sciences removes its ban on research in its records, this minor mystery may be dispelled. Bureau of Ordnance to Secretary of the Navy, 9 Jan. 1864, Bureau Letters, Jan.–April 1864 (RG 45); Welles to Bureau of Ordnance, 12 Jan. 1864, Department Letters No. 4, records of the Bureau of Ordnance (RG 74).

[45] Minutes 156, Reports Nos. 182 and 214, Permanent Commission to Ridgway, 9, 15 April, and 4 May 1864, Letters Sent, Ridgway to Permanent Commission, 20 May 1864, in Proposals Referred volume labeled "Supplemental Volume, 1862–1864," Ridgway note of 27 May 1863 attached to which is a letter of "G. S." of 27 March 1863, on Stimmers and Ericsson, in Letters Received — all of the Permanent Commission; Reports, Examining Board, 9 May 1862 and 13 June 1862; Ridgway to Fox, 25 July 1863, Miscellaneous Letters of the Secretary of the Navy, July, 1863, vol. III. All in RG 45. For Stimmers, see F. M. Bennett, *The Steam Navy of the United States*, 484–493, Pittsburgh, 1896.

Outside the Department there was considerable pressure to use new designs of vessels, other turrets, and even different steam engines. Under these circumstances it is understandable that the Navy Department would be sensitive about letting outsiders do research on ironclads and even more sensitive about having designs or devices certified as superior to the "objects of undoubted utility" on which so much money had been spent. And the Commission members, with the exception of Bache, were not disposed to challenge the status quo.

Since the amount expended on a function is often a touchstone of Executive and Congressional esteem, the status of scientific research and development in the Navy during the Civil War is roughly proportional to the availability of funds. The Navy Department's position was quite clear; unless there were specific appropriations for tests or experiments, it could commit funds only for objects of undoubted utility. Even when the Department decided on the construction of a novel vessel, as in the case of the *Monitor*, all the risk was borne by the contractors, who would receive not a penny if the ship did not meet performance requirements written into the contract.[46] The environment was markedly different from the present situation, when vast sums are expended for research and development as a matter of course.

Bache and Henry were well aware of the situation. In a letter of 21 August 1863, Henry, in discussing the possibility of further referrals to the National Academy, stated, "It will be necessary that the Heads of Departments be not frightened" by proposals involving expenditures. He went on to relate an experience of his. At the request of Dahlgren, who needed them in the assault on Charleston, Henry prepared lighting devices. In the absence of Fox, he had to present his bill for reimbursement to Welles who insisted on personally approving every expenditure of $100:

> Goodman, [sic] he has been so much in the habit of dealing with those who have no other aim than that of self advantage that he could not imagine that any one could have any other motive. Disgusted with being placed on so low a level I concluded to do nothing more with light and even not to make a charge for the expenses actually incurred.[47]

A trifling sum and a trifling incident but indicative of the Navy's parsimony toward research. In this environment the Navy Department could conceive of the Commission only as a negative device for rejecting inventions, not as a positive instrument for improving the fleet's capabilities.

Uppermost in the minds of the Secretary and his able Assistant Secretary, Gustavus V. Fox, was the embarrassment occasioned by the flood of inventions, not the desirability of coordinating the proposals with the research activities of the Department's bureaus (which were quite limited). Nor is there any evidence suggesting the conception of "mobilizing the nation's scientists," (the most significant professional groups being botanists and geologists, not physicists and chemists). What the Navy was violently criticized for during the Civil War was its lack of support of private inventors.[48] But the Army and the Navy knew from experience that most inventions submitted were not usable.[49] Inventions presented opportunities for financial bonanzas; and the resulting pressure on the military to adopt inventions raised the specter of

[46] Baxter, *op. cit.*, 261–262.

[47] Henry to Bache, 21 Aug. 1863, Henry Papers, Smithsonian Institution.

[48] For example, see "Imbecility in the Navy Department," *Scientific American*, 22 Mar. 1862, *6*: 185. In 1863 the magazine proposed formation of a "Board of Admiralty," composed of prominent inventors, to evaluate inventions, 28 Feb. 1863, *9*: 137–138.

[49] See Minutes, Naval Examining Board, 3 and No. 7 of Inventions Referred to Naval Examining Board for an extreme case.

misuse of public funds.[50] To winnow the few meritorious proposals from the dubious chaff was a serious problem, if only from the standpoint of placating public opinion impatient of the war's progress and eager to adopt promising devices.

Of course, the problem was not to be unique in American history. In World War I the Naval Consulting Board, headed by Thomas Edison, was set up to examine inventions. The National Inventors Council during World War II and up to the present has acted in a similar capacity. This remarkable solicitude for the individual inventor is the result of three American attitudes present today as well as in the Civil War. The first is pragmatic; in times of national emergency it would be folly to overlook any possible source of talent. The Wright brothers were, after all, only bicycle manufacturers when they became interested in aeronautics. The second attitude is the American distrust of experts, possibly because reliance on a few individuals violates egalitarian notions. Every major American war has been accompanied by criticism of the weapons used. That some of the criticisms were valid should not obscure the fact that most advances in military technology are the result of the labors of professional scientists and engineers, not of inspired amateurs. But encouraging private inventors seems an almost unconscious compensation for an irksome dependence on experts. The third attitude is a result of the prominent place occupied by inventors in the national folklore. Nurtured on tales of inventors whose discoveries transform the nation and reap wealth for themselves, the average American was, and still is, quite receptive to proposals to facilitate the realization of this small part of the "American dream."

[50] Corruption of public officials was a real threat. For an example of a bribe offered for a favorable opinion of an invention, see Maj. Gen. Totten to President Lincoln, 13 Dec. 1861, Totten Letterbooks, IX: 48–49, Records of the Office of the Chief of Engineers (RG 77).

The Conflict Between Pure and Applied Science in Nineteenth-Century Public Policy: The California State Geological Survey, 1860–1874

By Gerald D. Nash *

THROUGHOUT the first half of the nineteenth century both federal and state governments in America increased their activities to promote research and science. This brought government, science, and economic progress into a close relationship. In the wave of nationalism which swept over the United States after the Napoleonic Wars, most groups in American society gave their enthusiastic support to such endeavors. Gradually, however, a diversity of objectives became apparent which often threatened to undermine the rate of scientific progress. Unfortunately, some of the men who placed their faith in government research did not clearly articulate their differing objectives. Thus, they laid the basis for a conflict between advocates of pure science as opposed to applied science. Since the differences between research and applied technology, rarely easy to define, were but imperfectly understood, this blurring of aims was eventually to lead to open rupture. The goal of researchers desiring government support was to enhance the status of their newly emerging professional disciplines. On the other hand, economic or military interest groups desired research to gain immediate and practical ends. Such a divergence of goals developed in many spheres of federal- and state-sponsored scientific work during the nineteenth century. It occurred in the growth of the United States Coast Survey under Ferdinand Hassler; in the work of the Army's first Surgeon General, Joseph Lovell; in the efforts of Lieutenant Charles Wilkes of the United States Exploring Expedition; in the programs of Isaac Newton, the first Commissioner of the independent Department of Agriculture in 1862; and in the struggles of John Wesley Powell with the United States Geological Survey.[1]

* University of New Mexico.

[1] For specific aspects of these disputes and bibliographical references, see A. Hunter Dupree, *Science and the Federal Government: A History of Policies and Activities to 1940* (Cambridge, 1957), 29-33, 37, 52-55, 58-75, 152, 199-202.

ISIS, 1963, VOL. 54, PART 2, NO. 176.

Differences over the administration and goals of public programs in support of science often impeded the very progress of research.

A similar dichotomy developed in the states. This was true of the geological surveys which constituted one of the most important activities on behalf of science. At a time when state rather than federal encouragement of research was common, these new agencies reflected the real concern of Americans to utilize public aid to hasten the development of untapped resources. In the four decades after 1820, thirty-one states established such surveys; the degree of success which they attained varied. The purpose of the geologists seeking state support was generally to help them pursue their research at an accelerated pace. Economic interests, such as agriculturists and miners, desired creation of the surveys to facilitate the more rapid economic exploitation of regional resources. Their objectives were primarily practical; they had but scant respect for scientific advances except those that led to direct economic benefits. During the boom eras of the 1820's and the 1850's, both scientists and farm organizations supported the establishment of geological surveys for the attainment of their common aim — the discovery of new natural riches. The differing goals of their sponsors became increasingly clear when the surveys were put into operation. Within a few years, deep rifts and strong antagonisms resulted in a partial disruption of the scientific work.[2]

This conflict between the proponents of pure and applied science, common in many federal and state agencies during the nineteenth century, is well illustrated in the history of the California State Geological Survey between 1860 and 1874. The California Survey was among the most distinguished of all such scientific endeavors during this period. Under the leadership of Josiah Whitney, later Professor of Geology at Harvard and one of the outstanding scientists of his generation, it created many precedents, stimulated the work of scientists throughout the nation, and gained worldwide renown. At the same time, it served as a training school for several distinguished scientists who were to make their mark in succeeding years. Among them, William Brewer, later Professor of Agriculture at Yale University, was prominent. Another member of the Survey, Clarence King, was to become one of the nation's leading science writers. Charles F. Hoffman, one of the founders of American topography, and W. M. Gabb, a distinguished paleontologist, were also attracted by Whitney's reputation. The high scientific standards of the members of the California Survey were to make the dispute with the champions of practical research more clear-cut than elsewhere.

In its establishment, operation, and demise, the California Survey mirrored one of the characteristic problems created by government's increasing

[2] George P. Merrill (ed.), "Contributions to a History of American State and Geological and Natural History Surveys," in U. S. National Museum, *Bulletin* #109 (Washington, 1920), VII-XV: 27-40, 45-536; Walter B. Hendrickson, "Nineteenth Century State Geological Surveys: Early Government Support of Science," *Isis* (September, 1961), LII: 357-371; Josiah D. Whitney, "Geographical and Geological Surveys," *North American Review*, (1875), CXXI: 37-85, 270-314.

support for science in America. The common problems faced by the agriculturists and miners led them to unite in an effort to secure appropriations from the legislature for a geological survey. Agitation for such a research organization reflected itself early when on October 5, 1850, a leading newspaper, the *Alta California*, gave its endorsement to the project. Governor Burnett proposed it officially to the legislature in 1852. Two separate committees in both houses submitted favorable reports on the establishment of an agency to investigate the nature of the state's resources. Meanwhile, the lawmakers also enacted a resolution calling for immediate action on the part of Congress in establishing a federal geological survey " not only for the purpose of more speedily developing the mineral resources of the state, but to enable the agriculturist to predicate the success of his labors upon a sure data, and not to be entirely dependent upon rains to sustain and mature his crops." But Congress, enmeshed in sectional disputes over slavery, gave little attention to the request from the Californians, and left them to rely on their own resources.[3] Consequently, the burden of providing needed research activities fell back on state government for several years.

John B. Trask, a physician, had been making geological investigations on his own which he termed " geological surveys." At his behest, in 1851, the legislature granted him the honorary title of State Geologist and published his geological sketch of the eastern Sacramento Valley based on his personal observations. As Trask was popular with the lawmakers, they contributed a small sum of money for his work. A Joint Resolution of 1853, however, authorized him officially to undertake examination of the geological structure of the Sierras and coast ranges, " said report to comprise as near as possible the area of such lands . . . the facilities they offer, and requisitions necessary to insure their occupancy and improvement." Two thousand dollars were appropriated for the task.[4]

Trask made three separate reports concerning California's wealth which were published at state expense. Although generally not of a systematic nature, his studies had much useful information about the relative value of some of the mining areas in the state. The first pamphlet of 1853 discussed some rather general aspects of the geology of the Salinas Valley and parts of the San Joaquin and Sacramento Valleys. Trask also classified the rocks which he found there and made a list of old mines in the area. His observations on the second field trip were noted in a succeeding report which described the coastal mountains and northern California. This de-

[3] California, State Mineralogist, *Annual Report*, 1884 (Sacramento, 1884), 21-27, has a brief history of geological surveys in the state, 1838-1880, with bibliographical references; " Report of Committee on Joint Resolution to Congress on Geological Survey," California Senate, *Journal*, 1852 (San Jose, 1852), *Appendix*, 659-665; " Report on Need for Geological Survey," California Legislature, *Legislative Journal*, 1851 (San Jose, 1851), 1689-1702; Rodman Paul, *California Gold* (Cambridge, 1947), 302-304; for Burnett, see California, Governor, *Annual Message*, 1852, in Senate, *Journal*, 1852, 20.

[4] " Report of Committee on Public Expenditures Favoring Compensation for J. B. Trask," California Assembly, *Journal*, 1853 (San Jose, 1853), *Appendix* #56, 1-5; 1853 Stats. 144, 314, California Senate, *Journal*, 1852, 659-665.

scriptive material was accompanied by an analysis of the agricultural and mining possibilities of these regions. The third publication, which came out in 1856, was devoted to the coastal regions north of San Francisco. Altogether, there was much useful material of a general nature in these pioneering works.[5]

Meanwhile, the economic development of the state made the need for accurate information more acute than ever. The waning importance of the mining industry fostered increased demands by its spokesmen for extended state aid in the form of research and information. By 1859, the yields of the gold fields were steadily declining, and a general depression in the industrial and mountain regions was beginning to be felt. The increasing complexity of mining techniques created a demand for more and better knowledge about the industry. The large amounts of capital now required made investors especially eager for such information. Early panning and placer mining was giving way increasingly to expensive drilling techniques as less accessible deposits were approached. By 1860, the future possibilities of quartz mining were widely discussed and pondered. No wonder, therefore, that in every section of the state newspapers were clamoring for a state survey such as had existed in older areas for many years. As a journal in the mining regions summarized it aptly: "It is generally held that a thorough geological report, from a competent person, would place the immense resources of California in a more commanding position before the world's eye, besides resulting in a more systematic method of developing them." Concurrently, the gambling spirit aroused throughout California by the Comstock Lode re-emphasized the usefulness of precise scientific knowledge.[6]

Such problems in the economy were to be transmitted through political channels by a man whose contacts embraced both spheres. The actual negotiations for the creation of a survey in California were conducted by Stephen J. Field, a consummate politician and legal draftsman. In 1860, Field still exercised great influence over the legislators although he was a justice on the California Supreme Court. He was also a personal friend of many mining executives on the Pacific Coast. He was persuaded that the time had come for the establishment of a surveying corps in California. Thus he began a search throughout the East and Middle West for an appropriate scientist to head such an agency. His choice fell upon Josiah D. Whitney, the State Chemist of Iowa. Field's brother, the Reverend Henry M. Field of New York, was an old friend of the Whitney family. Whitney's brother-in-law, S. O. Putnam (secretary of the California Steam Navigation Company and a man of local prominence), had long wanted to see his relative chief of a California survey. He also exerted his influence.

Still in the early stages of his career, Whitney (1819-1896) was rapidly gaining prominence as one of the nation's abler young geologists. A native

[5] See California Senate, *Journal*, 1854 (San Jose, 1854), *Appendix*, #9; *ibid.*, 1855 (Sacramento, 1856), #14; *ibid.*, 1856 (Sacramento, 1857), #14.

[6] North San Juan *Hydraulic Press*, January 15, 1859, quoted in Paul, *op. cit.*, 304; see also *ibid.*, 116-170.

of Northampton, Massachusetts, and a graduate of George Bancroft's famous Round Hill School, he studied chemistry and mineralogy under the elder Silliman at Yale, from where he graduated in 1839. His first practical experience came during the following year when he served as an unpaid assistant in Charles T. Jackson's New Hampshire Survey. After further study in Germany he returned to the United States. In 1847 Jackson hired him to aid in a reconnaissance of mineral lands in northern Michigan. Whitney gained invaluable experience in this important scientific project and in 1850 collaborated with John W. Foster in writing a report on their findings. The reputation achieved through this work enabled Whitney to establish himself as a private consultant after he returned to his New England home. There, in 1854, he also published a book on American mining which became a standard work for several decades. Soon a new opportunity drew him west once more when in 1855 he was called to Iowa's state university and to its survey as chemist and mineralogist. Working closely with James Hall, he did pioneer research in classifying the physical geography, topography, climate, rock groups and rivers of the state. At the same time he lent his services to Amos H. Worthen and the Survey of Illinois where he explored lead and zinc deposits. He also advised the Wisconsin Survey. In 1860 Whitney was undoubtedly one of the men best qualified to explore California's relatively unknown resources. His reputation, Field's influence, and recommendations by Peirce, Agassiz, the Sillimans and other famous scientists, led the California Legislature in 1860 to appoint Whitney State Geologist to lead the newly established Geological Survey.[7]

The new California agency was closely patterned after similar organizations elsewhere. The organic act officially created a State Geologist who was to undertake a complete survey of the state, to furnish maps and diagrams, and to make scientific descriptions of rocks, minerals and fossils. In addition to annual progress reports, he was to submit a full and comprehensive review of results. The state guaranteed publication of the findings through its own printing office, and started the project off with a generous $20,000 appropriation.[8] As the *American Journal of Science* noted at the time: "No state geological survey was ever more auspiciously inaugurated, wisely provided for, or fraught with more interesting scientific and practical problems." [9]

[7] The above two paragraphs are based on Francis Farquhar (ed.), *Up and Down California in 1860-64, the Journal of William H. Brewer* (2d ed., Berkeley, 1949), xi-xii; Edwin T. Brewster, *Life and Letters of Josiah Dwight Whitney* (Boston, 1909), 183-191; Josiah D. Whitney, *The Metallic Wealth of the United States* (Philadelphia, 1854); Dumas Malone (ed.), *Dictionary of American Biography* (34 vols., New York, 1936), XX, 161-163, for a good biographical sketch by G. P. Merrill; see also James Hall and Josiah D. Whitney, *Report of the Geological Survey of the State of Iowa, 1855-57* (2 vols., n. p., 1858), I, 1-34, 259-472; for Jackson's work see "Catalogue of rocks, minerals and ores collected . . . 1847 and 1848, on the geological survey of the U. S. mineral lands in Michigan," in Smithsonian Institution, *Report*, 1854 (Washington, 1855), 338-357; for Hall, U. S. Army, Corps of Topographical Engineers, *Exploration and Survey of the Valley of the Great Salt Lake of Utah* (Washington, 1853), 399-414, and U. S. Geological Exploration of the Fortieth Parallel, *Report* (7 vols., Washington, 1870-1880), IV, part II.

[8] 1860 Stats. 225.

[9] *American Journal of Science* (1860), XXX, 157, quoted in Merrill, *Bulletin* #109, 31, n. 1;

Whitney conceived his task as a two-fold one. In the first place, he sought to explore extensively the general geological structure of the state, still unknown in 1860. He considered this the scientific aspect of his work. In the second place, he planned to examine particular areas intensively to provide information for entrepreneurs which would be of direct practical utility.[10] The Survey, then, was to concentrate on two aspects of its research function: direct investigation and the collection and dissemination of information.

As a devoted geologist, Whitney considered the pursuit of pure science to be most important. He wrote in 1861: "The first thing required . . . is a knowledge of the general geological structure of the state." With this end in view, Whitney divided the Survey into three divisions. First, he created a Topographical Section charged with making maps for the various portions of the state. A second division constituted the Geological group which was concerned with the investigation of the general geology, paleontology, and economic geology of California. The third section of the Survey was the Division of Natural History which studied California botany and zoology, then mostly unexplored.[11]

During the first five years of its existence, the Survey made admirable progress in its scientific work. Whitney landed in San Francisco on November 14, 1860, and began to gather his assistants. C. F. Hoffman started work as Topographer in March, 1861; W. Ashburner was sent to the Sheffield Scientific School at Yale to examine some of the California ores he collected. The others devoted most of the first year to exploration of the coastal ranges. Heavy storms and floods impeded the Survey's work during the winter and spring of 1861-1862, destroying most of the state's roads and bridges. Moreover, Brewer and three of his field assistants became ill with fever and were forced to suspend their work. Not until 1863 were Whitney and his corps able to venture into the Sierras to gather needed information. Once the data had been assembled the Topographical Section made a series of maps for California, the first substantial attempt on a systematic scale. The Geological Section made barometric observations. In 1864 it published the first of a long series of contemplated reports, a treatise on carboniferous, Jurassic, Triassic, and Cretaceous fossils. The Division of Natural History was not far behind in its own collections of plant and zoological species. Once this ten-year, general reconnaissance was completed, Whitney hoped to devote more time to the "practical" objectives of his agency.[12]

Brewster, *op. cit.*, 191-204; Thurman Wilkins, *Clarence King: A Biography* (New York, 1958), 55; Merrill, in *Bulletin* #109, 515-517; Farquhar (ed.), *op. cit.*, 3-54.

10 For Whitney's account, see California Academy of Sciences, *Proceedings*, 1863 (San Francisco, 1863), III, 23-29.

11 "Letter of the State Geologist Relative to the Progress of the State Geological Survey During the Years 1863-64," in California Legislature, *Appendices to the Journals of the Senate and the Assembly*, 16 sess. (Sacramento, 1866), III, #6, 1-14; quotation is from California, Geological Survey, *Geology* (Philadelphia, 1865-1882), xix.

12 "Letter . . . , 1861," in California Legislature, *Appendices*, 13 sess. (Sacramento, 1863), III, 5; California, State Geologist, *Annual Report*, 1862 (Sacramento, 1863), 5-9; *ibid.*, 1863 in California Legislature, *Appendices*, 15 sess. (Sacramento, 1865), I, 1-6; California Geological Survey, *Geology*, x-xvii, and *Paleontology* (2 vols., Philadelphia, 1864-1869).

Meanwhile, the legislators anxiously awaited immediate results. As the sub-committee on Mines and Mining Interests noted in 1863:

> We . . . expect that the reports will be of direct industrial value—particularly those portions relating to the Monte Diablo coal region, and the modes practiced in this state for extracting gold. The Botanist has devoted much attention to the grasses and clovers and is now engaged in extensive inquiries to ascertain what varieties are most palatable and nutritious to herbivorous animals.

The publication of the Survey's eagerly awaited first volume, on four kinds of fossils, was bound to prove a keen disappointment.[13]

Whitney's lack of tact and curt manner only deepened the frustration felt by the farmers and miners who had looked to the Survey for direct gains. Blunt and outspoken, Whitney saw little need to compromise his own scientific objectives with their own. His sharp tongue and biting sarcasm widened the gap between the advocates of theoretical research and those favoring applied science. As early as 1863, he told the legislators sharply that: " It is not the business of a geological surveying corps to act . . . as a prospecting party." Since the lawmakers were already engaged in reducing appropriations for his Survey, it was with doubtful propriety that Whitney, in a personal report before both houses, reminded them to their faces that: " We have escaped perils by flood and field, have evaded the friendly embrace of the grizzly, and now find ourselves in the jaws of the Legislature." In 1864, during the wild speculative boom promoted by war conditions and the Comstock Lode, he had the singular lack of judgment to choose the highly technical study of fossils for publication as the first volume of his series of reports. At the time, a story circulated which related that an opponent of appropriations for the Survey carried his point in the legislature by reading excerpts from this dry treatise. Certainly it is easy to imagine the feelings on the part of a group of intensely practical westerners, earnestly engaged in seeking a quick fortune, listening to a description of prehistoric reptiles in a book published at state expense.[14]

Whitney's relations with some of the governors were no better than with the legislators. John G. Downey (1859-1862) was especially antagonistic. Quite unlike the sensitive, devoted scholar at the head of the Survey, Downey expected results of a practical nature. At least one of Whitney's aides claimed that the governor wished the staid State Geologist to use his official influence to aid him in his mining stock speculations. When

[13] " Report of the Sub-Committee of the Committee on Mines and Mining Interests of the Assembly Concerning the State Geological Survey," in California Legislature, *Appendices*, 14 sess. (Sacramento, 1864), II, 3, and same for 1865 in *ibid.*, 15 sess., II, 5; 1863 Stats. 679; 1869-1870 Stats. 317; Wilkins, *op. cit.*, 101; Brewster, *op. cit.*, 208-290.

[14] " Lecture on Geology Delivered Before the Legislature of California at San Francisco, Thursday Evening, February 27, 1862," in California Legislature, *Appendices*, 13 sess., II; California, State Geologist, *Annual Report*, 1862, 8; quote is from " Letter of State Geologist . . . 1863-64," in California Legislature, *Appendices*, 16 sess., III, 8; Rossitter Raymond, " Biographical Sketch of J. D. Hague," American Institute of Mining Engineers, *Bulletin* #26 (February, 1909), 113 n.; Brewster, *op. cit.*, 300-301.

Whitney indignantly refused to convey the desired information, the governor [15] was said suddenly to have become very antagonistic to the Survey.

Another impediment to the Survey was the speculative oil mania that swept California at the close of the Civil War. Californians were in a frenetic mood as the state experienced a frenzied economic boom. Sparked by increasingly diversified exploitation of minerals and natural resources, and especially the riches of the recently discovered (1859) Comstock Lode, California was undergoing an intensive period of economic growth. The streets of San Francisco's financial quarter swarmed with hordes of speculators from all walks of life who milled about the three new stock exchanges. In such a tense and expectant atmosphere even vague rumors concerning the presence of petroleum deposits in southern California were enough to spark another wild speculative surge. In part, these rumors emanated from Tom Scott of the Pennsylvania Railroad, who had been laying plans for the building of a Pacific line with a terminus at San Diego. Since he had already made a great profit in the Pennsylvania oil fields between 1860 and 1865, he sought to investigate similar possibilities on some of his California lands. To ascertain their value, Scott sent the famed American geologist, Benjamin Silliman, Jr., to California to explore the prospects. Silliman made a rather hasty, superficial examination and submitted favorable reports based on inadequate samplings. He claimed that oil reserves abounded in southern California and that they could be exploited easily with rather small investments. Moreover, he stated that the quality compared favorably with products from the Pennsylvania wells. Such a mood of optimism, promoted by a man of Silliman's eminence, touched off scores of new prospecting enterprises and sparked a wild oil boom in California.[16]

Whitney doubted the presence of profitable oil deposits on the Pacific Coast. With his frank brusqueness, he resolutely opposed the promotion of all enterprises, corporate and individual, which claimed to exploit rich oil deposits. "You have no idea," he wrote to his good friend and co-worker, Brewer, "how the agents of these speculators are working in the sub-stratum of society . . . swindling the last dime out of the stupid women who are fools enough to listen to them." Moreover, he was no less vehement in expressing his opinion to the promoters directly and so earned their profound hostility. "Petroleum is what killed us," Whitney later noted with

15 Farquhar, *op. cit.*, 452; Brewster, *op. cit.*, 292-296; Josiah D. Whitney to William D. Whitney, March 3, 1874, in Brewster, *op. cit.*, 288-289.

16 Walter Stalder, "A Contribution to California Oil and Gas History," *California Oil World and Petroleum Industry* (November 12, 1941), 37; John Ise, *The United States Oil Policy* (New Haven, 1925), 85-92; California State Mining Bureau, *Annual Report*, 1884 (Sacramento, 1885), 293, and *Annual Report*, 1888 (Sacramento, 1889), 180-186; Benjamin Silliman [Jr.], "A Description of the Recently Discovered Petroleum Regions in California" (New York, 1865), pamphlet, and his article, "Examination of Petroleum from California," *American Journal of Science* (1865), XXXIX, 341-343; *Pacific Mining Journal*, October 23, 1865, lists sixty-four petroleum companies in California, with capitalization of over $35,000,000; see also Owen C. Coy, *The Humboldt Bay Region, 1850-1875* (Los Angeles, 1929), 234; Joseph L. King, *History of the San Francisco Mining and Stock Exchange Board* (San Francisco, 1910), 5 ff.

some exaggeration in a letter to his brother William. "By the word 'petroleum' understand the desire to sell worthless property for large sums, and the impolicy of having anybody around to interfere with the little game." With the wrath of speculators now upon him, the State Geologist faced mounting opposition.[17]

The conflict over policy between Whitney and the legislature, heightened by personal antagonism and the irrational mood provoked by a speculative frenzy, boded ill for the continuation of the Survey.

So much hostility had been aroused by Whitney that after 1866 the end seemed certain. The legislature adjourned in 1867 without making any appropriations for its work. Research was continued only because Whitney borrowed the needed funds, trusting to future compensation. After 1871, the legislature and Governor Newton Booth decided definitely on the abolition of the enterprise. Although $73,000 more was granted to the Survey between 1870 and 1874, this money was to compensate Whitney for his outlay, and to allow him to conclude his work. In 1874, the office was abolished and many of its incompleted projects were transferred to the University of California. Whitney left California to accept a chair in geology at Harvard.[18] The ambitious attempt of the California state government to promote agriculture and mining through scientific research came to an end.

Yet the scientific accomplishments of the Survey were considerable. Since the California legislature after 1874 was unwilling to provide funds for publication of results Whitney appealed to leading Californians for subsidies. Due to the generosity of railroad magnates such as Leland Stanford and Charles Crocker, and bankers like D. O. Mills and J. C. Flood, Whitney was able to issue at least six major reports about these findings. These included works on paleontology, geology, ornithology, and botany, in addition to special studies and maps.[19]

Earliest to appear were the volumes on paleontology. Much of this

[17] Josiah D. Whitney to William D. Whitney, April 13, 1868, in Brewster, *op. cit.*, 266-267; Merrill, *Bulletin* #109, 39; Josiah Whitney to William H. Brewer, May 12, 1865, in Josiah D. Whitney Papers, Bancroft Library, University of California; J. Ross Browne, *Report on the Mineral Resources of the States and Territories West of the Rocky Mountains* (Washington, 1868), 258-259.

[18] Total appropriations for the Survey, 1860-1874, were $245,600, in addition to $13,000 derived from sale of maps, a large sum in comparison to other states; see "Report of the State Geologist on the Condition of the Geological Survey of California, November 15, 1869," 4-5, in *Appendices*, 1871 (Sacramento, 1872), II; Josiah Whitney, "An Address on the Propriety of Continuing the State Geological Survey . . ." (San Francisco, 1868), pamphlet in University of California Library; "Statement Submitted by the State Geologist to the Committee Appointed to Investigate the Affairs of the Geological Survey," in *Appendices*, 20 Sess. (Sacramento, 1877), V, #16, 8; 1874 Stats. 694; The Harvard College Library contains a number of original maps made under Whitney's direction; *Mining and Scientific Press*, April 4, 1868; the California and Nevada delegations in Congress were able to secure creation of a national mining bureau in 1866, with establishment of U. S. Commissioner of Mining Statistics to collect information; J. Ross Browne, the famous Californian, was the first appointee; see Paul, *op. cit.*, 306-307.

[19] Some of these publications of the California Geological Survey include *Paleontology; Geology; Ornithology* (*Land Birds* ed. by S. F. Baird from MSS of J. G. Cooper, Cambridge, Mass., 1870); *Botany* (2 vols., Cambridge, Mass., 1876-1880); *Contributions to Barometric Hypsometry* (Cambridge, 1874).

research was done by Fielding B. Meek and William M. Gabb, who were to become outstanding in the field. They began their work with the collections of the California Academy of Natural Sciences, and went to Santa Barbara, San Bernardino, and Los Angeles during the winter of 1860-1861. In 1863, Whitney sent Gabb to Oregon and Washington to search for Tertiary fossils, while Brewer and King concentrated on the high Sierras. To secure perspective on the geological structure of the Pacific Coast from the south Whitney also sent A. Rémond to Sonora, Mexico on a similar quest. The results of their labors were contained in the two volumes on paleontology that appeared between 1864 and 1869. The first of these had a description of carboniferous and Jurassic fossils by Meek, and of Triassic and Cretaceous fossils of the Sierra Nevada by Gabb. King had originally discovered the Jurassic rocks in Mariposa County. In the second volume Gabb described Tertiary rocks on the Pacific Coast and made available new knowledge on lower members of the series. Not only did he describe the Tertiary Invertebrate and Cretaceous fossils which he found, but he also suggested a new classification of Pacific Cretaceous rocks. Prior to the Survey the presence of these fossils in California had been surmised; now it was proven. The importance of this study was to indicate the presence in the western United States of rocks equivalent in age to the Upper Trias of the Alps, and of widespread Cretaceous systems on the Pacific Coast.[20]

No less important were the publications on geology. So vast was the amount of data to be assembled that Whitney considered his report to be tentative only. In general, it contained a survey of formations in California's Monte Diablo range, the San Francisco peninsula and coast ranges, around the Bay of Monterey, and south of Los Angeles. The study also presented an account of the geology of the Sierra Nevada, east and west. Appended was a tabular statement of the operations of principal quartz mills in the state. Perhaps the main contribution of this work was to provide the first detailed scientific classification of the auriferous belt along the slopes of the Sierras, and a description of their fossils.[21]

Unfortunately, the publication plans for presentation of the results of zoological investigations never fully materialized. Whitney had hoped that the extensive researches of California's fauna by J. G. Cooper could form at least four volumes. Two of these were to be devoted to birds and mammals. With his talent for soliciting the aid of other scholars Whitney secured the services of Spencer F. Baird of the Smithsonian Institution, an eminent authority, to revise these volumes, and to take charge of the illustrations. New species in each genus were to be represented by at least one

[20] California, Geological Survey, *Paleontology*, I, xi-xvii, 3-16, 30-53 for Meek's work, and 19-35, 57-217 for Gabb; *ibid.*, II, vii-xiii; California, Geological Survey, *Geology*, xix-xxi; for other publications of Meek see his *Paleontology of Upper Missouri . . . 1855 and 1856* in Smithsonian Institution, *Contributions to Knowledge* (Washington, 1865), XIV, Art. V; U. S. Engineer Department, *Report of the Exploring Expedition from Santa Fe, New Mexico*, 1859 (Washington, 1876), and U. S. Geological Exploration of Fortieth Parallel, *Report*, IV, part I.

[21] California, Geological Survey, *Geology*, x-xvi, 8-166, 199-474, Appendix A.

illustration. A third book on fishes by Theodore Gill was also planned, in close cooperation with Cooper and the Smithsonian Institution. A final study was projected on shells since Cooper had assembled a large collection of Mollusca, including two hundred new species. Of this ambitious series only the volume on ornithology appeared. Its great contribution was to present the first systematic and scientific classification of birds in California.[22]

Some of the Survey's most distinguished work was performed in botany. Due to the aid of the many outstanding scientists whose help Whitney enlisted, the project became a national rather than merely a statewide endeavor. Although William H. Brewer was in charge of botanical investigations, he was greatly aided by Asa Gray. Brewer performed much of the labor of classification at the Harvard University herbarium. Gray agreed to work on the Compositae and to determine many of the species in the collection. He also secured one of his students, Professor Daniel C. Eaton of Yale, to describe ferns and the higher cryptogamic grasses, a field in which Eaton was an acknowledged specialist. For a description of grasses Whitney turned to another expert, George Thurber of New York and Cooper Union. Another of the distinguished authorities who contributed his special skills was George Engelman, the well-known founder of the Saint Louis Academy of Science, who described Cactaceae, for knowledge of which he was nationally known. Since a decade elapsed between the collection of data and publication of the report, much of it was revised and brought up to date by the increasingly eminent botanist, Sereno Watson. The first volume appeared in 1876 and had descriptions of Polypetalae by Brewer and Watson, and of Gamopetalae by Asa Gray. Watson also authored the second part of the work which contained lists of Lichenes by Edward Tuckerman, of Algae by W. G. Farlow, and of Fungi by H. W. Harckness and J. P. Moore. Here, too, were found Eaton's contributions on ferns, Thurber's on grasses, Engelman's on oaks, pines and Loranthaceae, and the English scientist M. S. Boott's on Carices. Altogether, the study classified sixteen hundred species of flowering plants and one hundred species of mosses. Brewer thought that these constituted 74 per cent of those surmised to exist in the state. Five per cent of those found were new to science, especially Polypetalae and Gamopetalae.[23]

The geographical and topographical work of the Survey was also important. Before Whitney began his work little accurate information about the state's geography was available, and no accurate map existed. Hoffman began his work by drawing fifty skeleton maps from materials he found

22 California, Geological Survey, *Ornithology*, passim; California, Geological Survey, *Geology*, xxii; Harvard University Museum of Comparative Zoology, *Memoirs*, XII (Boston, 1884); Cooper had made a reputation with his publication in U. S. War Department, *Report of Explorations and Surveys . . .* 1853-1856 (Washington, 1855-1860), III.

23 California, Geological Survey, *Botany*, I, 7 ff., II, vii-viii; California, Geological Survey, *Geology*, xxi-xxiii; American Academy of Arts and Sciences, *Proceedings*, 1862-1865 (Cambridge, 1865), VI, 519-556 and *ibid.*, 1868 (Cambridge, 1868), VII, 327; for works of Sereno Watson, see his *Bibliographic Index to American Botany* in Smithsonian Institution, *Miscellaneous Collections* (Washington, 1878), XV and with Asa Gray, *Manual of the Botany of the Northern United States* (6th ed., New York, 1889).

in the San Francisco office of the United States Surveyor-General. These maps could be used for topographical entries. Pioneering with the use of topographic mapping by triangulation, Hoffman also prepared maps of San Francisco Bay, Central California, Monte Diablo, and the coast ranges. Whitney's ultimate aim was to prepare the first detailed and accurate map of California; this was not completed until the close of the century.[24]

Despite such important scientific contributions, the Survey foundered. Apart from the clash of personalities, a prime reason for its abolition was the lack of a clear conception of its tasks by its various supporters. Whitney, as the foremost advocate of pure research, had scant appreciation of the necessity to produce practical results and to cultivate a friendly public. The lawmakers, deeply involved in a frontier economic boom, had neither patience, sympathy, nor understanding for achievements of long-range scientific value. For them, research was a means to an end only, a method to promote the more rapid exploitation of resources. The California State Geological Survey thus foundered upon the increasingly divergent conceptions of its task, as did so many state and federal research agencies of this period. Yet it was as a result of this hard-won experience in the nineteenth century that scientists and public policy-makers after 1900 gained a clearer recognition of the specific functions and objectives of government aid to science.

24 California, Geological Survey, *Geology*, xvi-xviii; California, Geological Survey, *Topographical Map of Central California together with Part of Nevada*, C. F. Hoffman, Principal Topographer (New York, 1873), and *Geological Map of the State of California* (Sacramento, 1873) and also *Map of California and Nevada* drawn by F. von Leicht and A. Craven (3d ed., rev. by Hoffman and Craven, San Francisco, 1880).

Rowland and the Nature of Electric Currents

*By John David Miller**

The mounted disk of ebonite
  Has whirled before, nor whirled in vain;
Rowland of Troy, that doughty knight,
  Convection currents did obtain
In such a disk, of power to wheedle,
  From its loved North the subtle needle.

'Twas when Sir Rowland, as a stage
  From Troy to Baltimore, took rest
In Berlin, there old Archimage,
  Armed him to follow up this quest;
Right glad to find himself Possessor
  Of the irrepressible Professor. . . .[1]

—JAMES CLERK MAXWELL, 1877

INTRODUCTION

THIS SERIO-COMIC VERSE from Scotland's eminent electrical theorist commemorates a delicate experiment performed in Berlin in 1876 by a young American civil engineer, Henry Augustus Rowland (1848–1901). Rowland is best remembered for his technical laboratory skill epitomized by his exacting determination of the mechanical equivalent of heat and his precisely ruled diffraction gratings, which were the prized tools of nineteenth-century spectroscopists.[2] Yet Rowland's electromagnetic researches have been largely overlooked, even though many of them were carried out in an involved theoretical context to demonstrate new actions of electric currents. In this latter work, which drew the keen interest of Maxwell and others, Rowland appears as something more than technician and mechanic. Indeed, he

*School of Education, University of California, Berkeley, California 94720. This paper was written through the assistance of a Berkeley faculty fellowship and is in part an extension of dissertation research carried out in 1967–1968 under a predoctoral fellowship at the Smithsonian Institution.

[1] Exerpted from the collection of Maxwell's serio-comic verse in Lewis Campbell and William Garnett, *The Life of James Clerk Maxwell* (London: Macmillan, 1882), p. 654.

[2] Published in a 125-page report in 1880, Rowland's studies on heat constituted the most exhaustive research on the subject following the work of James Prescott Joule at mid-century. (H. Rowland, "On the Mechanical Equivalent of Heat, with Subsidiary Researches on the Variation of the Mercurial from the Air-Thermometer, and on the Variation of the Specific Heat of Water," *Proceedings of the American Association for the Advancement of Science*, 1880, *15*:75–200.) By the late 1890s Rowland had distributed more than 100 of his gratings at cost to physicists throughout the world, one of special note going to Pieter Zeeman, who used it in making the important observation in 1897 of the magnetic splitting of the two D lines of the sodium spectrum (P. Zeeman, "On the Influence of Magnetism on the Nature of Light Emitted by a Substance," *Philosophical Magazine*, 1897, *43*:226–239).

emerges as an experimental physicist who could pursue his profession at a level equal to the best of his European colleagues and in doing so constitutes a novel figure in the exact sciences as then practiced in America. New documents, including two of Rowland's hitherto unknown scientific notebooks,[3] provide much of the basis for the present analysis, which focuses on the Berlin convection experiment and the series of generic researches which it spurred near the end of the century.

THE CONVECTION EXPERIMENT: INTERACTION OR AUTONOMOUS FLUID?

Accounts have been given elsewhere of how Rowland as the descendant of three generations of Protestant clergymen came to be enrolled in Amos Eaton's school of "practicle science," which was the Rensselaer Polytechnic Institute in Troy, New York.[4] It is only necessary here to note that Rowland's junior and senior years of 1868–1870 represented a period of intense scientific study and experimentation. This was carried out not so much in formal course work as through private reading and investigation within the confines of his boarding-house room. There he kept his large and rapidly growing scientific library as well as numerous pieces of experimental apparatus, much of which had been constructed by his own hands.

The theoretical notions of electricity which provided the context for his Berlin convection experiment were recorded by Rowland at this time in a two-volume notebook of 1868 in which he made numerous references to one distinctive source of electrical ideas. *The Experimental Researches in Electricity* of Michael Faraday[5] is the first entry in a list of Rowland's library books recorded at the end of the first volume. Also to be found in this volume along with more than a dozen similar citations are *Notes from "Faraday's Ex[perimental] Res[earches] in Electricity"* (p. 22) and the provocative title of the page (59) "*Thoughts Suggested* by the reading of the third volume of Far[aday's] Ex[perimental] Res[earches in Elec[tricity]." The discussion found directly under this citation concerned an analogy which Faraday drew between magnetism and galvanic currents,[6] but following immediately (p. 60) is Rowland's earliest written description of the convection experiment. As we shall see, the third volume of the *Researches* does contain ideas of Faraday's clearly fundamental to Rowland's theoretical notions of electricity, but it was in the first volume that Faraday was most explicit concerning convection action. There he wrote: "If a ball be electrified positively in the middle of a room and then be moved in any direction, effects will be produced, as if a 'current' in the same direction (to use the

[3] In all I recovered 37 of Rowland's notebooks from uncatalogued storage at the Johns Hopkins University in 1968. At that time Miss Harriette Rowland kindly consented to give approximately 300 family letters and other miscellaneous memorabilia collected by her father to the university. All of the foregoing materials as well as the Rowland-Gilman correspondence of the 1870s (once filed in the alumni records office) have been organized with the assistance of Miss Frieda C. Thies, archivist, in the Rowland manuscript collection at the Milton Eisenhower Library, Johns Hopkins University. Unless otherwise noted, citations refer to this consolidated collection. All of Rowland's technical papers cited below are also printed in *The Physical Papers of Henry Augustus Rowland* [collected for publication by a committee of the faculty of the university] (Baltimore: The Johns Hopkins Press, 1902).

[4] T. C. Mendenhall, "Henry A. Rowland Commemorative Address" in *The Physical Papers of Rowland*, pp. 1–17; S. Rezneck, "The Education of An American Scientist: H. A. Rowland, 1848–1901," *American Journal of Physics*, 1960, *28*:155–162.

[5] Michael Faraday, *Experimental Researches in Electricity*, 3 vols. (London, 1839–1855).

[6] *Ibid.*, Vol. III, p. 424. Rowland exploited this idea in deriving his magnetic analogy to Ohm's law. See n. 15 below.

convectional mode of expression) had existed . . . ."[7] And this statement was quoted by Rowland and one of his students in a report of convection experiments carried out later in 1889.[8]

In discussing Faraday's idea in an account of the Berlin work published for American readers Rowland observed that there seemed to be "no theoretical ground" for determining if an electrified body in motion produced magnetic effects, "seeing that the magnetic action of a conducted electric current may be ascribed to some mutual action between the conductor and the currrent."[9] The problem was this. Hans Oersted's famous experiments of 1820 had demonstrated the magnetic action of electricity conducted in copper wires,[10] but there was no reason to expect such action from electricity convected in space or apart from metallic conductors to interact with it.

Now, in Volume III of the *Researches* the idea of mutual action between electricity and its conductor is found in Faraday's "Thoughts on Ray Vibrations," which focus mainly on the description of radiation phenomena. There Faraday also wrote:

> Electricity is transmitted through a small metallic wire, and is often viewed as transmitted by vibrations also. That the electric transference depends on the forces or powers of the matter of the wire can hardly be doubted, when we consider the different conductibility of the various metallic and other bodies; the means of affecting it by heat or cold; the way in which conducting bodies by combination enter into the constitution of non-conducting substances, and the contrary. . . .[11]

Clearly taking up the idea of electricity interacting with its conductor, Rowland drew a sketch in the second volume of the 1868 notebook (p. 6), which is reproduced in Figure 1 and represents the vibratory nature of three electric currents of graduated intensities.

The varying conductivity of metallic bodies might imply the interaction of electricity

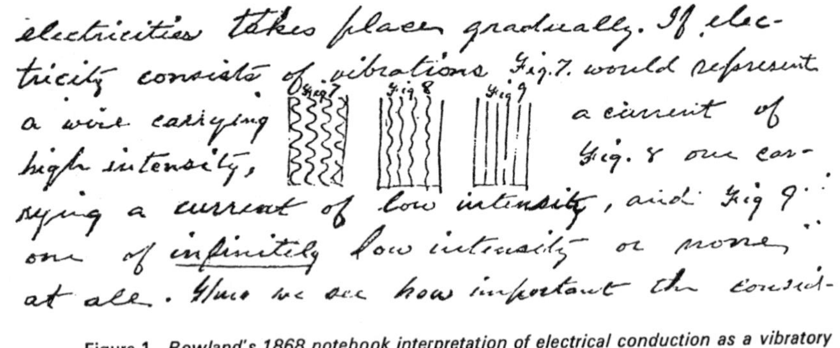
electricities takes place gradually. If elec-
tricity consists of vibrations Fig. 7 would represent
a wire carrying a current of
high intensity, Fig. 8 one carrying a current of low intensity, and Fig 9
one of infinitely low intensity or none
at all. Thus we see how important the consid-

Figure 1. *Rowland's 1868 notebook interpretation of electrical conduction as a vibratory interaction within a conductor. Currents of graduated decreasing intensities are represented from left to right.*

[7] *Ibid.*, Vol. I, p. 524.

[8] H. Rowland and C. Hutchinson, "On the Electromagnetic Effect of Convection-Currents," *Phil. Mag.*, 1889, *27*:445.

[9] H. Rowland, "On the Magnetic Effect of Electric Convection," *American Journal of Science and Arts*, 1878, *15*:30.

[10] H. Oersted, "Experimenta circa effectum conflictus electrici in acum magneticam," *Journal für Chemie und Physik*, 1820, *29*:275–281.

[11] Faraday, *Researches*, Vol. III, p. 448.

with its conductor, but did magnetic effects depend upon this interaction? Rowland was unsure and thus considered an alternative hypothesis that electricity behaved somewhat like the fluid substances postulated in eighteenth-century theories. In such theories an electrical fluid in motion produced effects which were independent of any kind of interaction between the fluid and the medium in which it resided; that is, the fluid existed in some free fashion, and its effects, magnetic or whatever, were attributes soley of its autonomous nature.[12]

The experiment which Rowland described in 1868 was organized in two sections in order to test each of the foregoing hypotheses. His plan is depicted in schematic form in Figure 2, which represents an electrically charged metal wheel containing a number of radial saw cuts. The wheel was to be rapidly revolved about an iron axis. The saw cuts were intended to prevent the electricity from circular "slippage" on the surface of the wheel, while the purpose of the iron axis was to concentrate magnetic action if it should occur. With the saw cuts reducing the possibility for interaction of the electricity with the metal of the wheel, evidence of magnetic action with this configuration would support a fluid-like model of electricity and the idea that this magnetism depended only on the mere motion of electricity in space.

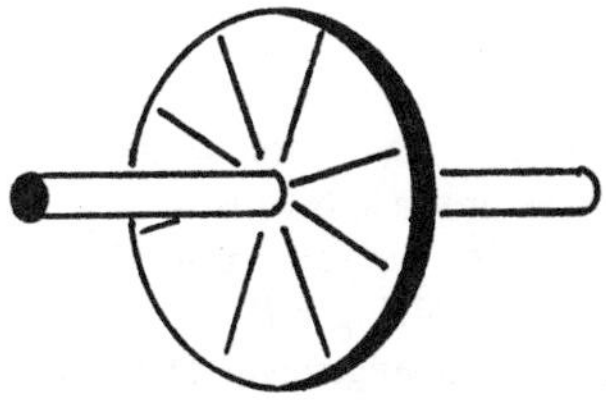

Figure 2. *Schematic representation of Rowland's proposed experiment of 1868 to test a fluid-like model of electricity.*

In order to observe such magnetism Rowland proposed using a sensitive astatic needle system, a popular instrument among electricians of the day. A typical system (one used later in 1889 by Rowland and one of his students) is displayed in the central and upper portions of the Frontispiece, a photograph of apparatus in the Smithsonian's Division of Electricity collection. The two thin horizontal rods are magnetized and may be adjusted vertically and axially on the glass tube supporting them to help neutralize the force of the earth's magnetism on a pair of tiny astatic needles suspended on a silk thread in the vertical hollow brass tube below. Although the needles are not visible in the photograph, one is located near the lower end of the vertical brass tube and the other at the junction of this tube and the horizontal brass section seen at the center right of the photograph. The two needles are aligned with axes parallel but with poles reversed in polarity in order to minimize any net torque on the system arising from components of the earth's magnetism escaping the effect of the long horizontal magnets above.

The lower needle was placed in the vicinity of the expected magnetism, while attached to the upper needle was a tiny mirror which reflected a beam of light through the open horizontal brass section onto a scale attached to a laboratory wall several feet away. The sensitivity to the system was enhanced not only by the arrangements to neutralize the earth's magnetism, but also by the path length of this reflected light beam which magnified by many times any motion of the astatic needles.

[12] See, e.g., Joseph Priestley, *The History and Present State of Electricity with Original Experiments* (2nd ed.; London: Dodsley, Johnson, and Cadell, 1769), pp. 444–450.

An astatic needle system also played a prominent role in the second section of Rowland's planned experiment as outlined in the 1868 notebooks. If no needle motion was observed at first with the wheel containing the radial saw cuts, it was to be replaced by one of solid metal as represented in Figure 3. The new wheel was to be charged with electricity and revolved rapidly in the vicinity of a stationary insulated ball of opposite electrical charge. Rowland's idea was that the stationary charge would attract the wheel's electricity, concentrating it in an area on the rim adjacent to the ball. Unlike the first part of the experiment, the charge on the wheel would remain stationary in space because of the ball's attraction. At the same time the wheel would be revolving relative to its own charge, an arrangement presumed to encourage interaction of the electricity with the circular surface of metal. Rowland was convinced that "the idea of a fluid will not suit electricity unless the first experiment [with saw cuts] succeeds in producing magnetism." On the other hand, under the theory that electricity interacted with the disk as a Faraday vibration, magnetism would be produced "in the second case [no saw cuts] but not in the first."[13]

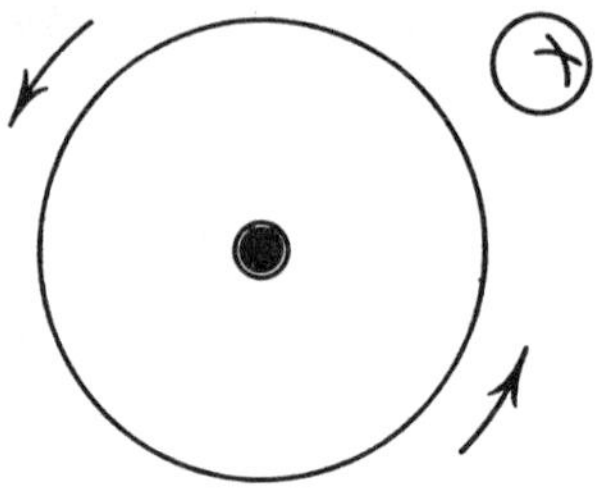

Figure 3. *Schematic representation of Rowland's plan of 1868 to test an interactive nature of electricity. A charged disk revolves near a stationary sphere of opposite electrical charge.*

IN SEARCH OF A LABORATORY

An opportunity for Rowland to carry out the convecton experiment did not present itself for seven years. Looking back some years later he noted, "as I recognized that the experiment would be an extremely delicate one, I did not attempt it until I could have every facility . . . ." The statement was consistent with his persistent complaints that facilities at Rensselaer Institute, where he had joined the faculty following graduation, were highly unsatisfactory for exact experimental research.[14]

The summer of 1875 offered a chance to leave Rensselaer. Daniel Coit Gilman (1831–1908), president of the recently endowed Johns Hopkins University at Baltimore, invited Rowland to go to Europe to inspect institutions of science as well as instrument shops and to make recommendations regarding suitable outfitting of a physical laboratory for the new university. Gilman had been impressed by letters which Rowland had shown him from Clerk Maxwell lauding some original experiments in magnetism carried out by Rowland in 1870–1873.[15]

[13] 1868 Notebooks, Vol. I, p. 61.

[14] H. Rowland, "Note on the Magnetic Effect of Electric Convection," *Phil. Mag.*, 1879, 7:442–443. Rowland had been appointed to the Rensselaer faculty in 1872, but this did not gain him much in the way of suitable laboratory space. During the winter of 1875 he told his family, "I have been working for the last week in a little shed attached to the institute with the thermometer down so far that my breath froze on my instruments" (Rowland to his mother, Feb. 23). See also Rowland to Edward Charles Pickering, Apr. 19, 1875 (by permission of the Harvard College Library).

[15] Maxwell to Rowland, July 9, 1873, and July 9, 1874. In these experiments (carried out mostly in his mother's Newark, New Jersey, home) Rowland gathered evidence supporting

After spending the summer in Scotland and England, which included a visit to Maxwell's estate in July, Rowland crossed to the Continent, arriving by himself in Berlin in late October. In terms of his primary mission he was not overly impressed by what he had so far seen. His standards were those of an experienced and exacting technician who had constructed most of his own instruments, including two elaborate electrical meters employed in the magnetic researches so highly praised by Maxwell. For Rowland's tastes, "the architect had got the best of the physicist" in too many European laboratories, and many of the instrument shops had "the appearance of museums of antiquity."[16]

It was largely through Gilman's insistence that Rowland had continued on to Germany at all.[17] Once there, however, he wrote enthusiastically to his employer:

> You were right when you said I would find nc lack of scientific spirit here and the apparatus shows it. In America we have apparatus for illustration, in England and France they have apparatus for illustration and experiment, but in Germany, they have *only* apparatus for experimental investigation. Our country is hardly ripe for the latter course though I should like to see it pursued to the best of our ability.[18]

Caught up in this spirit, Rowland applied for a general course of study in the university laboratory at Berlin under Hermann Helmholtz, but the distinguished German physicist's reply was prompt and negative, citing crowded conditions.[19] Disappointed but undismayed, Rowland waited a week and then composed a letter to Helmholtz suggesting more specific studies, namely an investigation of electric convection or an extension of the magnetic researches. Of the former he explained, "I have had this experiment in my notebook for years but have not yet found an opportunity to try it."[20] The convection experiment immediately interested Helmholtz, who, as Rowland informed Gilman, was "very polite this time and said that he would clear out a room for me downstairs where I could work. He says he has tried the experiment in another form but did not succeed. However, he likes my arrangement better and thinks it is a very promising experiment."[21]

---

his postulation of a magnetic analogue to Ohm's law for electric circuits. An account of the research was sent in desperation to Maxwell after repeated rejection by the editors of the *American Journal of Science* who did not understand it (J. D. Dana to Rowland, July 24, 1871, June 20, 1873, July 2, 1873). See also N. Reingold, *Science in Nineteenth Century America* (New York: Hill and Wang, 1964), pp. 262–275. Maxwell personally saw to the publication of Rowland's work in England (H. Rowland, "On Magnetic Permeability and the Maximum of Magnetism of Iron, Steel, and Nickel," *Phil. Mag.*, 1873, *46*:140–159).

[16] Gilman to the Johns Hopkins Trustees, Aug. 5, 1875; Rowland to Gilman Aug. 14 and Nov. 5.

[17] Gilman to Rowland, Aug. 16, 1875.

[18] Rowland to Gilman, Nov. 5, 1875.

[19] British electrician Arthur Schuster (1851–1924), who had performed experiments in the laboratory the year prior to Rowland's visit, reminisced at the close of the century about the "old laboratory of the great Helmholtz, in which we were about half a dozen students carrying on research work in a room in which each of us had to be satisfied with a table" (A. Schuster, "The New Physical Laboratory at Owens College, Manchester," *Nature*, 1898, *58*:621–622).

[20] Rowland to Helmholtz, Nov. 13, 1875. I am obliged to Dr. Christa Kirsten, Archiv Direktor, Deutsche Akademie der Wissenschaften, (East) Berlin, for providing me with a copy of this letter and permission to quote from it; 1868 Notebooks, Vol. I, pp. 60–61. Rowland also acknowledged the notebook record in an account of the Berlin researches published in the United States in 1878 (H. Rowland, "On the Magnetic Effect of Electric Convection," *Am. J. Sci.*, 1878, *15*:30).

[21] Rowland to Gilman, Nov. 19, 1875.

A TEST FOR OPEN CIRCUITS AND THE RATIO OF UNITS

In his letter to Helmholtz of November 1875 Rowland had reiterated the essential notions of his 1868 hypotheses:

> The question I first wish to take up is that of whether it is the mere motion of something through space which produces the magnetic effect of an electric current, or whether those effects are due to some change in the conducting body which, by affecting some medium around the body, produces the magnetic effects.[22]

Rowland favored the latter possibility. "The fact that it is the conducting *bodies* which attract or repel each other and *not* the electric currents leads one to believe the latter true."

Rowland's idea is clear. Before the Berlin experiments he believed that magnetic effects probably arose from some change in a conducting body brought on by the interaction of electricity with that body. Electricity could not produce magnetic action without its conductor, for if this could occur, what prevented electric currents in close proximity to each other from becoming detached from their conductors through mutual magnetic action? It was true that in the case of a spark there was the appearance that the electricity left the conductor. However, it was commonly known that large currents could be maintained in conductor elements in close proximity (electromagnets for instance) without such discharges, if the electric potential were kept low enough. If a physical conductor such as a copper wire was necessary for producing magnetic forces, the "mere motion of something through space," to use Rowland's words, would not produce similar action. Thus he favored the hypothesis of some change or interaction within the conducting body itself.

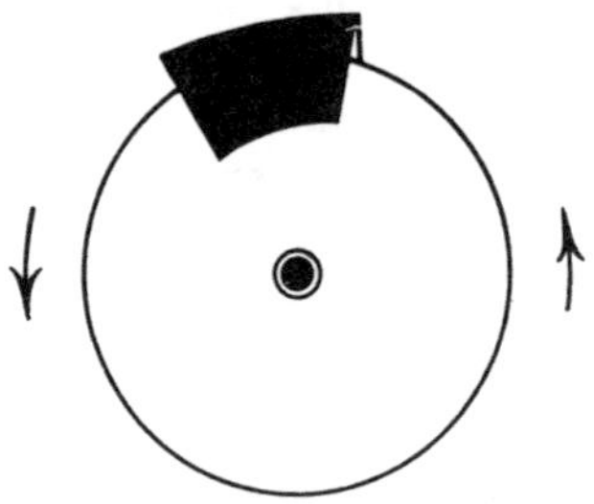

Figure 4. *Schematic representation of Rowland's investigation of 1876 testing the capacity of convected and conducted electricities to exchange identities within the same experimental configuration. A stationary charged metal sleeve encloses a sector of a revolving disk of opposite charge.*

The form of the experiment which Rowland proposed to Helmholtz was nearly identical to that described in the notebook of 1868. In the arrangement shown in Figure 4 a metal sleeve in the shape of a sector replaced the stationary ball (Fig. 3), but the anticipated action according to Rowland remained the same. In this experimental configuration there would be "no absolute motion of electricity yet the electricity is constantly conducted from one part of the disk to another." Alternatively the disk was to be covered with thin tinfoil having radial slits and the fixed metal plate eliminated. In this case Rowland expected "an electrified surface moving with a great velocity yet without conducting."[23] Once again for Rowland magnetic action in the latter case would provide evidence supporting an autonomous fluid-like nature for electricity, while such action from the former configuration would point more toward Faraday's theory of vibrations or interaction.

[22] Rowland to Helmholtz, Nov. 13, 1875.

[23] *Ibid.*

It appears that Rowland had discussed these ideas with Maxwell during their visit together in Scotland the previous summer of 1875, for Rowland noted to Helmholtz, "Maxwell assumes that the last case will produce magnetic effects although he has since told me he had no reason for the assumption." In his *Treatise on Electricity and Magnetism*, published in 1873, Maxwell indeed had indicated his belief that magnetic effects would be produced but did not give his reason for assuming so. "If the electric surface-density and the velocity can be made so great that the magnetic force is a measurable quantity, we may at least verify our supposition that a moving electrified body is equivalent to an electric current."[24] He had suggested further that the electrified surface might be that of a nonconducting disk and acknowledged that he was unaware of any attempts to perform such an experiment.

Rowland arrived in Germany at an opportune time. As he indicated in his November correspondence with Gilman, Helmholtz had been investigating the possibility of convection effects in the months prior to Rowland's arrival. Under study in Berlin was Helmholtz's extension of the potential theory of magnetic actions developed by Franz Ernst Neuman (1789–1895) in 1845.[25] By the introduction of a dielectric medium which could be indefinitely polarized, Helmholtz had extended this theory to include so-called "open circuits."[26]

Based upon his theory Helmholtz had computed a magnetic force to be expected from the "open" end of a wire from which an electrical discharge was directed upon a toroidal magnet, magnetized tangentially to its circular axis. An experiment in which an attempt was made to observe the action of this force had been carried out in the summer of 1875 by one of Helmholtz's students, but with negative results.[27] This outcome could be explained by supposing that there was some inadequacy in the theory or that the end of the wire was not "open" at all, in which case no force was predicted. It was possible that electricity was convected from the physical end of the wire by air which it repelled. If so, the negative effect could be easily explained by the current in the discharge which "completed" the circuit and accounted for the absence of any observable deflection of the toroidal ring. Helmholtz therefore was greatly interested in Rowland's proposal to investigate the possible magnetic properties of convected electricity.

In his first experiments in Berlin, Rowland employed a single gilded ebonite (vulcanite) disk, 21 centimeters in diameter, revolved about a vertical axis at the rate of 60 times a second. His method was to reverse the polarity of the electrification of the disk while at the same time observing the reflection of a beam of light from the mirror of his astatic needle system. After several weeks of trial (interrupted by the closing of the university for the Christmas holidays) he reported a distinct deflection of the beam several millimeters and wrote that this "qualitative effect, after once being obtained,

[24] J. Clerk Maxwell, *A Treatise on Electricity and Magnetism*, 2 vols. (Oxford: Clarendon Press, 1873), Vol. II, p. 370.

[25] F. Neumann, "Allgemeine Gesetze der induktierten elektrischen Ströme," *Schriften der Berliner Akademie der Wissenschaften*, 1845, pp. 1–88.

[26] H. Helmholtz, "Ueber die Theorie der Elektrodynamik," dritte Abhandlung: Die elektrodynamischen Kräfte in bewegtan Leitern, *Crelle's Journal für Mathematik*, 1874, *78*:273, 324.

[27] N. N. Schiller, "Elektromagnetische Eigenshaften ungeschlossener elektrischer Ströme," *Annalen der Physik und Chemie*, 1876–1877, *159*:456–473, 537–553; *160*:333–335. See also A. E. Woodruff, "The Contributions of Hermann von Helmholtz to Electrodynamics," *Isis*, 1968, *59*:300–311.

never failed." Rowland also reported a deflection when the gilding was scratched away in radial lines to prevent the possibility of annular currents, and numerous other tests were made to ascertain the effect of varying the velocity and electrification of the disk.[28]

In these experiments the indicated magnetic force was only about 1/50,000 of that of the earth's horizontal component in Berlin. But in spite of the minuteness of this

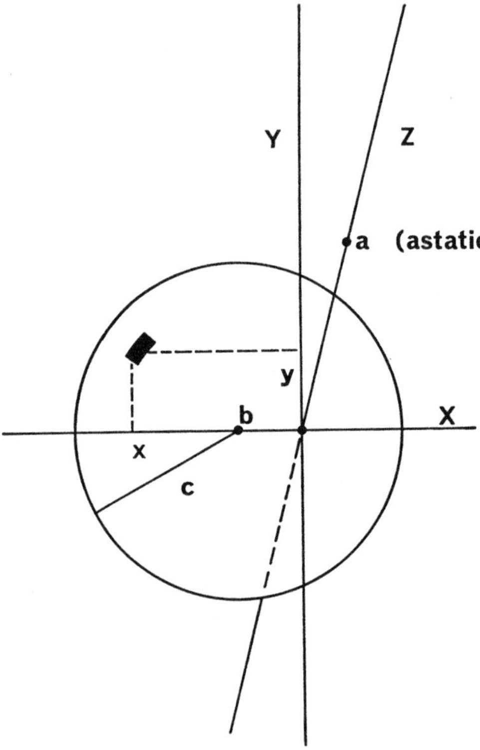

Figure 5. *The geometry of the Berlin convection experiment of 1876.*

action, Rowland gathered data to compare observed and computed increments of needle deflection. To do this he assumed a fluid-like model for electricity, with the magnetizing force due to the charge of any element of the disk's surface taken to be proportional to the "quantity of electricity" revolving in that element per unit of time. From the geometric configuration represented in Figure 5 Rowland derived the following expression for the total radial component of force (X) to be expected from

[28] H. Rowland, "On the Magnetic Effect of Electric Convection," *Am. J. Sci.*, 1878, *15*:31.

the entire surface (both sides) of his single disk acting in a plane parallel to this surface[29]:

$$X = \frac{8\pi N\sigma a}{v}\int_{-(C+b)}^{C-b}\int_{0}^{v'} \frac{(b+x)\,dx\,dy}{(a^2+x^2+y^2)^{\frac{3}{2}}} \tag{1}$$

Here N = the number of revolutions of the disk per second, σ = the density of the surface electrification, and a,b,c = geometric dimensions of the disk and needle position indicated in Figure 5.

The constant v represented the ratio of Maxwell's electrostatic to electromagnetic units and was necessary in the denominator of Equation (1) in order for the expected magnetic force to be computed in electromagnetic units, that is, in the same system in which the observed deflection was measured. The conversion factor was needed because in the laboratory the degree of surface electrification σ was most conveniently measured in electrostatic units.[30] The constant v retained the basic dimensions of length divided by time arising from the two sets of fundamental force and energy equations upon which the electrostatic and electromagnetic systems had been constructed.[31]

By the 1870s the value of v had been the subject of several investigations based on electrical methods. Much of the interest in the constant stemmed from Maxwell's hypothesis in the 1873 *Treatise* that the "agreement or disagreement of the values of V [velocity of light] and of v [ratio of units] furnishes a test of electromagnetic theory of light."[32] At that time he gave the following tabular comparison of the two velocities:

Velocity of Light (meters per second)		Ratio of Electric Units	
Fizeau	314000000	Weber	31740000
Aberration, etc., and Sun's Parallax	308000000	Maxwell	28800000
Foucault	298360000	Thomson	28200000

Now in order to compute the expected radial horizontal component of magnetic force given by Equation (1) above, Rowland had to assume some value for v. He chose the value given by Maxwell himself in the above table, that is, 288 million meters per second. Correspondingly, the measured magnitude of this force component was obtained by determining the oscillation period of the astatic needle system under the magnetic influence of the whirling disk.[33] Rowland's averages for these measured and computed values of deflection are shown below for three series of

[29] *Ibid.*, p. 34.

[30] The electrification was inferred from the length of a sample spark drawn from Leyden jars supplying the system after calibration data worked out by William Thomson in 1859–1860. In *Reprints of Papers on Electrostatics and Magnetism* (London: Macmillan, 1872), pp. 247–259.

[31] The clearly dimensionless integrand of Eq. (1) does not enter in here. The dimensions of electrical quantities in these systems were given by Maxwell and Henry C. Fleeming Jenkin in a report to the British Association in 1863. *Reports on Electrical Standards* (Cambridge: Cambridge Univ. Press, 1913), pp. 126–135.

[32] Vol. II, p. 387.

[33] After a method developed by Coulomb in which the indicated force was nearly proportional to the square of the vibration frequency ("Second mémoire su l'électricité et le magnétisme," *Histoire de l'Académie Royal des Sciences*, 1788, pp. 578–612).

experiments compiled from a total of sixty-two readings of individual deflection positions.[34]

	Series I	Series II	Series III
Measured magnetic force[35] (horizontal component)	0.00000327	0.00000317	0.00000339
Computed magnetic force (horizontal component computed using Maxwell's value of v)	0.00000337	0.00000349	0.00000355

Rowland noted that the difference between expected and measured values of force was 3, 10, and 4 per cent respectively using Maxwell's value for v. However, he observed, "The value $v = 300{,}000{,}000$ meters per second, satisfies the first and last series of the experiments best."[36]

The second form of the convection experiment (Fig. 4) which employed the stationary metal plate to induce electrostatic action was also tried by Rowland in Berlin. At first it might seem curious that he went to this additional trouble, since magnetic action had been reported from the radially scratched disks. Helmholtz, however, suggested using the second form as a test that convected and conducted electricities could exchange identities within the same experimental configuration. Briefly, the idea was to consider that electricity induced under the sector was carried forward as a convection current by the disk until it was released at the edge of the sector, where it might return as conduction currents through other parts of the disk. Tests with this arrangement were in accord with previous results, leading Rowland to the conclusion that "electricity produces nearly if not quite the same magnetic effect in the case of convection as of conduction, provided the same quantity of electricity passes a given point in the convection stream as in the conduction stream."[37] A new action was claimed for the electric fluid.

[34] The electrification was normally made positive, negative, and then positive again, giving three readings, although on occasion Rowland performed a third reversal, accounting for the even number of readings.

[35] Possessing the fundamental dimensions of $(\text{length})^{-1/2}$ $(\text{mass})^{-1/2}$ $(\text{time})^{-1}$ in the electromagnetic system in which Rowland employed the units of centimeters, grams, and seconds respectively, that is, the dimensions of the absolute electromagnetic system as given by Maxwell in the *Treatise* (Vol. II, p. 242).

[36] H. Rowland, "On the Magnetic Effect of Electric Convection," *Am. J. Sci.*, 1878, *15*:38. Since v occurs throughout these computations as a simple factor, it is not difficult to compute the exact value of the constant corresponding to the mean values of Rowland's first and last series of experiments. From the expression

$$v/288\ (327 + 339) = (337 + 355) \times 10^6 \text{ m/sec}$$

results 299 million m/sec while a similar computation including the second series gives 305 million m/sec. The variability of Rowland's data is examined in greater detail below.

[37] *Ibid.*, p. 33. By scratching fine concentric circles through the gilding of the disk a conduction configuration was obtained which could be easily characterized mathematically. When a current C left the vicinity of the sector covering $1/n$ of the circumference of a ring thus formed, it was expected to divide according to Ohm's law into a conduction current returning under the sector equal to $-C(n-1)/n$ and one outside the vicinity of the sector, equal to $+C/n$. Thus if the convection and conduction currents exchange identities without net currents being

THEORIES OF THE NEEDLE ACTION

In a letter to Gilman of early February 1876 Rowland assessed his efforts in Berlin:

> I have been highly successful and have not only detected the magnetic action of the revolving electrified disc, but have also measured it with tolerable accuracy. At first I obtained no result, but by working at it, avoiding all sources of disturbance, and giving the highest delicacy to the apparatus, I have at last succeeded in obtaining a result. The force measured is 1/50,000 of the earth's magnetism and that within half an inch of a disc revolving sixty times in a second. I am very pleased at the result of two or three months labor. I think it may have some effect on the theory of electricity.[38]

Rowland also had written to his family explaining that the experiment was not a thing which would "produce a great stir because there will be too few to see its importance."[39]

Helmholtz for one had followed the work closely and was pleased with Rowland's results, reimbursing him for his expenses (however modest) and personally seeing to the presentation of the researches to the Berlin Academy of Science.[40] Here he endorsed Rowland's findings: "Mr. Rowland has now carried out a series of direct experiments, in the physical laboratory of the University here, which give positive proof that the motion of electrified ponderable substances is also electromagnetically operative [effective]."[41] Nonetheless Helmholtz believed that the needle action could be explained equally well by a number of contemporary electrodynamic theories.

He cited as one example the theory of Wilhelm Eduard Weber (1804–1891) in which convected electricity acted like a fluid in motion with which was associated a

---

created or destroyed,

$$+C \quad - \quad \frac{C(n-1)}{n} \quad = \quad \frac{+C}{n}$$

(convection current under sectors)	(conduction current under sectors)	(conduction current outside sectors)

It followed that the magnetic effect to be expected near the sectors would correspond to a current of approximately

$$\frac{+C-C(n-1)}{n} \quad \text{or} \quad \frac{C}{n}$$

This was $1/n$th of the action that would be expected by electrifying the entire disk, and this magnitude of effect was reported by Rowland.

[38] Rowland to Gilman Feb. 2, 1876.

[39] Rowland to Mary Rowland, Jan. 26, 1876.

[40] The cost was about $80 Rowland to Anna Rowland Feb. 21 1876. Helmholtz followed the research closely throughout its duration, Rowland reporting to Gilman, "He is very quiet and dignified and pays very little attention to those working under him and so I feel quite pleased at his condescension in coming in to see how I am getting along every day" (Feb 2 1876) Nevertheless Rowland went on to remark candidly

> Prof. Helmholtz' character interests me much. Although the foremost physicist on the continent he is by no means brilliant and in some of his suggestions I can sometimes fancy there is a trifle of stupidity. His wonderful genius lies in his power of concentration. When he thinks about a thing of any complexity his whole mind is on it. I have sometimes thought that one might knock him down without his feeling or knowing it. It is in this way that he brings forth his great results but if he does not think thus deeply his thoughts are almost as other men. As an experimenter, he is quite poor. In these aspects he reminds one of Newton who, to all appearances, was not very "smart" but yet had a wonderful power of concentration like Helmholtz. . . .

[41] As translated in the account published in the *Phil. Mag.*, 1876, *2*:233–237; Helmholtz's account to the Academy read: "Hr. Rowland hat nun eine Reihe directer Versuche im Physikalischen Laboratorium der hiesigen Universität ausgeführt, welche den positiven Beweis geben, dass auch die Bewegung elektrisirter ponderabler Körper elektromagnetisch wirksam ist" (H. Helmholtz, "Versuche über die elektromagnetische Wirkung elektrischer Convection, ausgeführt von Hrn. Henry A. Rowland . . . ," *Akademie der Wissenschaften Monatsberichte*, 1876 *16*:211–217).

magnetic action at a distance. The magnitude of this action was proportional to the velocity of the fluid; in Rowland's experiment, it was proportional to the speed of the disk.[42] Alternatively Helmholtz thought that the needle action could be described equally well by Maxwell's displacement theory or Helmholtz's own potential theory, which employed the concept of dielectric polarization. In these theories Helmholtz had written: "The volume-elements of the stratum of air situated between the resting and moved plates suffer continual displacements in the direction of a rotation [a]round radially directed rotation-axes." Helmholtz believed that "the arising and disappearing components of this polarization would constitute the current which is indicated by the astatic pair of needles."[43]

Thus under Helmholtz's interpretations the experiment did not signal a shift in attention to any particular theory. True, Rowland's work nicely explained the negative results of the open-circuit experiments preceding his in Berlin, but it said nothing for such experiments if the electric potential was kept low enough to prevent an electrical discharge into the air. Even in 1881, when J. J. Thomson computed the magnetic force of a moving charged sphere showing that Rowland's observations were in agreement with Maxwell's theory of electric displacement, G. F. Fitzgerald was quick to respond that it had never been actually demonstrated "that open circuits, such as Leyden jar discharges, produce exactly the same effects as closed circuits, and until some such effect of displacement currents is observed, the whole theory of them will be open to question."[44]

Moreover, Rowland's computation of the ratio of units from the Berlin work was too rough to constitute substantial new evidence for Maxwell's theory or against any other. Certainly Rowland himself did not consider it as such, spending considerable effort later at Johns Hopkins to make a precise determination of the constant.[45] Indeed, Thomson concluded as much in 1885 in an exhaustive report to the British Association, which was based on a survey of major electrodynamic theories and experiments drawn mostly from the Berlin work of the seventies. Evidence to date, Thomson said, told "nothing as to whether any special form of the dielectric theory such as Maxwell's or Helmholtz's is true or not."[46]

Indeed, if Rowland had reported no action, the situation would have been as unsettled, for such results could be most easily written off to the difficulty of the experiment. As for Maxwell, he reserved his usual reticence except to compliment as he did in correspondence and verse the skill of his young protégé in measuring a previously unobserved action of electricity.[47]

[42] W. Weber, "Elektrodynamische Maasbestimmungen," *Ann. Phys.*, 1848, *73*:193–240.

[43] Helmholtz, *Phil. Mag.*, 1876, p. 237.

[44] J. J. Thomson, "Effects Produced by the Motion of Electrified Bodies," *Phil. Mag.*, 1881, *11*:229–249; G. F. Fitzgerald, "Note on Mr. J. J. Thomson's Investigation of the Electromagnetic Action of a Moving Electrified Sphere," *Scientific Proceedings of the Royal Dublin Society*, 1883 (read 1881), *3*:254.

[45] H. Rowland, "On the Ratio of the Electromagnetic to the Electrostatic Unit of Electricity," *Am. J. Sci.*, 1889, *38*:289–298.

[46] J. J. Thomson, "Report on Electrical Theories," *Report of the Fifty-fifth Meeting of the British Association for the Advancement of Science*, 1885, p. 149.

[47] Maxwell to Rowland, Mar. 7, 1876. See also T. K. Simpson, "Maxwell and the Direct Experimental Test of His Electromagnetic Theory," *Isis*, 1966, *57*:428.

TESTING FOR NEW ELECTRIC ACTIONS IN BALTIMORE

But wouldst thou twirl that disk once more,
Then follow in Childe Rowland's train,
To where in busy Baltimore
He brews the bantlings of his brain. . . .

In the spring of 1876 Rowland returned to America carrying numerous receipts for equipment to outfit the new physical laboratory at Baltimore. He had been authorized by Gilman to spend about $6,000, but by mid-November of 1877 bills drawn against the physics department solely for apparatus very nearly exceeded twice this amount.[48] Rowland had been emphatic that money be spent only on research apparatus and "not for illustration of lectures," telling Gilman that though it was

> . . . *sometimes* possible to produce good work with poor apparatus just as it is possible to cut down a tree with a penknife yet there is other work which *cannot* be done by any possibility without calling to our aid all the resources of mechanics . . . [and] to this last class belong many of the higher questions in mathematical physics.[49]

In spite of unfavorable reaction from American colleagues such as mathematician-astronomer Simon Newcomb (1835–1909) and geophysicist-logician Charles Sanders Peirce (1839–1914),[50] Rowland's chief intention for the elaborate equipment was to carry out extensive absolute measurements of physical constants such as the ratio of units. "Particularly in magnetism and electricity," he told Gilman, "there is a wider field open in this direction than any other and it can only be entered by having the most perfect instruments."[51] By 1879, according to lists compiled at Harvard University, the Johns Hopkins collection of scientific apparatus greatly exceeded in quantity and quality that of any other American institution and for that matter most European laboratories.[52]

Rowland was soon vindicated in his purchases, but not exactly in the way he had anticipated. At Johns Hopkins in the late seventies Edwin Herbert Hall (1855–1938) found two advantages unique in America for carrying out electrical investigations: he gained access to instruments of precision and to the counsel of Henry Rowland. Following an idea and experimental configuration suggested specifically by Rowland and putting the delicate instruments to a use for which they had not been primarily intended, Hall reported another new magnetic action of electricity in 1879.[53] His measurements indicated that a powerful magnet modified the distribution of electricity being conducted in a thin strip of gold foil; that is, he reported the generation of a new electromotive force acting mutually perpendicular to the longitudinal axis of the foil and to the magnetic lines of force, as represented in Figure 6.

From Hall's laboratory notebook records the substantial role played in this discovery by a fluid-like model of electricity is clearly evident. Hall wrote:

[48] Johns Hopkins University, Physics Department Account Ledger, Nov. 15, 1877.

[49] Rowland to Gilman, Jan. 6, 1876.

[50] Newcomb to Gilman, Feb. 17, 1876; Peirce to Gilman, Jan. 13, 1878.

[51] Rowland to Gilman, Mar. 25, 1876.

[52] "List of Scientific Apparatus," *Harvard University Library Bulletin*, 1879, Nos. 11–12, pp. 302–304, 350–354. Rowland's student Edwin Hall wrote from Europe in 1881 that the Johns Hopkins University would be the loser if it exchanged its apparatus, for example, with the Cavendish Laboratory, especially if "what belongs to Prof. Rowland personally" was included (Hall to Gilman, July 31, 1881).

[53] E. Hall, "On a New Action of the Magnet on Electric Currents," *American Journal of Mathematics*, 1879, *2*:287–292.

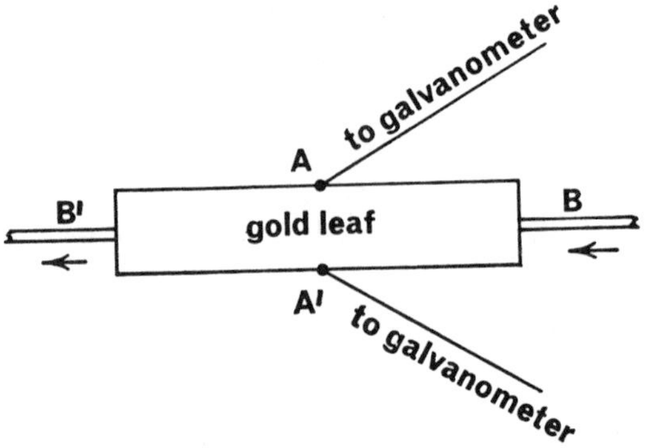

Figure 6. *Schematic representation of the experimental configuration suggested by Rowland to Edwin Hall. Lines of strong magnetic force may be pictured as acting in a direction perpendicular to the page.*

> I reasoned thus: If electricity were an incompressible fluid it might be acted on in a particular direction without moving in that direction. I took an example about like this. Suppose a stream of water flowing in a perfectly smooth pipe which is however loosely filled with gravel. The water will meet with resistance from the gravel but none from the pipe at least no frictional resistance. Suppose now some body brought near the pipe which has the power of attracting a *stream* of water. The water would evidently be pressed against the side of the pipe but being no[t] compressible and with the gravel, completely filling the pipe, it would not move in the direction of the pressure and the result would simply be a state of stress without any actual change of course by the stream. It is evident however that if a hole were made traversing through the pipe in the direction of the pressure and the two orifices thus made were connected by a second pipe, water would flow *out toward* the attracting object and *in* at the opposite orifice. . . .[54]

The two orifices in Hall's analogy correspond to the electrical junctions A and A' in Figure 6, while the central stream of water of which he speaks represents the flow of electricity from B to B'. Moreover, Hall's ideas were evidently amenable to Rowland, who, it seems, had tried the experiment before Hall. Hall noted in his laboratory records, "I set something of the above reasoning before Prof. R[owland] and he advised me to try the experiment as he had not made the trial very carefully himself."[55]

But there is more to suggest Rowland's substantive role in these experiments. Not only is he mentioned by name in numerous references to his ideas in Hall's laboratory records, but there is also an interesting statement concerning Hall's work made by Rowland more than a decade later in a letter to G. F. Fitzgerald. There Rowland revealed that the Berlin charge convection experiment,

> . . . together with that of Mr. Hall (Hall effect) which was really my experiment also, were made to find the nature of electric conduction. Indeed I had already obtained the Hall effect on a small scale before I made Mr. Hall try it with a gold leaf which gave a

[54] E. Hall, "Notebook of Physics [Johns Hopkins University, 1877–1880]," Hall papers (bms Am 1734–2), pp. 92–93 (by permission of the Harvard College Library).

[55] *Ibid.*, p. 95.

larger effect. My plate was copper or brass and I only obtained 1 mm. deflection. Mr. Hall simply repeated my experiment, according to my direction, with gold leaf.[56]

Physicist Joseph Ames, a close colleague of Rowland's at Johns Hopkins, might well have been thinking of Hall's investigations when describing the role which Rowland took in regard to his students' researches:

> Rowland's fundamental idea of a student was that he was responsible for his own investigation. The idea embodied in it might have been suggested to the student, and he might have received some help at various critical moments, but Rowland never regarded himself as responsible in any way for the investigation; that is, for its performance.[57]

In fact, concerning the publication of student researches Ames noted, "There have been several striking cases where it might have seemed to an impartial observer that Rowland's name should have appeared on the title page."

Although the fluid-like model of electric action had been extremely helpful in the convection experiment and Hall's measurements, the importance of the concept to the fundamental nature of electric currents had considerably dwindled for Rowland by the mid-1880s, if not sooner. Addressing the Electrical Conference at Philadelphia in 1884, he observed that "no hypothesis as to the nature of electricity" rested at the base of the "beautiful fabric of mathematical electricity" provided by the theories of Helmholtz, Thomson, Maxwell, and others; indeed these theories seemed independent of the "fate of the so-called electric fluid." He saw the "idea that electricity is liquid" as a useful concept but believed that there was only one conclusion to be drawn regarding its fundamental nature: "Electricity is a property of *matter*. . . . When we know what matter is, then the theories of light and heat will also be perfect; then and only then, shall we know what is electricity and what is magnetism."[58]

The limitations of the fluid analogy were stressed again in an address of 1888 to the Electrical Club in New York,[59] and the following year Rowland told the members of the American Institute of Electrical Engineers:

> The notion that electricity was a subtle fluid which could flow along metal wires as water flows along a tube, . . . remains in force today among all except leaders in scientific thought. This notion still serves a very important purpose in science. But for many years, it has been recognized that it includes only a very small portion of the truth. . . .[60]

Rowland went on to cite the recent experiments of Heinrich Hertz, observing in this context, "The electric current is an unsolved mystery . . . we know that we must look outside of the wire at the disturbance in the medium before we can understand it." The ether in which Hertz's waves propagated was "a much more important factor in

[56] Rowland to G. F. Fitzgerald, May 12, 1894. I am obliged to Mr. Desmond Clarke, Librarian/Secretary to the Royal Dublin Society for furnishing me a copy of this letter and permission to quote from it.

[57] J. Ames, "The Work of the Physical Laboratory," *The Johns Hopkins University Alumni Magazine*, 1917, *5*:158.

[58] H. Rowland, "Address as President of the Electrical Conference at Philadelphia," Sept. 8, 1884, in *Report of the Electrical Conference at Philadelphia* (Washington, D.C.: GPO, 1886), pp. 20 ff.

[59] H. Rowland, "The Electrical and Magnetic Discoveries of Faraday" (Address at the Opening of the Electrical Club House of New York City, 1888), in *Physical Papers*, pp. 638–652.

[60] "On Modern Views with Respect to Electric Currents" (Address before the American Institute of Electrical Engineers, New York, May 22, 1889), *Transactions of the American Institute of Electrical Engineers*, 1889, *6*:342–357.

science than the air we breathe" and constituted one of the "most modern views held by physicists with respect to electric currents."

By demonstrating electromagnetic wave action in the space surrounding a wire Hertz's experiments had greatly elevated Maxwell's electromagnetic theory of light for physicists, and Rowland saw at this time an opportunity for making a contribution in the same direction by repeating the convection experiment and determining the ratio *v*. Since the Berlin researches, apparently only three physicists had attempted the delicate experiment: in Vienna Ernst Lecher (1856–1926) had reported negative results in a paper of 1883 which described experiments with charged disks of brass and pasteboard covered with graphite; on the other hand, Wilhelm Röntgen and Franz Himstedt (1852–1933) reported positive action in experiments of 1888–1889 carried out in conjunction with tests for Maxwell's displacement current.[61]

Rowland and his student Gary Talcot Hutchinson (1866–1939) knew of these conflicting reports when they began new experiments at Baltimore in 1889. However, their objective was not so much to produce further evidence for the evanescent effect as it was to make a precision measurement of the ratio of units.[62] Rowland's measurements of *v* in Berlin had been rough, and shortly after his return to Baltimore he had attempted a precise determination of the constant by a different experimental plan. At first he had been optimistic concerning his results, writing to Maxwell in 1879, "I believe the experiment is a link in the proof of your [electromagnetic] theory seeing that the result is, by the first rough calculation, 299,000,000 meters per second, though the corrections may amount to 1/2 per ct. or so."[63] When corrections were computed, however, a value of 297,900,000 meters per second resulted, which was too far from reported velocities of light to satisfy Rowland. Thus he postponed publishing these results until 1889, when the new convection experiments were completed and other direct measurements were simultaneously carried out by one of his students, Edward Bennett Rosa (1861–1921).[64]

The elaborate Rowland-Hutchinson convection experiments of 1889 did not live up to expectations. A photograph of the 1889 apparatus is reproduced as the Frontispiece. The two larger gear systems seen at the left and right were intended to be used in accurately determining the speed of revolution, but they served in addition to create spurious capacitance effects with the electrified disks at the center of the apparatus.[65] Enormous difficulties were encountered with insulation, because the winter had been

[61] E. Lecher, "Einige elektrische Versuche mit negativen Resultaten," *Repertorium der Physik*, 1884, *20*:151–153; W. Röntgen, "Elektrodynamische Wirkung bewegter Dielektrika," *Ann. Phys.*, 1888, *40*: 93–108; F. Himstedt, "Elektromagnetische Wirkung der Convection," *Ann. Phys.*, 1889, *38*: 560–573.

[62] H. Rowland and C. T. Hutchinson, "On the Electromagnetic Effect of Convection-Current," *Phil. Mag.*, 1889, *27*:445.

[63] Rowland to Maxwell, Apr. 7, 1879, Cambridge University Library, Cambridge, England. James Clerk Maxwell papers (microfilm).

[64] In the 1879 experiments Rowland used a precision spherical condenser discharged through a galvanometer. The magnitude of the charge of the former provided the electrostatic measure, while the deflection of the latter provided the electromagnetic. The condenser is extant in the Smithsonian Division of Electricity collection (Catalogue No. 318,513). "On the Ratio of the Electro-static to the Electro-magnetic Unit of Electricity," *Phil. Mag.*, 1889, *28*:304–315; E. B. Rosa, "Determination of *v*, the Ratio of the Electro-magnetic to the Electro-static Unit," *Am. J. Sci.*, 1889, *38*:298–312.

[65] This apparatus is in the Smithsonian Division of Electricity collection (Catalogue No. 318, 515). I measured this gear-disk capacitance in 1968 with a General Radio Bridge (type 1608-A) and found it to represent about 30 per cent of the plate-disk capacitance.

unusually damp and electrical leakage was "abundant."[66] But in addition to this, the men encountered a mysterious and intermittant magnetic interference which was never identified or controlled.

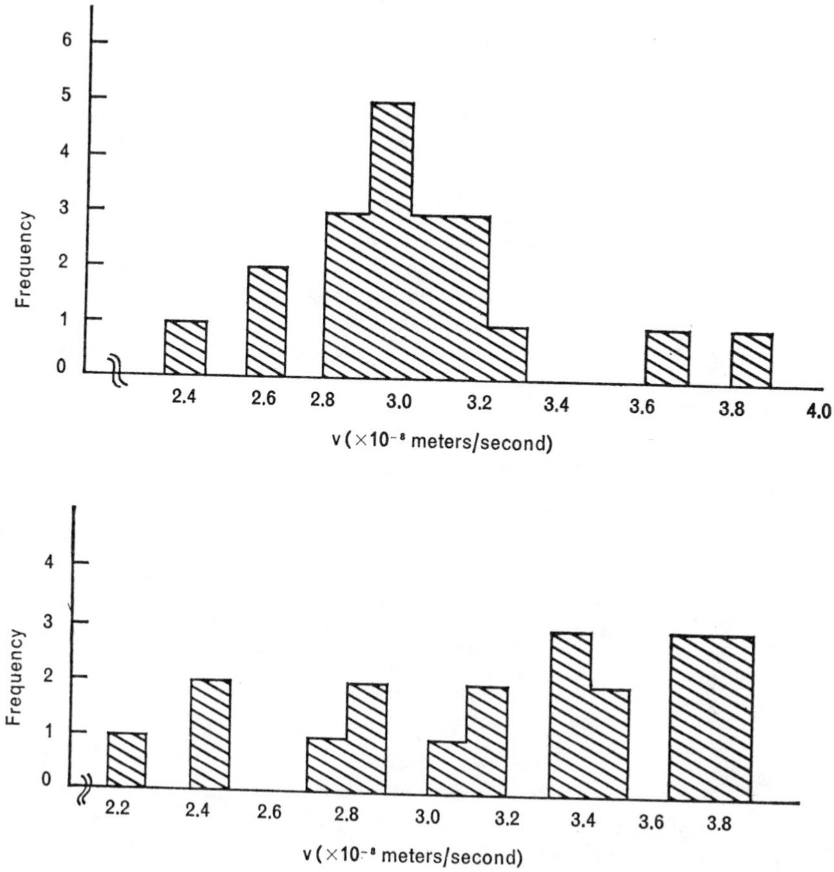

Figure 7. *Histograms constructed from raw data of the 1876 Berlin convection experiment* (*upper*) *and the Baltimore experiments of 1889* (*lower*).

A comparison of the shapes of two histograms shown in Figure 7 which I have constructed from the raw data of Rowland's experiments of 1876 and 1889 is suggestive of spurious interference during the work in Baltimore. The graphs display the

[66] *The Baltimore Sun Almanac* of 1890 (p. 142) recorded the rainfall for the months of November 1888 through April 1889 as 27.46 in. compared with 19.97 in. for that period of the previous year.

incidence of values of v for each set of experiments, and a normal distribution is more closely approximated by the (upper) Berlin histogram.[67] The Baltimore interference may have come from commercial electric power or trolley lines which criss-crossed the city by the late 1880s, but in any event there was an apparent advantage in searching for the delicate action with less sophisticated equipment in an electrically quieter decade.

AN ETHER MODEL OF THE CONVECTION EFFECT

Through the tragic circumstances of an incurable illness Rowland was distracted from his laboratory work in fundamental physics during much of the 1890s, devoting a large portion of his time to commercial electrical enterprises.[68] Nevertheless he continued to reflect upon the puzzle of electric conduction in correspondence, papers, and speeches. Thus to the inquiry of Fitzgerald in 1894, Rowland characterized the convection experiments and the measurements of Hall as showing that "magnetic action is due to the so-called moving charge." Yet he was perplexed that magnetic attraction between wires seemed to indicate that the force acted on the wires and not on the charge, while in Hall's experiments there was evidence that the charge itself was affected. Rowland could only reiterate to Fitzgerald the hypothesis under which the Berlin experiments originally had been proposed to Helmholtz in 1876: "I was in the hope of finding that the magnetic action of a current in the wire was due to some change in the wire caused by the current. This seems is not so. And yet the forces are between the wires! This is still a puzzle to me."

Fitzgerald viewed the effect measured by Hall as a "secondary action" of electric convection but went on to observe:

> It is of course very remarkable that the action is on the matter on which the charge is moving and not on the charge but I am not sure that we ought to expect that for it is a mechanical force and mechanical force is always exerted on matter and not on the ether notwithstanding (?) Faraday's and Maxwell's stress on the ether. What we call an electric charge is a quality of the surface of the matter and if this is moving, on the matter or with the matter makes no difference, if this is moving relatively to the ether in its neighborhood there is a different force on the matter than there would be if this surface quality were not moving.
>
> I suppose of one could quite clearly understand all these puzzles one could frame a satisfactory theory as to the structure of ether.[69]

A paper written by Rowland in the months following this correspondence was very much preoccupied with the structure of an ether and its relation to electric conduction. He began by disposing once again of the electrical fluid:

---

[67] Approximately the same number of needle deflection entries was published in each set of experiments, facilitating the comparison made here. The central value of v which I have used is 3.0×10^6 m/sec, although any other values tabulated by Maxwell in the *Treatise* might have been used, since doing so would not appreciably affect the shape of these histograms.

[68] Rowland was married in 1890 at age 42, and at that time learned through an insurance medical examination that he had diabetes (then incurable) and was given by his doctors about ten years to live. As his children were born he turned toward commercial pursuits to provide for their future, but ironically most of his inventions were too complicated in actual construction to ever realize substantial practical value.

[69] Rowland to Fitzgerald, May 12 (Royal Dublin Society Library); Fitzgerald to Rowland, Apr. 13 and May 26.

> We can now answer with certainty that electricity no longer exists. Electrical phenomena, electrostatic actions, electromagnetic action, electrical waves,—these still exist and require explanation; but electricity, which, according to the old theory, is a viscous fluid throwing out little amoeba-like arms that stick to neighboring light substances and, contracting, draw them to the electrified body, electricity as a self-repellent fluid or as two kinds of fluid, positive and negative, attracting each other repelling themselves,—this electricity no longer exists. For the name electricity, as used up to the present time, signifies at once that a substance is meant, and there is nothing more certain to-day than that electricity is not a fluid.[70]

Replacing these theories, Rowland saw the ether as accounting for "all ordinary electrical and magnetic actions" as well as light, the earth's magnetism, and gravitation. "Moreover Maxwell has connected the theories of electricity and of light, and no theory of one can be complete without the other. Indeed they must both rest upon the properties of the same medium which fills all space—the ether."

Needless to say, the reports of the important discoveries of Röntgen, Zeeman, and others during the last five years of the decade were studied as carefully in Baltimore as in physical laboratories elsewhere in the world. And, as reflected in his speech delivered as president of the American Physical Society in the fall of 1899, for Rowland the ether had become as complex as for any other physicist: "What is matter; what is gravitation; what is ether and the radiation through it; what is electricity and magnetism; how are these connected together and what is their relation to heat?" Electricity the "subtle spirit of the amber" and the "fluid" had "vanished, thrown on the waste heap of our discarded theories to be replaced by a far nobler and exalted one of action in the ether of space." All actions, "feeble and mighty" had "their being in this wonderous ether."[71]

But there is more here than speculative rhetoric, for Rowland believed that the convection experiment conceivably demonstrated a relationship between electricity and the ether. When matter was electrified it had "sufficient hold on the ether to communicate its motion to the ether." Zeeman's observation of the widening of the lines of the sodium spectrum in a strong magnetic field was evidence for Rowland of this kind of interaction at the molecular level. Vibrating, electrified "matter" of the molecule gripped the ether, generating magnetism, and moreover, "The experiment on the magnetic action of electric convection shows the same thing. By electrifying a disc in motion it appears as if the disc holds fast to the ether and drags it with it, thus setting up the perculiar ethereal motions known as magnetism."[72]

These ideas provided the basis for a series of interesting investigations carried out by Rowland and several of his students in 1899–1900. In one experiment a coil containing several miles of wire was revolved at great speed about its central axis. Rowland's idea was that the ether moving relative to the coil's circumference might constitute a current which in turn would induce a secondary current in the wire. However, after nearly eight months of testing, no consistent effect was observed.[73]

But further complicating the picture at this time were the negative results reported from a series of elaborate convection experiments carried out by Victor Crémieu, a

[70] H. Rowland, "Modern Theories as to Electricity," *The Engineering Magazine*, 1895, *8*:589.

[71] H. Rowland, "The Highest Aim of the Physicist," *Am. J. Sci.*, 1899, *8*:403–405.

[72] *Ibid.*, p. 406.

[73] N. E. Gilbert, "Some Experiments upon the Relations between Aether, Matter, and Electricity," *Phil. Mag.*, 1902, *3*:361–380.

doctoral student in physics at the University of Paris. At first Crémieu had attempted to measure an inverse convection effect—that is, to observe the electrification of a body due to a changing magnetic force. Failing at this he had tried to repeat Rowland's experiment and was equally unsuccessful in several attempts.[74] Rowland reacted by setting one of his own students, Harold Pender (1879–1959), to work on a series of convection experiments with new apparatus which had been constructed in 1889 but never put to use.[75] New experiments were also undertaken at Harvard University, and while at the Cavendish Laboratory in England, J. J. Thomson's student Harold A. Wilson published a highly critical analysis of Crémieu's findings, noting that they conflicted with other important phenomena of which Rowland's convection effect afforded "a simple explanation." Wilson cited not only the action measured by Hall but also the deflection of cathode rays in a vacuum which had been reported by Thomson in 1897.[76] Both of these experiments, Wilson believed, substantiated the magnetic action of convected current.

Nonetheless, Fitzgerald, who had speculated on the effect six years earlier in correspondence with Rowland, was not as anxious as Wilson to write off Crémieu's work. In a paper to the British Association in 1900 he maintained that too little was known of the "theory of etheral effects of a charge of electricity forced to move by mechanical actions" to be certain that both Rowland's and Crémieu's observations were not correct.[77] Fitzgerald thought that the difference in experimental results might have something to do with Crémieu's method of measuring magnetic action. Instead of the standard astatic needle system which Rowland and his students used, Crémieu employed an induction coil and galvanometer to measure magnetic action from revolving disks which were alternately charged and discharged. The anticipated alternating magnetism arising from the varying electric charge was supposed to induce a current in the induction coil–galvanometer circuit. Fitzgerald postulated however that some interaction of the moving charge and the ether neutralized induction forces in Crémieu's coil, precluding any current indication of the galvanometer.

By Christmas of 1900 the newest of Rowland's double-disk apparatus was working in Baltimore and, meeting Fitzgerald's objection, it employed an induction coil rather than an astatic needle. Rowland could not directly oversee the experiments, for he had become seriously ill, indeed bedridden with diabetes, and never returned to the laboratory. After months of tests, Pender concluded that his work showed "beyond any doubt that *electric convection does produce magnetic action.*" The written account

[74] V. Crémieu, "Recherches sur l'existence du champ magnétique produit par le mouvement d'un corps électrisé," *Comptes rendus des Séances de l'Académie des Sciences, Paris*, 1900, *130*:1544–1549; "Recherches sur l'effet inverse du champ magnétique que devrait produire le mouvement d'un corps électrisé," *Compt. rend.*, 1900, *131*:578–581; "Sur les expériences de M. Rowland relatives à l'effet magnétique de la convection électrique," *Compt. rend.*, 1900, *131*:797–800; "Nouvelles recherches sur la convection électrique," *Compt. rend.*, 1901, *132*:327–330.

[75] This apparatus is extant in the Smithsonian Division of Electricity collection (Catalogue No. 318,514). H. Pender, "On the Magnetic Effect of Electrical Convection," *Phil. Mag.*, 1901, *2*:179–208.

[76] E. P. Adams, "On the Electromagnetic Effects of Moving Charged Spheres," *Phil. Mag.*, 1901, *12*:285–299; H. A. Wilson, "The Magnetic Effect of Electric Convection," *Journal de Physique*, 1901, *2*:319–320; and "On the Magnetic Effect of Electric Convection and on Rowland's and Crémieu's Experiments," *Phil. Mag.*, 1901, *2*:144–150.

[77] G. F. Fitzgerald, "Note on M. Crémieu's Experiment," in *Report of the Seventieth Meeting of the British Association, Bradford, 1900* (London: John Murray, 1901), p. 628.

of the experiments was not completed until June 10, 1901, but news of the action was communicated to Rowland shortly before his death on April 16, 1901.[78]

Crémieu did not give up easily, believing enough in his results to plan a new series of experiments to test for the possibility of open circuits, such as those postulated in the potential theories a quarter of a century earlier. At the same time he attributed Pender's results to electrical leakage and spurious magnetic interference from electric trolley lines which the latter had admitted ran near the Baltimore laboratory.[79] Pender responded by moving his entire apparatus into the countryside, twelve miles from the center of Baltimore and two miles from the nearest electric line. Under these conditions the system sensitivity was half again as great as it could be made in Baltimore, while much attention was given to the quality of insulation in various parts of the system. All values of the ratio of units from numerous trials in this work fell within 3 per cent of their arithmetic mean, which Pender reported as 3.00×10^{10} centimeters per second.[80]

In view of the repeated conflict of these experimental findings Crémieu reviewed the history and present state of the convection experiments in a paper of 1902. There he wrote, "La seule conclusion possible de l'ensemble de faits que je viens de résumer, c'est qu'il faudra encore beaucoup d'expériences. . . ."[81] At the suggestion of Sir William Thomson and under arrangements made by Jules Poincaré, just such a repetition of experiments was begun by Pender and Crémieu at the University of Paris in January 1903. Working side by side for three months the two physicists struggled to account for the constant disagreement of their results. Eventually further analysis of Crémieu's experimental schemes indicated that they always differed in at least one detail from the configurations used in Baltimore. Crémieu's moving charge surfaces and condensing plates were nearly always covered with a thin layer of some dielectric, usually with caoutchouc or India rubber, but often with sheets of mica.[82] Moreover, when used on Pender's apparatus, these coatings, especially mica, considerably diminished the previously observed magnetic effects. The results of one such series of measurements are given below.

	Galvanometer deflection, mm.
Disks bare, condensing plates bare	140
Disks covered with mica, condensing plates bare	100
Disks and condensing plates covered with mica	15

Although the dielectric clearly had the effect of reducing the deflection, the men did not have enough time remaining at their disposal in Paris to investigate the specific mechanism of this action, which was to remain mysterious. They were satisfied that the dielectric accounted for Crémieu's negative findings, but they remained noncommittal regarding the significance of their work to theories of electric action.

[78] H. Pender, "On the Magnetic Effect of Electrical Convection," *Phil. Mag.*, 1901, *2*:207–208.

[79] "Convection électrique et courante ouverts," *J. Phys.*, 1901, *10*:453–471; "Sur l'existence des courants ouverts," *Compt. rend.*, 1901, *132*: 453–471.

[80] H. Pender, "On the Magnetic Effect of Electrical Convection, II," *Physical Review*, 1902, *15*:299.

[81] "État actuel de la question de la convection électrique," *J. Phys.*, 1902, *1*:771.

[82] H. Pender and V. Crémieu, "On the Magnetic Effect of Electric Convection," *Phys. Rev.*, 1903, *17*:385–409.

> It is not for us to say if these magnetic effects are really due to electric convection in the sense in which Faraday and Maxwell understood this expression, nor to decide if they are in accord with the fundamental hypotheses of the present theories.[83]

As for Rowland, his premature death denied him the privilege of analysis of these experiments as well as an opportunity to search further for an interaction of electricity and an ether.

The convection experiment of 1876 and the work of Edwin Hall three years later marked the last productive exploitation of the electrical fluid concept. Such theories had been useful for more than a century and a half, yet Rowland himself had not remained unalterably attached to this model for electromagnetism and eventually had abandoned it in favor of electric interactions with an ether. Ether theories in turn provided the basis for new experimental configurations at Baltimore as well as a repetition of the convection experiment carried out successfully by Harold Pender near the end of the century.

Thus the involved theoretical context of Rowland's electromagnetic work is inescapable. Indeed, in view of his capacity to translate abstract theories and models of electromagnetic action into the concrete physics of the laboratory, Maxwell's doughty knight may have himself best described his place in nineteenth-century American physics when he wrote to Gilman from Berlin that summer of 1875: "I . . . believe I see more clearly the path in which I hope to excell in the future. It lies midway between the purely mathematical physicist and the purely experimental, and in a place where few are working."[84]

[83] *Ibid.*, p. 409.

[84] Rowland to Gilman, Aug. 14, 1875.

Rowland's Magnetic Analogy to Ohm's Law

*By John David Miller**

IN AN EARLIER PAPER I have described the work of the American physicist Henry Augustus Rowland (1848–1901) concerning the nature of electric currents.[1] The following analysis attempts to construct the hitherto-unknown account of Rowland's early experimental and theoretical researches in magnetism.[2] These investigations, carried out largely in his mid-twenties, again reveal his capacity to make productive use of models and analogies, elaborating them mathematically to describe physical phenomena. This skill was practiced by few of Rowland's American contemporaries.

The series of experiments which Rowland began in 1870 at his mother's Newark home had been designed originally to determine the distribution of magnetism in several iron and steel bars. He soon found it difficult, however, to interpret his measurements. In the early 1870s little was accurately known of the effects of various media and geometric configurations on the transmission of magnetic forces. By mid-century the experiments of investigators such as Harris, Sturgeon, Joule, Lenz, Jacobi, Dub, and Henry had drawn attention to several specific configurational factors affecting the behavior of electromagnets.[3] However, their experiments had produced no single model of magnetic action which simultaneously took the shape of the core, its material composition, and the arrangement of the energizing coil into account. In the winter of 1870, as an ardent home experimenter and recent graduate in civil engineering of Rensselaer Institute, Rowland was well aware of the researches of these established electricians.[4]

Received July 1973: revised/accepted Aug. 1974.

*School of Education and Lawrence Hall of Science, University of California, Berkeley, California 94720. A draft of this paper was presented as part of the Joseph Henry Symposium, Washington, D.C., Dec. 1972.

[1] "Rowland and the Nature of Electric Currents," *Isis*, 1972, *63*:5–27.

[2] Much of this account has been constructed from Rowland's laboratory records and correspondence in the Rowland manuscript collection at the Milton Eisenhower Library, The Johns Hopkins University. Unless otherwise noted, citations refer to this archive.

[3] W. S. Harris, "On the Power of Masses of Iron to Control the Attractive Force of a Magnet," *Philosophical Transactions of the Royal Society of London*, 1831, *121*:501–506; W. Sturgeon, *Scientific Researches* (London: Thomas Crompton, 1850), p. 300; J. Joule, *Scientific Papers* (London: Physical Society of London, 1884), p. 8; H. Lenz and K. Jacobi, "Ueber die Gesetze der Elektromagnete," *Annalen der Physik und Chemie*, 1839, *47*:225–270; C. Dub, "Anziehende Wirkung der Elektromagnete," *Ann. Phys. Chem.*, 1850, *81*:46–72; J. Henry, "On the Application of the Principle of the Galvanic Multiplier to Electromagnetic Power in Soft Iron, with a Small Galvanic Element," *American Journal of Science and Arts*, 1831, *19*:400–408.

[4] Notebook, 1868–1871 "[Student records kept while at Rensselaer Institute]," Vol. II, pp. 120,

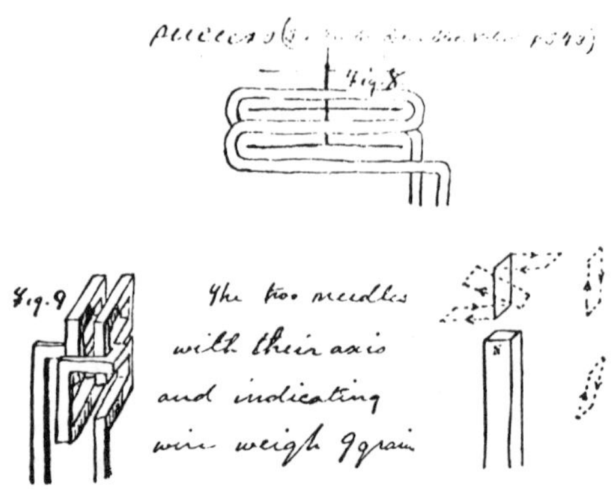

Figure 1a. Sketches of Rowland's copied from Michael Faraday's twenty-eighth series of researches.

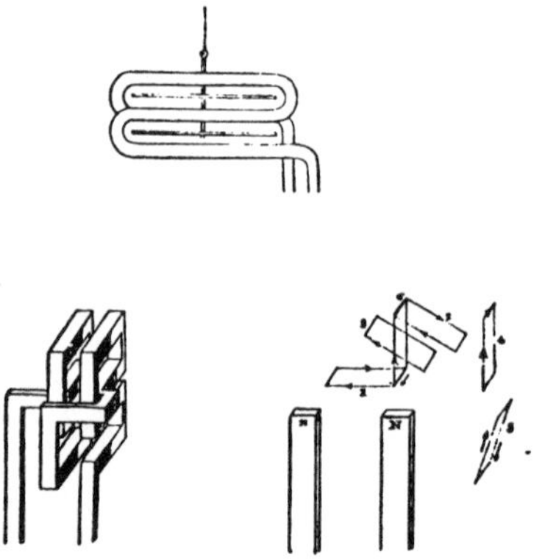

Figure 1b. Selected sketches from Michael Faraday's twenty-eighth series of researches.

His large and growing private collection of scientific texts, as listed in his laboratory notebooks,[5] provided a convenient resource, but he also had access to scientific periodicals in the library at West Point, where his uncle, John Forseyth (1810–1886), was Academy Chaplain.[6] Yet, in addition, Rowland emphasized, "I had also read Faraday's *Researches* which last, has been my guide ever since."[7] That Rowland was well acquainted with the sections of Michael Faraday's *Experimental Researches in Electricity* which dealt with magnetic action is evident because many of Faraday's concepts appeared in Rowland's notebook of 1868–1871, including "diamagnetism" (Vol. I, pp. 29–31), "induction" (Vol. I, pp. 34, 57–58), "electromagnetic analogies" (Vol. I, pp. 59), and "intensity of electricity and magnetism" (Vol. II, pp. 3–7).[8] Rowland had even copied in studious detail many of Faraday's drawings, notably those from Faraday's twenty-eighth series, "On lines of magnetic force; their definite character; and their distribution within a magnet and through space." A selection of Rowland's sketches, reproduced in Figure 1a, displays a striking resemblance to those of Faraday's, reproduced as Figure 1b. Later in the 1880s Rowland, as a mature and internationally recognized physicist, revealed publicly his near discipleship to Faraday. "Who will follow in his [Faraday's] footsteps and live such a life that the thought of it almost fills one with reverence?"[9] "We can

122. Some of Rowland's references to these researches are also contained in fragmented notes scattered in loose-leaf fashion throughout this volume.

[5] *Ibid.*, Vol. I, pp. 125–126.

[6] Mary Rowland to her brother, Nov. 18, 1872; Rowland to Maxwell [1875], Cambridge University Library, Cambridge, England. James Clerk Maxwell Papers.

[7] H. Rowland to E. C. Pickering, Apr. 19, 1875 (by permission of the Harvard College Library).

[8] Notebook, 1868–1871 "[Student records]," pp. 34–35; Michael Faraday, *Experimental Researches in Electricity*, Vol. III (London: Bernard Quaritch, 1850), pp. 336–350.

[9] H. Rowland, "Address as President of the Electrical Conference at Philadelphia," Sept. 8,

only receive with gratitude what Faraday has given freely to us, and speak his name with the reverence due."[10] Faraday was no ephemeral influence, particularly in shaping Rowland's view of magnetism.

In his experiments of 1870 Rowland attempted to make precise numerical measurements of the strength and direction of Faraday's lines of force in several kinds of electromagnets constructed with a variety of core metals. His chief sensor, which he had also constructed, was a tiny coil of wire connected to a delicate galvanometer. Through magnetic induction, glavanometer deflections of varying intensity occurred when the coil was moved suddenly from each test position.[11]

The difficulty Rowland experienced in interpreting these measurements was that the configuration of pattern of Faraday's lines appeared to shift as the current in the electromagnets was varied. Rowland had observed a similar phenomenon in 1868 when he recorded an account of pole shifting in a horseshoe magnet as an armature was placed between its ends.[12] In his experiments of 1870 the magnetic configuration shifted only a few percent as the current was varied. But this was enough to upset Rowland's venerated sense of precision and arouse his interest in seeking an exact mathematical description of the lines of force.

As a starting point Rowland turned to an analogy of magnetism which Faraday had provided in his *Researches.* As is well known, Faraday did not depend on explicit mathematical formalism in his theories, and Rowland, some years later, in discussing his own early magnetic studies, admitted, "When I wrote my first paper I knew very little about the mathematical *theory* of the subject."[13] However, it was not the formal language of mathematics with which Rowland was unfamiliar in 1870, for his mathematics courses at Rensselaer had taken him through the calculus of variations;[14] rather, he was ignorant of the *applications* of mathematics to magnetism.

The magnetic analogy Faraday described in his *Researches* began as follows:

> The magnet, with its surrounding sphondyloid of power, may be considered as analogous in its condition to a voltaic battery immersed in water or any other

1884 (*Report of the Conference*, p. 18, Washington D.C., 1886); reprinted in *The Physical Papers of Henry Augustus Rowland* (Baltimore: The Johns Hopkins Press, 1902), p. 626.

[10] H. Rowland, "The Electrical and Magnetic Discoveries of Faraday" (Address at the Opening of the Electrical Club House of New York City, 1888); *Electrical Review*, New York, Feb. 4, 1888; reprinted in *Physical Papers*, pp. 649–650.

[11] Notebook, 1870, 1871 "[Distribution of Magnetism: Studies of Magnetic Permeability]," p. 17. Rowland realized that the degree of galvanometer deflection for his nearly frictionless instruments was almost independent of the speed of test coil movement. This well-known ballistic technique depended on the sensor coil being displaced in a fraction of the natural period of the galvanometer and on the magnetic field being approximately uniform. (H. Rowland, "On Magnetic Permeability, and the Maximum of Magnetism of Iron, Steel, and Nickel," *Philosophical Magazine*, 1873, *46*:140–159; see p. 149.)

[12] Notebook, 1868–1871, Vol. I, p. 26.

[13] Rowland to Pickering, Apr. 19, 1875.

[14] Notebook, 1871–1875 "[Private Library Booklist]," p. 3. Rowland lists "Woodhouse, Calculus of Variations," presumably referring to Robert Woodhouse (1773–1827), *The Principles of Analytic Calculation* (Cambridge: Cambridge University Press, 1803).

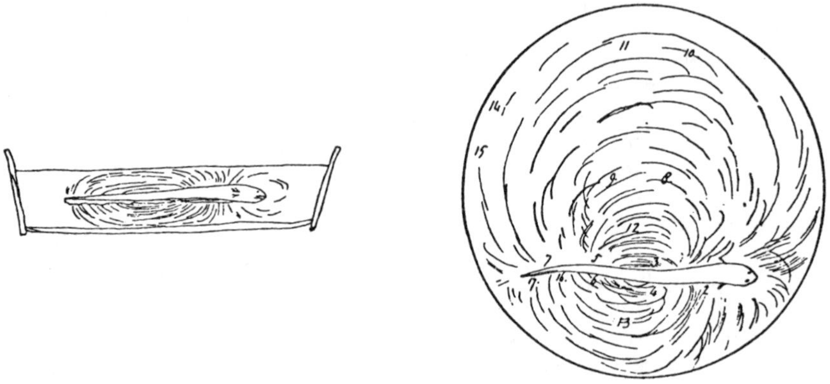

Figure 2. Michael Faraday's sketches of the electric action of an eel, 1838* (Faraday's Diary, *Vol. III, London: G. Bell and Sons, 1933, p. 354): "Now with reference to the current in the water all round the fish. Every part of the water is for the moment filled with its influence and conducts, and the fish causes on all sides of him, above and below him, a current of Electricity from his anterior to his posterior parts. The lines in the diagram may generally represent this, but those which would accurately represent lines of equal force would doubtless differ somewhat in form from them."

> electrolyte; or to a gymnotus or torpedo, . . . at the moment when these creatures, at their own will, fill the surround fluid with lines of electric force.[15]

During the fall of 1838 Faraday had carried out numerous investigations of the peculiar electrical action of *Gymnotus electricus,* an eel having electric organs distributed along its body and tail. Faraday had studied the distribution of electricity surrounding the eel at its moments of discharge, sketching the electric lines (reproduced in Fig. 2) while describing their action. After watching the eel bend itself into a semicircle to surround and electrocute small fish before consuming them, Faraday concluded, "a bent voltaic battery in its surrounding medium . . . or a gymnotus curved at the moment of its peculiar action . . . present exactly the like results."[16] He explained,

> I think the analogy with the voltaic battery so placed, is closer than with a case of static electric induction, because in the former instance, the physical lines of electric force may be traced both through the battery and its surrounding medium, for they form continuous curves like those I have imagined within and without the magnet.[17]

Faraday had coined the word *sphondyloid* to denote tubes of force or regions of power.[18] For example, when a magnet "in place of being a bar, is made

[15] Faraday, *Researches in Electricity,* Vol. III, p. 424.
[16] *Ibid.,* p. 428.
[17] *Ibid.,* p. 425.
[18] See *sphondyloid* in *A New English Dictionary,* ed. J. A. Murry (Oxford: Clarendon Press, 1888), p. 588.

into a horseshoe form, we see at once that the lines of force and sphondyloids are greatly distorted or removed from their former regularity (symmetry)." Thus, "A line of maximum force from pole to pole grows up as the horseshoe form is more completely given; . . . the power gathers in, or accumulates about this line, just because the badly conducting medium, i.e., the space or air between the poles is shortened."[19] Rowland made a note of this analogy in his notebook of 1868 in a section entitled "*Thoughts Suggested by the reading of the third volume of Far[aday's] Ex[perimental] Res[earches] in electricity,*" and he acknowledged later in a paper of 1875 that his own work could be "considered simply as a development of Faraday's idea of the analogy between a magnet and a voltaic battery immersed in water."[20]

In the same notebook Rowland had also recorded some experiments designed to test the advantages of the continuous geometries emphasized by Faraday. For instance, Rowland had observed a more than proportional increase in lifting power of a horseshoe magnet on its armature when the armature made contact with both poles rather than with a single pole.[21] In another investigation he aligned a pair of thin bar magnets vertically to the earth with mutually opposing magnetic poles and observed a greater attraction between one set of opposing poles when an armature was placed between the other set.[22] Finally, Rowland's notebook contained a galvanic analogy of Rowland's own design to explain this experiment.[23]

His ideas are represented in Figure 3, where two cups, A and B, containing slightly conductive liquid and strips of platinum, and two galvanic batteries, C and D, were arranged as shown. Rowland explained that "the liquid in A diminished the intensity of the electricity circulating in the wires and therefore there will be less intensity at B." But if the wires at A were joined, the intensity at B would be much increased. The open platinum plates were analogous to the poles of magnets, the magnets being represented by the batteries C and D. Connecting the platinum plates together at A was physically analogous to connecting one set of opposing poles of two bar magnets with an armature, yielding an increase in magnetic intensity at the other set, that is, at B in the analogy.

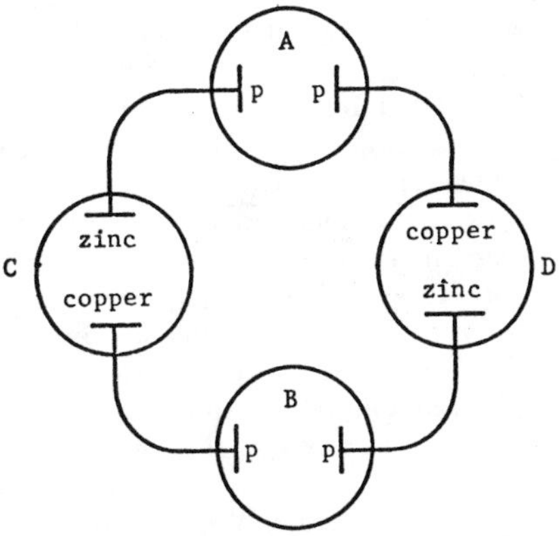

Figure 3. Schematic representation of Rowland's earliest galvanic-magnetic analogy.

[19] *Researches in Electricity,* Vol. III, p. 428.
[20] Notebook, 1868–1871, Vol. I, p. 59; H. Rowland, "Studies on Magnetic Distribution," *Phil. Mag.,* 1875, *50*:257.
[21] Notebook, 1868–1871, Vol. I, p. 69.
[22] *Ibid.,* Vol. II, p. 70.
[23] *Ibid.*

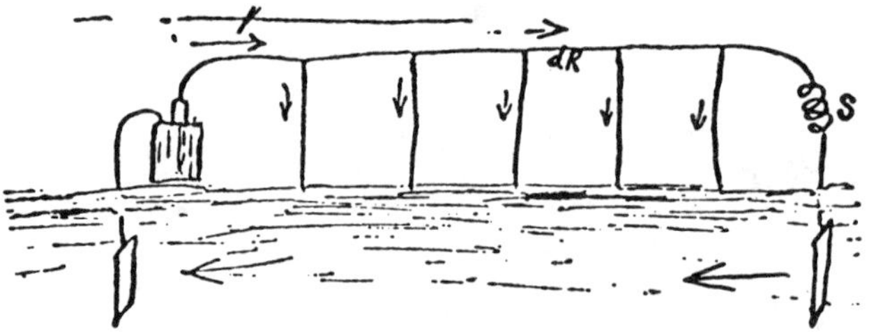

Figure 4. Faraday's galvanic analogy as interpreted by Rowland.

In his account of his magnetic researches published in 1873,[24] Rowland did not provide his derivation for translating Faraday's voltaic battery analogy into explicit mathematical form. However, he had recorded it in the early pages of a laboratory notebook of 1872–1876.[25] This work is entitled "Ohm's Law Applied to Magnetism" and is organized around a neatly drawn diagram with enumerated equations. However, since several complicated algebraic details of the derivation are not included, it suggests that this work was copied from an earlier, less polished study for the purpose of preserving a permanent reference record.

Rowland's derivation was divided into two parts. First, using Ohm's law he studied the circuit represented in Figure 4 to determine under this analogy how a single pair of hypothetical magnetic poles might distribute lines of force both through the magnet itself and through distributed elements of the medium surrounding it. Now for Faraday the lines of force in the medium surrounding a magnet were greatly affected by the medium's composition. Indeed, Faraday viewed this medium as "essential as the magnet itself, being a part of the true and complete magnet system."[26]

In Rowland's figure, the vertical elements represent the lines of magnetic force in the surrounding medium upon which Faraday laid so much stress. Rowland called the elements "insulators," giving them an electrical resistance per unit length corresponding to the imperfect insulators of a telegraph line.[27] The galvanic cell at the left of Figure 4 represents the source of the lines of force, while the horizontal path represents the conduction of the lines within the magnetic material itself. The coil *S* at the far right represents the resistance

[24] "On Magnetic Permeability" (cited in n. 11).

[25] Notebook, 1872–1876 "[Magnetic Theory]," pp. 31–37.

[26] *Researches in Electricity*, Vol. III, p. 426. Faraday used the term "magnetic conduction" to describe the aptitude of a medium to effect this relation. Later, in 1872, William Thomson suggested the noun "permeability," which was to be the term adopted by Rowland in his earliest paper on the subject (W. Thomson, *Reprints of Papers on Electrostatics and Magnetism*, London: Macmillan, 1872, p. 484). In 1882 Robert Holford Bosanquet (1841–1912), an English physicist, replaced Rowland's terminology by employing "magneto-motive force" to designate the force driving the lines of induction and "reluctance" to denote the resistance to these lines. (H. Bosanquet, "Preliminary Paper on an Uniform Rotation Machine and on the Theory of Electromagnetic Tuning Forks," *Proceedings of the Royal Society of London*, 1882, *34*:445–447.)

[27] Notebook, 1872–1876, p. 31.

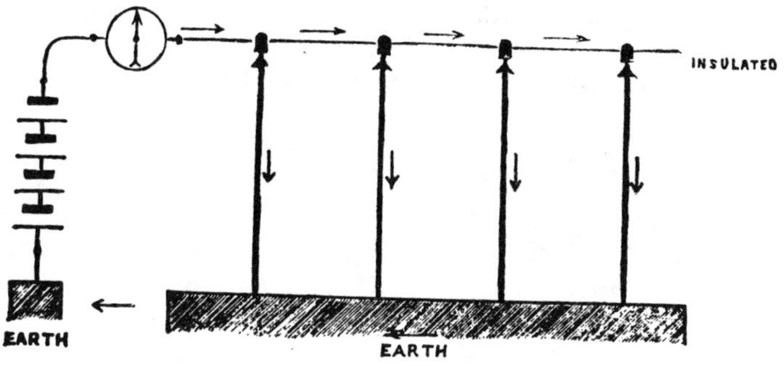

Figure 5. Drawing of a telegraph circuit appearing in The Telegraph Journal, *1873, No. 2, p. 32.*

to the lines of force issuing from the end of the magnetic pole. Rowland's circuit was new, but only in this special analogy. Figure 5 represents the circuit (without termination resistance) as it appeared in *The Telegraph Journal* in 1873.

In his mathematical analysis Rowland considered an elemental section of the circuit of length dl, as represented in Figure 6. Parameters included R, the total horizontal resistance of the line; R', the total leakage resistance of the insulators; ρ, the composite resistance of these two beyond a given point; and S, the terminal resistance represented by the coil S at the right of Figure 4.

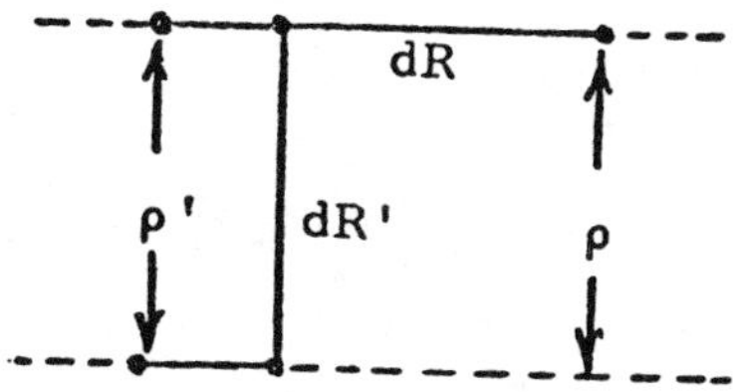

Figure 6. An element of Rowland's Ohmic analogy.

From Ohm's law[28] he wrote:

$$\rho' = \frac{1}{\dfrac{1}{\rho + dR} + \dfrac{1}{dR'}}$$

Since in practice dR' was much greater than dR,

$$\rho' = (\rho + dR)\left(1 - \frac{\rho + dR}{dR'}\right)$$

[28]George Simon Ohm (1787–1854) had described the flow of electricity in a wire as being proportional to the tension (*Spannung*) and inversely proportional to the resistance of the wire. G. Ohm, "Die galvanishe Kette," mathematisch Bearbeitet (Berlin: T. H. Riemann, 1827), p. 50. Although greeted with ridicule when first published, the law was widely accepted by electricians of the 1870s.

with Rowland not bothering to use an approximation sign. By using the relations $dR' = R'/dl$ and $dR = Rdl$ (R' and R are obviously not dimensionally equivalent), and by neglecting second-order terms, he arrived at

$$\rho' - \rho = d\rho = \frac{R'R - \rho^2}{R'} dl$$

The integration of this equation,[29] followed by a series of algebraic manipulations, produced a formula which gave ρ as a function of position on the line, the end resistance, and the line resistance parameters, that is $\rho = \rho(l, R, R', S)$.

With the distribution of line resistance for a single voltaic element established, Rowland imagined a whole series of such elements representing a string of small bar magnets placed end to end (with alternating poles) in the fashion

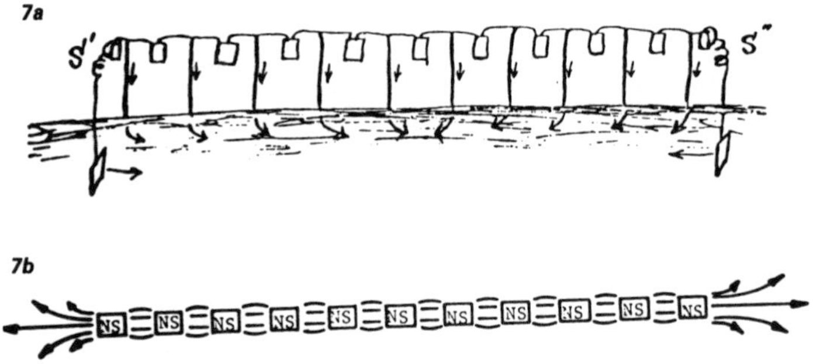

Figure 7a. The extended interpretation of Faraday's analogy as reproduced from Rowland's notebook. 7b. A magnetic interpretation.

shown in Figure 7.[30] This is an extension of the scheme shown in Figure 4 and symbolizes a long, thin bar magnet. The galvanic cells are seen in Figure 7a to be deployed along the telegraph line in a fashion suggestive of the distribution of the electric organs in Faraday's *gymnotus*. Summing the effects of the electric driving force associated with each cell and employing the function previously derived for ρ, Rowland obtained equations which described the distribution of Faraday's lines of force in and near a long, thin magnetic bar. The individual driving force of each cell corresponded in the analogy to the magnetic force issuing from the poles of the small hypothetical magnets represented in Figure 7b. The function describing ρ accounted for the aggregate resistance to the lines of force combined with resistances from within and without the magnet.

[29] Notebook, 1872–1876, p. 31.
[30] *Ibid.*, p. 33.

Rowland's general equations as published in 1873[31] were

$$Q_\epsilon = \frac{M1}{2\sqrt{RR'}} \frac{1 - A}{(A\epsilon^{rb} - 1)} (\epsilon^{rx} - \epsilon^{r(b-x)})$$

$$Q' = \frac{\epsilon^{rb} - 1}{(A\epsilon^{rb} - 1)(\sqrt{RR'} - s')} \frac{M}{r} - \frac{M}{2R} \frac{1 - A}{A\epsilon^{rb} - 1} \times (\epsilon^{rb} + 1 - \epsilon^{rx} - \epsilon^{r(b-x)})$$

in which

$$r = \sqrt{\frac{R}{R'}}$$

and

$$A = \frac{\sqrt{RR'} + s'}{\sqrt{RR'} - s'}$$

where

R = resistance to lines of force of 1 meter of length of bar
R' = resistance of medium along 1 meter of length of bar
Q' = lines of force *in* bar at any point
Q_ϵ = lines of force passing from bar along small distance I
ϵ = base of Napierian system of logarithms
x = distance from one end of helix
b = total length of helix
s' = resistance at end of helix of the rest of bar and medium
M = magnetizing force of helix

Now these complex exponential forms were obviously far from the simple proportionality law of Ohm. But Rowland went on to study the predictions of his equations for the center of a long, thin magnet. Here, from his observations of the lines in the media surrounding the magnet and from symmetry considerations, he expected the lines to assume homogeneous paths and his equations to reflect this simplification.[32] Rowland did not publish the details of the algebraic transformation of these equations under these limiting conditions. But clearly at the center of a magnet of any length $x = b/2$ the number of lines Q_ϵ passing along the bar at some small distance vanishes, since $\epsilon^{rb/2} - e^{r(b-b/2)} = 0$. But the number of lines in the bar, Q', does not vanish, having at this position the value

$$Q' = \frac{1 - \epsilon^{-rb}}{(A - \epsilon^{-rb})(\sqrt{RR'} - s')} \frac{M}{r} - \frac{M}{2R} \frac{1 - A}{\mathrm{A} - \epsilon^{-rb}} (\epsilon^{-rb} + 1 - 2^{-rb/2})$$

[31] Rowland, "On Magnetic Permeability," p. 145.
[32] *Ibid.*, p. 142.

For the case of the infinitely long bar, Rowland let $b \to \infty$. Then

$$\lim_{b\to\infty} Q' = \frac{M}{r(\sqrt{RR'} + s')} - \frac{M}{2R}\frac{\sqrt{RR'} - s'}{\sqrt{RR'} + s'} + \frac{M}{2R}$$

Collecting terms, this is

$$Q' = \frac{M(R + rs')}{Rr(\sqrt{RR'} + s')}$$

Which, upon substituting for r, readily reduces to

$$Q' = \frac{M}{R}$$

Thus in the center of the magnet the number of lines passing through the magnetic medium was proportional to M, the magnetizing force in the magnet, and inversely proportional to R, the resistance to these lines of force. He had therefore found a magnetic analogy to Ohm's law for electric circuits. Or, as Rowland himself wrote: "throughout their whole course they [the lines] obey a law similar to Ohm's law, and the number of lines passing in any direction between two points is equal to the difference of magnetic potential of those points divided by the resistance to the line."[33]

We recall that Rowland's original intention in his magnetic researches was to make numerical determinations of the intensity of Faraday's lines of force in various magnetic metals. The Ohmic analogy allowed him to study the changing permeability of different materials in a convenient fashion if they were forged in the shape of long, thin magnets. But this was difficult to do in the metal-working technology of the 1870s while maintaining a uniform composition in the material. Moreover, even when such bars could be constructed, their lengths necessarily remained finite, and hence a small error was always introduced into the computation of their permeabilities.

Rowland therefore devised the strategy of using ring or toroidal shaped magnets, which could be forged more uniformly and which conducted the lines of force uniformly, very closely approximating the conditions at the center of a long, thin bar magnet. This was the obvious limiting case in which Faraday's horseshoe magnets went over to the "more completely given" geometry of the circle. One such ring sketched by Rowland in January 1871 is reproduced in Figure 8.[34]

Through the expedient of the ring geometry and using precision galvanometers of his own construction, Rowland was able to chart in detail the permeability

[33] Rowland, "On Magnetic Distribution," p. 258.

[34] Notebook, 1870–1875, p. 17. Three or four months following Rowland's experiments of Jan. 1871 the ring configuration was also employed by Aleksander Grigorievich Stoletow (1839–1896), a professor of physics at the University of Moscow. Stoletow used a toroidal configuration to study what he called the "magnetizing-function" of iron. (A. Stoletow, "On the Magnetizing-Function of Soft Iron, Especially with Weaker Decomposing Powers," *Phil. Mag.*, 1873, *45*:40).

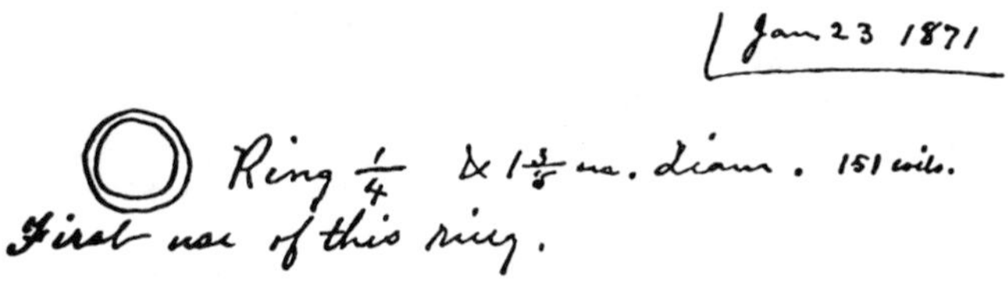

Figure 8. A magnetic ring tested in 1871 by Rowland.

of numerous magnetic materials. He plotted his data (magnetizing force as abscissa and resulting magnetization as ordinate) so that the variation in the permeability was apparent both by the concavity of the curve near its origin and by the leveling off of the magnetization for large magnetizing forces (saturation effect).[35] Summing up, Rowland characterized his work as the first "hitherto made on this subject in which the results are expressed and the reasoning carried out in the language of Faraday's theory of magnetic force."[36]

It was not until after mid-April 1873 that copies of John Clerk Maxwell's *Treatise on Electricity and Magnetism* arrived in America.[37] Rowland was refining and writing his past three years of work and was among the first to add a copy of Maxwell's *Treatise* to his scientific library.[38] As is well known, Maxwell's *Treatise* set forth, in formal mathematics, a comprehensive theory of magnetic action based on Faraday's lines of force. It appears that the major effect on Rowland of this late contact with Maxwell's work was to extend his awareness of the theoretical studies on magnetism undertaken by other physicists of the nineteenth century. Although he made no references to these studies in his notebooks of 1868–1872, his paper of 1873 cited those on magnetism carried out by such theorists as Simeon Poisson (1781–1840), Karl Neumann (1832–1895), and George Green (1793–1841).[39] Rowland admitted that he knew little of the mathematical theory of magnetism in 1870 and, because of his ignorance, had already invented his own units of measurement by the time Maxwell's *Treatise* arrived. He did not bother to rewrite his entire paper, however; he simply published conversion constants and equations relating his mathematical formulations to those of these theorists.[40]

Poisson and Neumann had calculated the permeabilities by analyzing the Newtonian forces between molecules subjected to magnetic influences. Rowland's computation, while molecular in the sense that it employed elemental galvanic cells, as we have seen, depended fundamentally on the laws of electrical circuits.

[35] Rowland, "On Magnetic Permeability," p. 158.
[36] *Ibid.*, pp. 142–143.
[37] J. Clerk Maxwell, *A Treatise on Electricity and Magnetism*, 2 vols. (Oxford: Clarendon Press, 1873).
[38] Scientific Books' bills of sale, 9 receipts, 1871–1873.
[39] Rowland, "On Magnetic Permeability," p. 141.
[40] *Ibid.*

This contrast illustrates the novelty of Rowland's use of Faraday's battery analogy.

In his *Treatise* Maxwell had postulated that an analogy existed between the conduction of lines of magnetic force and of electric current: "In isotropic media the magnetic induction depends on the magnetic force in a manner which exactly corresponds with that in which the electric current depends on the electromagnetic force." He saw this as representing Faraday's view that "Magnetic induction is a directed quantity of the nature of a flux and it satisfies the conditions of continuity as electric currents and other fluxes do."[41] Since both Maxwell and Rowland had begun with Faraday's analogy of electric continuity, one would expect considerable congruity in their equations. In fact, this was precisely what Rowland had demonstrated in his paper of 1873 by deriving his simplified formula, or Ohmic law, for computing permeability from Maxwell's general theory.[42]

Rowland's 1873 paper was the product of several earlier accounts of the experiments he had begun three years earlier. The editors of the *American Journal of Science* had repeatedly rejected these earlier papers, finally admitting that they did not understand Rowland's exacting mathematical methods.[43] Giving up on the *Journal,* Rowland sent the 1873 version of his work directly to Maxwell himself, who quickly recognized the validity of Rowland's analogy and definitions in magnetic conduction paralleling Faraday's hypotheses. Maxwell arranged for the immediate publication of Rowland's work in *The Philosophical Magazine.*[44]

Later, in 1875, Rowland became despondent over the lack of laboratory facilities for carrying out exact experimental research at Rensselaer Institute, where he had been teaching for three years.[45] That year he met Daniel Coit Gilman, who was visiting Peter S. Michie, professor of astronomy at West Point and a neighbor of Rowland's uncle. Gilman saw Maxwell's letters to Rowland, thought them "worth more than a whole stack of recommendations," and hired the young Rensselaer engineer to organize the physics department in the newly endowed Johns Hopkins University.[46] There models and analogies maintained visibility throughout the next two decades, notably in the work of Rowland's student Edwin Hall, in Rowland's continuing convection experiments, and in his unsuccessful searches for electromagnetic action of an ether.

[41] *Treatise on Electricity,* Vol. II, p. 51.
[42] "On Magnetic Permeability," p. 145.
[43] James D. Dana to Rowland, May 18 and July 24, 1871; June 20, 1873.
[44] Maxwell to Rowland, July 9, 1873.
[45] Rowland to his mother, Feb. 23, Mar. 22, 1875; Rowland to E. C. Pickering, Apr. 19, 1875.
[46] Peter S. Michie to Rowland, June 1875.

The Genesis of American Neo-Lamarckism

By Edward J. Pfeifer *

TODAY LITTLE IS heard of the neo-Lamarckian evolutionists who flourished in the United States toward the end of the last century.[1] This neglect results largely from the shortcomings of neo-Lamarckism itself which faded in the twentieth century as a viable form of evolution. The neo-Lamarckians nevertheless were of major importance in American science. Their researches enlarged biology, paleontology, and geology, and they did make a respectable contribution to evolutionary theory.

The first significant expression of American neo-Lamarckism occurred at the Boston Society of Natural History one evening in 1866, when Alpheus Hyatt, Louis Agassiz's brilliant student, read his paper, "On the Parallelism between the Different Stages of Life in the Individual and Those in the Entire Group of the Molluscous Order Tetrabranchiata." [2] In this formidably titled work Hyatt argued that genera of fossil shellfish had passed through stages corresponding to those in a single life. Generic youth was shown by the multiplication of subordinate species and by the richness of their ornamentation. After the full flowering of maturity, generic old age was indicated by a decline in numbers of species and ultimately by the contortions of shells, which foretold extinction.

Hyatt attributed this cycle to a process of acceleration and retardation. By acceleration he meant that characteristics belonging to adults of a maturing species became embryonic in the next higher species. Therefore later groups enjoyed more advanced traits at an earlier age and were free to make still other gains, which then became hereditary. When the vital powers of the group began to wane, the process reversed itself and increasingly degraded characteristics were inherited until the genus passed away completely.

* Saint Michael's College, Winooski, Vermont.

[1] Neo-Lamarckism is discussed briefly by Stow Persons in his *American Minds, A History of Ideas* (New York: Holt, 1958), pp. 242–244. See also *Evolutionary Thought in America*, ed. Stow Persons (New Haven: Yale University Press, 1950), especially Robert Scoon, "The Rise and Impact of Evolutionary Ideas," pp. 4–42. Alpheus S. Packard, Jr., devoted a chapter to neo-Lamarckism in his *Lamarck, the Founder of Evolution: His Life and Work* (New York: Longmans, Green and Co., 1901). Additional information will be found in Philip Fothergill, *Historical Aspects of Organic Evolution* (London: Hollis and Carter, 1952), pp. 160–166. See also George W. Stocking, Jr., "Lamarckianism in American Social Science: 1890–1915," *Journal of the History of Ideas*, 1962, *23*:239–256.

[2] *Memoirs of the Boston Society of Natural History*, 1866–1867, *I*, Part 1:193–209.

Hyatt was thus the first of Agassiz's students to identify himself as an evolutionist. Painful as this defection must have been to the master, Agassiz might yet see his own influence in Hyatt's work. By comparing a fossil series with a single life Hyatt followed Agassiz's procedure in the embryonic recapitulation theory that he taught. Moreover, Agassiz, apparently following the French paleontologist Charles d'Orbigny, had once remarked that the contorted shells of fossil ammonites conveyed the impression of a death struggle. Edward S. Morse, one of Hyatt's fellow students, thought that Hyatt began these investigations because he was unable to see that mollusks follow a single structural pattern as Agassiz insisted.[3] In so doing, Hyatt was true to his teacher, though he did fall into heresy. Hyatt himself said that he became an evolutionist and admirer of Lamarck during his first year with Agassiz, an act of some independence in view of Agassiz's characterization of Lamarck as "an absurd egoist."[4]

While Hyatt was formulating these views, Edward Drinker Cope, the equally brilliant young paleontologist from Pennsylvania, was developing similar ones independently. From researches on fossil amphibians, completed in 1865 but published later, he too concluded that acceleration and retardation determined the course of evolution.[5] Thus the neo-Lamarckians, conceding nothing to Darwinians, could boast that their theory had its own coincidence at birth.

From these beginnings neo-Lamarckism spread so rapidly that within fifteen years a number of distinguished naturalists professed this view of evolution. The leaders remained Hyatt and Cope, their continued work in paleontology gaining distinction for American science. The entomologist Alpheus S. Packard, Jr., Hyatt's fellow student and longtime professor of biology at Brown University, assumed a leading role too and became the group's most ardent publicist. Among the lesser figures were the paleontologist and explorer William H. Dall, the Philadelphia botanist Thomas Meehan, and the ornithologist Joel A. Allen. Among geologists Clarence King and Joseph Le Conte were to be counted as neo-Lamarckians. Eventually Cope's student Henry Fairfield Osborne took his place among the leaders. After 1867 they had the *American Naturalist* in which to publish their views, an opportunity which Hyatt, Packard, and Cope, as editors, exploited fully.

In essence, neo-Lamarckism was an attack upon natural selection as the primary factor in evolution. The Americans charged Darwin with exaggerating the efficacy of natural selection and overlooking the more fundamental question of why variations occurred at all. Natural selection could not originate anything, they insisted, and they pointed out that Darwin had

[3] *Proceedings of the Boston Society of Natural History*, 1901–1902, *30*:418.

[4] A. G. Mayer, "Alpheus Hyatt, 1838–1902," *Popular Science Monthly*, Feb., 1911, *78*:133.

[5] Edward Drinker Cope, in *The Primary Factors of Organic Evolution* (Chicago, 1896), p. 8, says that Hyatt's paper probably appeared shortly before his own first expression of these views at a meeting of the American Philosophical Society. Cope apparently expanded this paper and published it considerably later. For a convenient collection of Cope's early writings, see his *The Origin of the Fittest* (New York, 1887).

simply taken for granted the tendency to vary. Cope expressed these strictures clearly when he wrote:

> It has seemed to the author [Cope] so clear from the first as to require no demonstration, that natural selection includes no *actively* progressive principle whatever; that it must first wait for the development of variation, and then, after securing the survival of the best, wait again for the best to project its own variations for selection.[6]

Hyatt had similar reservations. He voiced them at the Boston Society in reference to a new species of deer, with spike horns rather than antlers, supposedly identified in the Adirondacks. Assuming that such a species actually existed and that natural selection had shaped the new head gear, Hyatt yet denied that it could account for the initial appearance of horns of any shape.[7]

At the same time Hyatt voiced his dissatisfaction with Saint George Mivart, the English naturalist, whose recent *Genesis of Species* (1871) was a widely read attack upon natural selection. Hyatt, who was aware that American distrust of natural selection coincided with a similar feeling in England, felt that his own work and Cope's strengthened Mivart's case. He was irked that Mivart did not recognize their contribution. In 1867 Richard Owen had clarified his position toward Darwin by openly accepting evolution but with decided reservations about natural selection. About the same time Alfred R. Wallace, whose thoughts on natural selection commanded attention, expressed his doubts that it could account for certain human features. And in 1871 Mivart published the sharpest attack of all. The difference between the movements was that Darwin's British opposition did not construct a rival theory.

Even without these criticisms Darwin had known that his theory was incomplete without an explanation of why variations occur and had tried to fill the gap with his theory of pangenesis, offered in 1868.[8] He thus presented a decidedly Lamarckian doctrine, as if in anticipation of the neo-Lamarckians, who, not greatly impressed, continued to accuse him of assuming the tendency to vary.

The downgrading of natural selection led to the denial that evolution must proceed at a creeping Darwinian pace. Dall concluded that "missing links" should sometimes be attributed to evolution by "leaps, gaps, saltations, or whatever they may be called."[9] He was positive that "leaps" had occurred and that they would be accounted for when the process of evolution was fully understood. He drew the analogy of a puddle of water which appeared permanent but was continually exerting pressure on its mud barriers. These might eventually give way and free the water to run off into another puddle. In this analogy Dall displayed the neo-Lamarckian

[6] Cope, *The Origin of the Fittest*, p. 175.

[7] *Proc. Boston Soc. nat. Hist.*, 1870–1871, *14*:141–145.

[8] Charles Darwin, *The Variation of Animals and Plants under Domestication* (New York, 1868), pp. 428–483.

[9] William H. Dall, "On a Provisional Hypothesis of Saltatory Evolution," *American Naturalist*, March, 1877, *11*:135–137.

disposition to think of evolution as a struggle between forces promoting change and others maintaining equilibrium. As long as the forces remained in balance there would be organic stability; but when the evolutionary force had overcome restraints it might then produce rapid change before coming again into equilibrium. In Cope's writings this doctrine took the form that organisms might change genus without changing species. This would happen if generic traits were modified while specific ones were not. Evolutionary change would then be more abrupt than Darwin admitted. In support of this view Cope cited fossils which combined change in generic characters with constancy in specific ones and referred to animals that did the same in prenatal development. He was inclined to admit that natural selection governed the modification of species but he attributed greater changes to a more important factor.

The neo-Lamarckians believed that Othniel C. Marsh, who usually confined his attention to fossils, had in fact observed a salamander undergo a change of generic magnitude. Marsh had transported an axolotl from its habitat in the Rockies and was surprised that it quickly lost its gills and assumed the form of a mature salamander. To the neo-Lamarckians this was evidence that abrupt change did occur in nature.[10] Today however the axolotl is known as a larval stage which either passes its life and, demonstrating the phenomenon of neoteny, breeds in that condition or metamorphoses into an adult. Marsh had simply witnessed for the first time the second pattern of development.

By espousing saltatory evolution the neo-Lamarckians evaded the charge, often made against Darwinians, of requiring more time than Lord Kelvin's physics and astronomy allowed. Huxley in his famous New York lectures of 1876 had left the age of the earth for others to determine.[11] Biologists, he said, would go about their own business in confidence that evolution had operated within the time eventually granted. To this Packard retorted that Huxley would need no subterfuge if he would believe in rapid evolution.

In the campaign against natural selection the neo-Lamarckians construed the concept narrowly, as relating mostly to struggle among living forms. Since they regarded it as of secondary importance at best, they had to offer something in its place. The true cause of hereditary change, they thought, lay in the wider relation of organism to environment, as the axolotl seemed to teach. Allen held that geography and climate were the essential factors in modifying birds and tried to formulate laws governing change on this basis. Packard thought that creatures living in caves became blind through disuse of their eyes and then produced blind offspring. Cope offered the theory that environment governed animal movement, which led in turn to structural modification. Stimulated by Cope, John A. Ryder expressed

10 For Hyatt's discussion of the axolotl, see his "Modern Ideas of Derivation," *Amer. Nat.*, June, 1870, *4*:230–237. For contemporary views see Sherman C. Bishop, *Handbook of Salamanders* (Ithaca: Comstock Publishing Co., 1943), p. 12; also Gavin de Beer, *Embryos and Ancestors* (rev. ed.; Oxford: Clarendon Press, 1951), p. 53.

11 These lectures will be found in Thomas H. Huxley's *Science and Hebrew Tradition* (New York: D. Appleton, 1900), pp. 46–138.

similar views. Eventually both he and Cope tried to account for the evolution of teeth on these principles. From the foregoing it is obvious that neo-Lamarckian environmentalism was a very complicated relationship, which involved the action and reaction of organisms and environment, the habits and instincts of animals, use and disuse of organs, the struggle for food, and other conditions of existence.

As far as Cope was concerned, the environmental relationship produced change by relocating an organism's "growth force." [12] He did not pretend to understand this entity completely but felt that Joseph Henry had demonstrated its existence. Henry, experimenting with a potato, had compared the weight of its starch with the weight of its sprout. Since the sprout was lighter, Cope concluded that some of the starch was not converted into growth but had escaped as carbonic acid and water. Transformation of the starch, he thought, produced chemical force, "Chemism," which was in turn converted into growth force. Cope noted that some substances could change chemical force into light, heat, or electricity, but that only protoplasm, "the material out of which structures and tissues are made," could convert it into growth. Henry's experiment assumed greater significance to Cope than to anyone else, but then, there was only one Cope.

Cope reasoned that no animal could generate growth force beyond a fixed capacity. By using a particular member, however, the animal might concentrate the force therein, with division of cells the result. This part of the animal would then become enlarged, though another might be stunted. If the animal possessed no organ capable of development by use, effort would still draw the growth force and originate an entirely new structure, though this would appear only in the animal's offspring.

Characteristics acquired in this way would then become hereditary through the neo-Lamarckian process of acceleration. Acceleration simply meant that an animal's growth speeded up, so that at reproduction he passed along more advanced traits than his forebears bequeathed to him. Similarly, the neo-Lamarckians believed in retardation, by which an animal's growth would be slowed down with transmission of degraded characteristics the result. Hyatt, as has been seen, wove acceleration and retardation into his theory of generic youth and old age. Cope thought that acceleration proceeded according to exact or inexact parallelism. By exact parallelism he meant that certain genera were identical to embryonic stages of higher, related genera. In inexact parallelism there was no such identity. Cope attributed the phenomena of parallelism to unequal rates of acceleration and retardation. Should they proceed at one pace, he held, there would be no such thing as inexact parallelism.

When these ideas were woven together they formed a system more complete than Darwin's. Neo-Lamarckism supplied answers for those who demanded that evolutionists account for variation and for the gaps in the fossil record. Unfortunately the answers became untenable to later genera-

[12] Cope, *The Origin of the Fittest*, pp. 17–18.

tions of scientists. Modern genetics disproved the heredity of acquired characteristics and restored natural selection to favor. Radiology, by establishing the age of the earth, proved that evolution need not have moved by jumps. Though it might still have done so, most paleontologists and biologists do not now believe in saltatory evolution, nor in cyclical interpretations of the fossil record, nor in vital forces.[13]

To recount the neo-Lamarckians' failure, however, is to obscure two facts which enhance their stature. George Gaylord Simpson states that the neo-Lamarckian emphasis upon environmental factors was a major contribution to evolutionary thought.[14] One should also remember that the neo-Lamarckians were original researchers who developed their theories scientifically. Simpson also testifies to this fact. Even though he specifically rules against Hyatt's cyclical theory, he still concedes that Hyatt's work upon the ammonites did provide a basis for such a view.[15]

The neo-Lamarckians themselves claimed to have developed these ideas without being conscious disciples of Lamarck. Cope testified that as late as 1871 all he knew of Lamarck came from Darwin's preface to the *Origin* and from Chamber's *Encyclopaedia*.[16] Packard stated that Hyatt in his formative period had not read Lamarck either.[17] It is impossible, of course, to claim that Hyatt and Cope knew nothing of Lamarck in those early years. Hyatt had certainly learned of him from Agassiz, and Cope must have acquired some of Lamarck's ideas even without reading his works. Nevertheless, the genesis of the term " neo-Lamarckism " makes clear that Cope and Hyatt had formed their theories before comprehending their relationship with Lamarck. From Packard's original estimate of Lamarck's hypothesis as being merely a non-Darwinian version of evolution [18] they moved to an awareness that they themselves entertained a " modified Lamarckian philosophy of animal differentiation." [19] Packard coined the term " neo-Lamarckianism " in 1884 to describe their view, though he ultimately recommended the shorter form of the word.[20]

For another reason it is likewise clear that neo-Lamarckism was not a conscious discipleship. The conclusion is inescapable that the strongest influence on the American neo-Lamarckians was Agassiz. Hyatt, Packard, Dall, and Le Conte, whose address at the 1877 meeting of the National Academy of Sciences was a high spot of neo-Lamarckism, all studied under Agassiz. It was from the museum at Cambridge that Hyatt and Packard, along with Morse and Frederick W. Putnam, went to the Peabody Institute at Salem, where they began to publish the *American Naturalist*. Le Conte even claimed that his master by stressing the progressive, though interrupted, char-

13 For a contemporary evaluation of neo-Lamarckian principles, see George Gaylord Simpson, *The Major Features of Evolution* (New York: Columbia University Press, 1953), pp. 266–270, 291–293, 359–376.

14 *Ibid.*, p. 266.

15 *Ibid.*, pp. 291–292.

16 Cope, *The Origin of the Fittest*, p. 423.

17 Packard, *op. cit.*, p. 386.

18 " 'The Development of Chloeon (Ephemera) Dimidiatum.' By Sir John Lubbock," *Amer. Nat.*, Oct., 1867, *1*:428–431.

19 John A. Ryder, " The Gemmule vs. the Plastidule as the Ultimate Physical Unit of Living Matter," *Amer. Nat.*, Jan., 1879, *13*: 12–20.

20 Packard, *op. cit.*, pp. 396–397.

acter of the fossil record was actually the father of modern theories of evolution — a statement which applied naturally to Le Conte's own neo-Lamarckism.[21]

Agassiz's influence was also apparent from attempts to reconcile his doctrines with Darwinism. The first would-be mediator was Theophilus Parsons, professor of law at Harvard. He suggested, during the debates over Darwinism at the American Academy of Arts and Sciences in 1860, that variations need not always be minute. Another who tried to effect a compromise along the same lines was Thomas Meehan, the Philadelphia botanist, who was continually publishing facts which bore on evolution. Sometimes these told in Darwin's favor, sometimes not. Although Meehan had gone along with the majority of naturalists in accepting evolution, his activity caused doubt as to his opinions. In 1876, after a paper at the American Association for the Advancement of Science, he found it necessary to set the record straight. In a letter to the *Popular Science Monthly,* he protested that he was indeed an evolutionist, despite the contrary reputation that he seemed to be acquiring.[22] His paper was supposed to show that evolution sometimes took place by leaps and that the search for missing links was futile. Meehan thought that he thus provided " a plank " on which Agassiz and Darwin might stand together. As Meehan shows, the beauty of neo-Lamarckism to those still swayed by Agassiz was that it would interpret his separate creations as rapid evolution. Then Agassiz's name would be even more glorious, and his synthetic forms and prophetic types could be granted their true significance.

Although the American neo-Lamarckians developed independently, once they saw that Lamarck had anticipated them, they publicized him as a patron saint. Cope discussed the relationship in his *Primary Factors of Organic Evolution* and paid tribute to Lamarck as the first advocate of evolution. Packard, sharing this interest, took leave from Brown to write Lamarck's biography, a chapter of which dealt with neo-Lamarckism. In 1888 the *American Naturalist* printed in translation the seventh chapter of the *Philosophie zoologique,* from which anyone could see that Lamarck had stressed environmental relationships and the inheritance of acquired characteristics.[23] These concepts became so widely accepted that by 1880 neo-Lamarckian views possibly predominated over Darwinian ones among American biologists. As Darwin's own outlook in successive editions of the

21 Joseph Le Conte, " Man's Place in Nature," *Princeton Review,* Nov., 1878, *2,* 4th Series:776–803. Le Conte made the same claim in other places too. Edward Lurie in *Louis Agassiz, A Life in Science* (Chicago: University of Chicago Press, 1960), p. 287, denies the validity of Le Conte's contention, at least as based upon Agassiz's work in embryology. On Agassiz's relation to evolutionary thought, one should also see Ernst Mayr, " Agassiz, Darwin, and Evolution," *Harvard Library Bulletin,* 1959, *13*:165–194.

22 Thomas Meehan, " Getting Right on the Record," *Pop. Sci. Mon.,* Nov., 1876, *10*:102–103. Meehan actually protested that he was a Darwinian. His views, however, certainly differed from Darwin's.

23 J. P. B. A. Lamarck, " On the Influence of Circumstances on the Action and Habits of Animals, and that of the Actions and Habits of Living Bodies, as Causes which Modify their Organization," *Amer. Nat.,* Nov., Dec., 1888, *22*:960–972, 1054–1066.

Origin had become increasingly Lamarckian, this American trend could almost be said to have his blessing. At least neo-Lamarckians were pleased to note the drift in Darwin's opinions.

They also spied a Lamarckian trend among European thinkers. Their most prominent authorities in this connection were Owen, Mivart, Ernst Haeckel, and Herbert Spencer. Owen, as Ryder pointed out, had long professed Lamarckian inclinations [24] and Mivart's attacks on natural selection had heartened neo-Lamarckian reviewers. Haeckel supported them too by emphasizing embryology as the key to understanding evolution. Ryder, however, regarded Spencer as offering the most comprehensive principle of evolution. And truly much in his *Principles of Biology* did square with neo-Lamarckism. He too saw organisms as subject to opposing forces promoting change and stability and his opinions on adaptation and the formation and development of organs were close to theirs. He also questioned the universal efficacy of natural selection.

There were still significant disagreements between Spencer and the neo-Lamarckians. Spencer's forces were physical and governed by the law of conservation of energy, which he continually discussed in relation to biology. The neo-Lamarckians on the other hand were unconcerned with physics and their vital principle was a more mystical, vitalistic concept. Furthermore Spencer was less skeptical than they about natural selection. He felt that at most a small group of phenomena escaped its governance, while the neo-Lamarckians expected it ultimately to assume secondary importance in the wider evolutionary theory which they believed was under formulation.

Ready as the neo-Lamarckians were to cite European thinkers, they received little notice in return. Hyatt's displeasure at Mivart indicated as much, and neither Haeckel nor Spencer cited them in the works which cheered. Darwin seems to have been long unaware of them too. At least when Packard visited Down, Asa Gray's introductory letter characterized him as one of the Hyatt-Cope school, who thought that they had an evolutionary law.[25] Whether this was news to Darwin or not, Darwin had the opportunity to learn about neo-Lamarckism from the pamphlet which Packard left behind. Darwin did not understand the pamphlet, he advised Gray. Perhaps he really meant that he disagreed, but Hyatt and Cope would probably accept the statement at face value. The *American Naturalist* at any rate complained editorially that Darwin misrepresented the doctrine of acceleration in the sixth edition of the *Origin*.[26]

Long afterward in the biography of Lamarck, Packard catalogued some forty European scientists as supporting the American school. Since Packard wrote to prove that Lamarck was truly the father of evolution, he was happy to show that Haeckel shared this opinion. Packard, however, tended to see in Lamarck's work only what would support his own positions. Though there was much of this, he consequently failed to analyze Lamarck's thought as searchingly as Charles C. Gillispie has done. Nor did Packard

[24] Ryder, *loc. cit.*

[25] *Letters of Asa Gray*, ed. Jane Loring Gray (Boston, 1893), Vol. 2, p. 624.

[26] "Error in Darwin's Origin of Species," *Amer. Nat.*, May, 1872, *6*:307–308.

come to grips with the considerations from which Professor Gillispie denies that Lamarck was philosophically a predecessor of Darwin.[27]

Packard also felt that Lamarck's attitude toward science and religion, had it prevailed in 1859, would have allayed theological controversy over Darwinism. This claim, however, is extremely doubtful, for the Deistic passages quoted, as replete with Lamarck's wisdom, resemble those which Darwin placed in the *Origin* in the futile hope of preventing strife.[28] Nor did Packard emphasize Lamarck's materialism, although Haeckel did so. Gillispie stresses this aspect of Lamarck too and points out that he provided no " escape into transcendentalism." [29]

Neo-Lamarckism in contrast did, as a cluster of scientific addresses delivered in the middle eighteen-seventies shows. The first of these was the work of John W. Dawson, principal of McGill and a prominent figure in Canadian science. He spoke strictly as a paleontologist, he said, to show why some students of fossils hesitated to accept the theory of evolution.[30] After stating a number of well-worn objections against Darwinism, he observed that a new group of laws was apparently breaking into the subject, among which were the major planks of neo-Lamarckism. Not that Dawson gave his allegiance to them. He merely introduced them as a hopeful sign that biology and theology might yet reconcile their differences. In his closing paragraphs he warned that life could not be explained on strictly material grounds.

Clearly Dawson had not surrendered to evolution in any form. Yet he would flirt with neo-Lamarckism and had already stated at McGill that Cope's was the most promising of these views, a fact which spoke volumes for the religious implications of neo-Lamarckism. There was no such ambiguity about the position of Le Conte, the California geologist whose address to the National Academy of Sciences demonstrated some of the same concerns and motivations that moved Dawson.[31] Le Conte spoke of critical periods in the history of the earth, one of which supposedly preceded each great division of geological time. During critical periods, Le Conte said, environmental change caused rapid organic change and produced living forms that could survive and perhaps change gradually until the next critical period, when again evolution might proceed " by a few decided steps." This was good neo-Lamarckism, of course, and Le Conte duly noted that he thus obviated appeal to the imperfection of the fossil record and demand for time that physicists would not grant. Ultimately Le Conte explained that the " Present " should be considered a major era separate from the Cenozoic with the Quaternary as its critical period and man as the distinctive form of life evolved therein.

Le Conte, like Dawson, was speaking as a scientist. He emphasized there-

[27] Charles C. Gillispie, " Lamarck and Darwin in the History of Science," in *Forerunners of Darwin: 1745–1859*, ed. Bentley Glass *et al.* (Baltimore: Johns Hopkins Press, 1959), pp. 265–291.

[28] Packard, *op. cit.*, p. 374.

[29] Gillispie, *op. cit.*, p. 275.

[30] " Address of J. W. Dawson," *Amer. Nat.*, Oct., 1875, *9*:529–522.

[31] Joseph Le Conte, " On Critical Periods in the History of the Earth and their Relation to Evolution: and on the Quaternary as such a Period," *American Journal of Science*, August, 1877, *114*:99–114.

fore the "geological importance of the appearance of man" as lying in human power to modify "the whole fauna and flora of the earth." But Le Conte referred in passing to man's "transcendent dignity" and thus gave *carte blanche* to theological interpretations of his paper. Not that Le Conte was likely to object, for he explained elsewhere that human nature resulted from the "introduction of some new principle characteristic of *humanity* as distinguished from *animality*, of *reason* as distinguished from *instinct*, of *spirit* as distinguished from *matter*."[32] He thus provided the escape which Gillispie shows that Lamarck did not.

A much more striking address but with marked similarities to Le Conte's was Clarence King's "Catastrophism and Evolution" delivered at Yale in 1877.[33] King, the spectacular young man who at twenty-four had led a major government survey, began by remarking that the earliest geological induction of man was the occurrence of terrestrial catastrophe. Though this doctrine had gone into disfavor, King was not at all sure that it was dead, since men were by nature either catastrophists or uniformitarians. Some "build arks straight through their natural lives, ready for the first sprinkle," while others do not "even own an umbrella." In geology each type simply gravitated into its appropriate school.

King reduced the difference between the two schools to a disagreement over the rate at which energy was expended. The same force stops a train, he noted, whether the engine coasts to a stop or crashes into an obstacle. He rejected "sweeping catastrophism" as an error of the past, but he also rejected "radical uniformitarianism," or the "harmless, undestructive rate of today prolonged backward into the deep past," as inadequate to account for some features of American geology. The canyons of the Cordilleras "could never have been carved by the pigmy rivers of this climate to the end of infinite time." Soon King charged the uniformitarians with accepting the crust changes to which he referred "with unruffled calmness" and scolded them for denying accelerated change in the past from the analogy of the present.

His words, coming from a scientist, must have been music to orthodox ears as he assailed the logic of the uniformitarians:

> In plain language, they start with a gratuitous assumption (vast time), fortify it by an analogy of unknown relevancy (the present rate), and serenely appeal to the absence of evidence against them as proof in their favor. The courage of opinion has rarely exceeded this specimen of logic. If such a piece of reasoning were uttered from a pulpit against evolution, biology would at once take to her favorite sport of knuckle-rapping the clergy in the manner we are all of us accustomed to witness.[34]

---

32 Joseph Le Conte, "Scientific Relation of Sociology to Biology," *Pop. Sci. Mon.*, Feb., 1879, *14*:430.

33 Clarence King, "Catastrophism and Evolution," *Amer. Nat.*, August, 1877, *11*:449–470. For recent works on King, see Thurman Wilkins, *Clarence King* (New York: Macmillan, 1958) and Richard A. Bartlett, *Great Surveys of the American West* (Norman: University of Oklahoma Press, 1962). An excellent treatment of King will also be found in William H. Jordy, *Henry Adams, Scientific Historian* (New Haven: Yale University Press, 1952).

34 King, *op. cit.*, p. 462.

Though King was speaking of geology, he was thus far in agreement with the neo-Lamarckians and their rapid evolutionary pace. And he soon turned his fire against the Darwinians by charging that the error of the uniformitarians had "been confidently built in as one of the corner-stones of the imposing structure of evolution." Biologists, he asserted, generally denied catastrophism to save evolution, but he felt that the crumbling nature of their uniformitarian foundation would cause the ruin of everything resting upon it. King obviously had no confidence in uniformitarian natural selection.

These observations provided King a transition to the causes of evolution. On this subject he felt that biologists were too concerned with the strictly biological environment and, therefore, "the favorite child, Natural Selection, has been fed into a plethoric, overgrown monster." And he took another slap at the logic which made man "what we are because this brain and this body form the most effective fighting-machine the dice box of ages has thrown." In spite of these criticisms, he drew encouragement from the fact that in America, "more than in Europe," biologists were beginning to study the physical environment as a factor in evolution. He stated his own belief that rapid geological change would prove to be the primary cause.

Accelerated change, he continued, would confront organisms with the alternatives of adapting or perishing. Plasticity would then be the key to survival. Natural selection might account for changes during uniformitarian times, but King held that the ages of greatest organic change coincided with ages of geological catastrophe. Then he turned his attention to "that strange procession of fossil horse skeletons, among whose captivating splint-bones and general anatomy may be described the profiles of Huxley and Marsh." Though they considered the bones proof of evolution, "the fossil jaws are utterly silent as to what the cause of the evolution may have been." King noted that he had studied the region from which these discoveries came. His own work proved that each new modification followed a geological catastrophe:

> And the almost universality of such coincidence is to my mind warrant for the anticipation that not very far in the future it may be seen that the evolution of environment has been the major cause of the evolution of life; that a mere Malthusian struggle was not the author and finisher of evolution; but that He who brought to bear that mysterious energy we call life upon primeval matter bestowed at the same time a power of development by change, arranging that the interaction of energy and matter which make up environment should, from time to time, burst in upon the current of life and sweep it onward and upward to ever higher and better manifestation. Moments of great catastrophe, thus translated into the language of life, become moments of creation, when out of plastic organisms something newer and nobler is called into being.[35]

---

[35] *Ibid.*, p. 470.

As clearly as King demonstrated the religious possibilities of neo-Lamarckism, he also illuminated scientific aspects of it. His assault upon uniformitarianism suggests the impulse behind his later study of the age of the earth, in which he out-Kelvined Kelvin by offering a figure of twenty-four million years.[36] King's antipathy to uniformitarians also made him critical of Lamarck, whom he placed by name among his target group. Although such behavior in a neo-Lamarckian may seem strange, in fact it was not. As Packard explained, Lamarck's major shortcoming as an evolutionist was his ignorance of the geological succession of living forms. Yet Packard was forgiving because only modern geology, he said, had shown that periods of geological crisis following periods of quiet had destroyed some living forms and had modified others.[37] King's criticism of Lamarck on this point was thus good neo-Lamarckism.

King's address also revealed something about tensions which had developed in science. He surely did *lese majesty* to Othniel C. Marsh, Yale's great paleontologist, whose star had risen on Huxley's praise and upon Huxley's use of Marsh's horse fossils in the New York lectures. By interpreting Marsh's fossils in a non-Darwinian fashion King had challenged the authority upon whose praise Marsh's fame had grown. And far from least, in view of Marsh's quick temper, King had in effect flaunted the banner of Cope, with whom Marsh was feuding, although not over evolution.

Thus the neo-Lamarckians continued to press their cause, but in doing so they displayed greater hostility toward Darwinians than to Darwin himself. Packard indeed closed the biography of Lamarck with praise of Darwin, who by "patience, industry, and rare genius for observation and experiment, and his powers of lucid exposition, convinced the world of the truth of evolution, with the result it has transformed the philosophy of our day."[38] Though differences might continue over the explanation of evolution, in the acceptance of the fact all were Darwinians, Packard said. This tribute was an act of saving grace for Packard, because by the time his book was published the rediscovery of Mendelian genetics, which would ultimately rehabilitate natural selection and discredit the key doctrines of neo-Lamarckism, had already occurred.

[36] Clarence King, "The Age of the Earth," *Amer. J. Sci.*, Jan., 1893, *145*:1–20.

[37] For contemporary attitudes toward geological change and evolution, see Simpson, *op. cit.*, pp. 241–244.

[38] Packard, *op. cit.*, p. 424.

Applied Microscopy and American Pork Diplomacy: Charles Wardell Stiles in Germany 1898-1899

*By James H. Cassedy**

AS THE NINETEENTH CENTURY drew to a close, the most pressing problem for the American Ambassador to Germany, Andrew D. White, was in large part a scientific matter. Not even the Spanish-American War nor the question of Samoa seemed to disturb United States-German relations as much as the problem of German restrictions on imports of American pork.[1] Despite their best efforts, the embassy's career diplomats had not been able to cope with the scientific rationale which the Germans used to justify these restrictions. And, from a distance of three thousand miles, experts in the Department of Agriculture were not in a position to help as much as was needed. What was required, White wrote to Secretary of State John Sherman, was that "the Agricultural Department send over here one or more thoroughly trained experts fit to deal with the whole subject, and especially accustomed to the use of the Microscope."[2] In a rapid response to White's suggestion, Agriculture Department officials agreed to detail Charles Wardell Stiles, a young zoologist, to work on the matter. The State Department in turn commissioned Stiles as Agricultural and Scientific Attaché to the United States Embassy in Berlin. Stiles reported there for duty in early April 1898.[3]

Stiles' assignment came at a stage of history when both the United States and Germany were thrusting with new-found strength, and sometimes arrogance, into the world arena. Diplomatically, the assignment was an episode in the growing economic

* History of Medicine Division, National Library of Medicine, Bethesda, Maryland 20014.

[1] In late 1898 White wrote to Washington that pork was "the most serious question visible above the horizon from this post at this time." White to Secretary of State John Hay, Oct. 29, 1898, Despatches from U.S. Ministers to German States and Germany, 1799–1906, Microcopy No. 44 of material in National Archives, Washington, Roll No. 83. Subsequent citations are to date and roll number only. The very large continuing volume of diplomatic correspondence from the Berlin embassy on this subject throughout 1898 and 1899 is additional evidence of its high priority.

[2] White to Secretary of State John Sherman, Dec. 27, 1897, Roll 83.

[3] White to Sherman, Apr. 12, 1898, Roll 84. Stiles lived from 1867 to 1941. The fullest account of his family background and early years is F. G. Brooks, "Charles Wardell Stiles, Intrepid Scientist," *Bios*, 1947, *18*: 139–169. General short summaries of his career include those in *National Cyclopaedia of American Biography*, Current Vol. D, 1934, pp. 62–63, and in *Who Was Who in America, 1897–1942*. There has been no full biographical study.

For general background see Thomas N. Bonner, *American Doctors and German Universities, A Chapter in International Intellectual*

confrontation of the two nations. Scientifically, it came at a time when America was just beginning to adopt the fruitful new laboratory techniques of the age of bacteriology, while the Germans were enjoying, sometimes flaunting, the undisputed eminence and scientific prestige which their brilliant uses of these techniques had brought them. Stiles' activities in Germany marked a new complexity in America's international scientific affairs as well as a new sophistication in handling them. Stiles' presence also constituted a somewhat audacious American challenge to the authority of a segment of the German biomedical establishment.

The United States had had a problem with its pork exports for nearly twenty years before Stiles arrived in Berlin. It was a problem born of affluence, to be sure, but nonetheless a problem. The vigorous post-Civil War growth in the American stock-raising and meat-packing industries had made the United States the world's leading exporter of pork and pork products by the 1870s. This success led in turn to protective measures by European countries whose own meat interests were affected. By 1881 all of the leading European importing nations—Great Britain, France, Greece, Turkey, Italy, Austria, and Germany—had placed restrictions or outright prohibitions on American pork.

Restrictions based on economic arguments were understandable and simple to deal with. Unfortunately, the occurrence of trichinosis in Europe at this time complicated the situation by making it also a scientific and sanitary matter. Many Europeans willingly blamed the presence of trichinosis upon the American pork, since restrictive measures could then be justified in the name of hygiene. The American government, they asserted, has shown itself to be impotent in ridding its exported pork of the parasites which caused the disease.

The Europeans had long since developed scientific controls as well as economic sanctions to defend themselves against trichinosis. The key measure was the microscopic inspection of pork—a procedure which had been introduced during the rapid expansion of laboratory methods across mid-nineteenth-century Europe. Some parts of Germany began using microscopes for pork inspection as early as 1863, and the practice was compulsory in Prussia by 1877. Compulsory microscopic inspection spread to other German states and other European countries soon afterward.[4]

Together the various measures effectively shut off the flow of most American pork to Europe throughout the 1880s. In response there was a spate of official American investigations and reports. Several of these mustered good scientific arguments against the European allegations, but to little effect.[5] While all of the restrictions hurt, the

---

Relations, 1870–1914 (Lincoln: Univ. of Nebraska Press, 1963); and William D. Foster, *A History of Parasitology* (Edinburgh/London: Livingstone, 1965), Chap. 5.

Despite Stiles' carefulness in zoological terminology and his accuracy as a bibliographer, he was oddly inconsistent in designating his own name on his scientific publications and correspondence. His two most common usages were C. W. Stiles and Ch. Wardell Stiles, but he also used Charles W. Stiles, C. Wardell Stiles, and Charles Wardell Stiles on occasion. To his friends and peers in the scientific world, however, he was and is known best by his unabbreviated full name.

[4] A general review of these and later developments is given in Ch. Wardell Stiles, *Trichinosis in Germany* (Washington: GPO, 1901), pp. 9–34, Bull. No. 30, Bureau of Animal Industry.

[5] Among the principal reports were one by a Marine Hospital Service official, W. C. W. Glazier, *Report on Trichinae and Trichinosis* (Washington: GPO, 1881), Exec. Doc. 9, 46th Cong., 3rd Sess., Senate 592; one conducted by the Department of State (Michael Scanlan, ed.), *American Pork: Result of an Investigation Made*

German ban was by far the most offensive to Americans. This was partly because the Germans were harsh in applying their restrictions, but also because the German market was one of the largest. Since the United States in turn had its own protective tariffs against some German goods, considerable ill feeling existed between the two countries throughout the 1880s, though it did not seem to affect the flow of American medical and science students to German universities. Despite the best efforts of diplomats during this time, American hog raisers and meat packers received no relief from the German restrictions until congressional legislation of 1890 and 1891. Laws passed at that time provided for the inspection of meats for export, including microscopic examination of pork for trichinosis. Most countries thereupon eased their restrictions. However, in the case of Bismarckian Germany, it took the added threat of an American duty on sugar before the Germans, in the Saratoga Agreement of August 1891, removed their ban on pork.[6]

Responsibility for organizing and carrying out the new microscopic inspection of exported pork was assigned to the Bureau of Animal Industry of the Department of Agriculture. By vigorous action the bureau recruited and put to work a corps of some two hundred microscopists by 1895.[7] Organization of such a force was no small task for any American agency in the early 1890s. Since there had not previously been any large-scale microscopy projects in the United States, relatively few trained operators were available.[8] The Department of Agriculture, however, with a substantial scientific establishment, was as well prepared as any agency to undertake this task. The Bureau of Animal Industry alone had such able scientists as Daniel E. Salmon, Theobald

---

under Authorization of the Department of State of the United States (Washington: GPO, 1881); and, the most thorough, one by a commission of scientists and businessmen appointed by President Chester A. Arthur. The scientific parts of the latter report were prepared by Daniel E. Salmon of the Department of Agriculture; Salmon circulated his findings in his article "Trichiniasis," in *First Annual Report of the Bureau of Animal Industry for the year 1884* (Washington: GPO, 1885), pp. 475–491.

[6] Microscopic inspection of American hogs began in June 1891, about three months after the authorizing legislation was passed. The first inspection took place at Chicago abattoirs of the Morris, Armour, and Swift packing companies. Representatives of several foreign governments observed the new inspection procedures. *Eighth and Ninth Annual Reports of the Bureau of Animal Industry, for 1891 and 1892* (Washington: GPO, 1893), pp. 39–40.

General brief accounts of the diplomatic aspects of the pork dispute are found in Alexander Deconde, *A History of American Foreign Policy* (New York: Scribner's, 1963), pp. 308–309, and Thomas A. Bailey, *A Diplomatic History of the American People* (7th ed., New York: Appleton-Century-Crofts, 1964), p. 419. For greater detail, see John L. Gignilliat, "Pigs, Politics, and Protection: The European Boycott of American Pork, 1879–1891," *Agricultural History*, 1961, *35*: 3–12, and Louis L. Snyder, "The American German Pork Dispute, 1879–1891," *Journal of Modern History*, 1945, *17*: 16–28.

[7] This number grew modestly to about 265 in 1900 and then fluctuated some from year to year depending on the export trade. See *Annual Reports of the Department of Agriculture* for the various years. The mushrooming in the use of microscopy in the Department as a whole during the 1880s and early 1890s was such that over 500 instruments were owned by the various divisions by 1895. This situation led the secretary that year to abolish the archaic Division of Microscopy, with its single microscope. See *Yearbook of the United States Department of Agriculture, 1895* (Washington: GPO, 1896), p. 57.

[8] City health departments were just beginning to establish their first laboratories, Providence with a laboratory for water analysis in 1888 and New York with its diagnostic laboratory in 1892. In many other respects American microscopy had advanced more rapidly in the post-Civil War years. This upsurge had, by the 1890s, seen the introduction of the microscope into many colleges and universities, the establishment and growth of native American optical firms, and the establishment of many local and national microscopic societies. For further details, see Samuel Henry Gage, "Microscopy in America (1830–1945)," *Transactions of the American Microscopical Society*, 1964, *83*, No. 4 (Supplement): 1–125.

Smith, Emile de Schweinitz, and Charles Wardell Stiles. For these men, the bureau provided a productive milieu for a wide range of research activities.

Stiles, as the bureau's zoologist, had no special connection with the routine aspects of the pork microscopic inspection system; rather, his duties covered a wide range of the bureau's practical concerns with animal parasites. Stiles had brought excellent credentials to his job. Prominent among these was a Leipzig Ph.D. gained in 1890 under Rudolf Leuckart, who was probably the most eminent helminthologist of the day.[9] Stiles had shown his competence in laboratory research early by correcting Leuckart's description of the Ascaris embryo.[10] Subsequently, after joining the Bureau of Animal Industry in 1891, Stiles spent much time in systematically working over the principal American helminthological collections, including those of the bureau, of the United States National Museum, and of the late Joseph Leidy, as well as new specimens sent in by the bureau's inspectors.[11] Some of Stiles' publications resulting from this work came to the attention of the American Microscopic Society in 1894. There, a reviewer, Henry B. Ward, applauded this "introduction of modern microscopical methods into another department of goverment work." Ward also encouraged microscopists to use the extensive card catalogue of helminthological literature that Stiles and his associate, Albert Hassall, had been developing.[12] However, throughout the 1890s, as Stiles observed, there were "but few men in America who are giving special attention to the subject of animal parasites."[13] This situation by itself gave Stiles a position of authority when the question of trichinae in America's exported pork products came up in the bureau.[14]

Authors of standard American diplomatic history texts normally terminate their accounts of the German-American pork dispute with the Saratoga Agreement and the

[9] Stiles earlier studied medicine and science at the University of Berlin under Hermann von Helmholtz in physics, Emil DuBois-Reymond in physiology, F. E. Schultze in zoology, and Heinrich W. Waldeyer in anatomy. After receiving his Ph.D. in 1890 he spent another year in scientific study at Robert Koch's laboratory in Berlin, at the Austrian Zoological Station in Trieste, at the Collège de France, and at the Pasteur Institute in Paris with Pasteur himself, Elie Metchnikov, Edouard Balbiani, and the parasitologist Raphael Blanchard. For details on Stiles' European student experiences, see Brooks, "Stiles," pp. 144–151.

[10] Leuckart and others had identified a certain appendage of the embryo as a perforated tooth. Stiles, using large microscopic magnification, found the appendage was instead the three lips of the adult Ascaris. Charles W. Stiles, "Notes sur les parasites. I. Sur la dent des embryons d'Ascaris," *Bulletin de la Société Zoologique de France*, 1891, *16*: 162–163.

[11] See, e.g., Ch. Wardell Stiles, "A Revision of the Adult Tapeworms of Hares and Rabbits," *Proceedings of the United States National Museum*, 1896, *19*: 145–235. Many of these studies were undertaken, as Stiles wrote, "entirely from a standpoint of economic and practical helminthology." Ch. Wardell Stiles and Albert Hassall, "Notes on Parasites, 48.— An Inventory of the Genera and Subgenera of the Trematode Family *Fasciolidae*," *Archives de Parasitologie*, 1898, *1*, No. 1: 81 ff.

[12] Henry B. Ward, "American Work on Cestodes," *Proceedings of the American Microscopical Society*, 1894, *15*: 183–188. Ward felt that two of Stiles' papers had made an "important advance in the morphological and systematic knowledge of the Cestoda. . . . They command attention not only since they emanate from an American worker, but also by reason of the methods employed." Ward was impressed that Stiles had been able to study at firsthand some material published by the zoologist Karl Rudolphi 85 years earlier and that he gave "the first accurate description of the latter which has been published."

[13] C. W. Stiles, "Notes on Parasites—No. 11. Distoma magnum Bassi, 1875," *Journal of Comparative Medicine and Veterinary Archives*, Aug. 1892, *13*: 466.

[14] During this same decade Stiles introduced medical zoology into the American medical school curriculum with courses at Georgetown University in 1892, the Army Medical School in 1894, and the Johns Hopkins University in 1897.

beginnings of America's microscopic inspection.[15] Actually the controversy continued nearly ten years longer. As America's pork exports revived in the early 1890s, many felt that American microscopic inspection would ensure the continuing untroubled expansion of the European market. Unfortunately, the Saratoga Agreement failed to reckon with the considerable autonomy of the individual German states, and the United States by 1897 found sale of its pork barred or harassed at this lower governmental level, where a host of local regulations had sprung up, notably to require the microscopic re-inspection of American pork. Fees required for such re-inspection naturally threatened to price United States pork out of the German market. American producers, officials, and diplomats were understandably upset by these new barriers, and some began to speak about retaliatory action.[16] They were also alarmed by renewed German claims that trichinosis was being traced to American pork, and they feared the loss of other European markets if the allegations could not be effectively refuted. Bureau of Animal Industry officials were particularly angered in this matter by the affront implicit in the Germans' rejection of the results of the new American microscopic inspection system.

Ambassador White realized that most of the allegations against American pork were simply part of a smokescreen raised by German meat interests and Agrarians to keep the cheaper American meat out of the country. However, the Agrarian propaganda was effective in poisoning the opinion of the public as well as the officials of the German Central Government. White was convinced that the United States Government would have to go to great lengths with the Germans to demonstrate its good faith before it could hope to gain official and public acceptance of American pork. The sending of a scientific attaché would be an important first step. For the long run, however, White thought that the United States would have to greatly enlarge its microscopic inspection system. Whatever its effectiveness, he wrote, the current American system did not inspire confidence abroad. Accordingly, it should "be conformed, just as far as possible, to the German [system], in its methods and in its extent." While the costs of such a system would obviously be considerable, "the expense of the largest force required and of the most perfect methods of inspection would be small compared with the ultimate gain."[17]

White's propositions met a mixed reception from Daniel E. Salmon, Chief of the Bureau of Animal Industry. Salmon had been closely involved with the pork problem ever since he investigated it in the early 1880s. He knew all the German arguments against United States pork. Along the way he had discovered, from bitter personal experience, that the German scientific establishment was made up of two highly disparate types of individuals:

> For the modest, untiring, devoted, generous German scientists, of whom there are so many, I have the greatest respect and admiration; but for that other class of critical, bumptious, arrogant, narrow-minded individuals who have made themselves offensive by

[15] See Deconde, *American Foreign Policy*, pp. 308–309, Bailey, *Diplomatic History*, p. 419.

[16] In 1895 the Secretary of Agriculture wrote: "Reciprocal certification of the chemical purity of wines exported from those countries to the United States may some time be demanded from the German and French Governments as a sanitary shield to American consumers, for certified American meats are as wholesome as foreign wines." *Yearbook of the United States Department of Agriculture, 1895* (Washington: GPO, 1896), p. 11.

[17] White to Sherman, Dec. 27, 1897, Roll 83.

> their effusive and unscrupulous efforts to convince the world that our inspection is a farce, and our science a delusion, I have nothing of a complimentary character to say.[18]

Salmon also knew something of what it would cost to emulate the German system of microscopic inspection. Prussia alone, with "the only really effective inspection" in Europe, had some 26,000 inspectors by the mid-1890s. This was greater than the entire enlisted ranks of the United States Army on the eve of the Spanish-American War. Salmon had come to believe, however, that the elaborate and expensive German system was not all that its advocates claimed.[19]

Salmon was not about to recommend such a system for the United States, but he needed more current information about its workings in order to effectively answer White and others who saw it as desirable. It was also essential to find out if there really were flaws in the American microscopic inspection and if trichinous American pork was being shipped to Europe after all. The time had come, he wrote, "first, to correct the error, if it existed, in the system of microscopic inspection conducted by the bureau, or, second, to deny authoritatively the charge."[20] Implicit in these matters was the need to justify a key assumption behind American scientific thinking on pork since the 1880s: that the curing process effectively neutralized any trichinae which may have been present in the meat. White's suggestion about an agricultural and scientific attaché thus presented Salmon with the opportunity both to obtain definite scientific information and to present the American scientific viewpoint on preventing trichinosis.

Stiles went to Berlin in early 1898 with the full confidence of Salmon and his other superiors in the Agriculture Department. He was familiar with the pork problem, since he had for some time past helped Salmon prepare technical data for White's diplomatic endeavors. At the same time Stiles had an easy entrée into the European zoological and biomedical community. As early as 1892 the French Société de Biologie had chosen him to succeed Leidy as its American correspondent, and he had similar connections with other European scientific societies.[21] For several years he had also been the United States delegate to the International Zoological Congress. Most important, he had stature as member and secretary of the newly organized five-man International Commission on Zoological Nomenclature. Stiles' appointment to this commission had been partly due to his talent as a mediator between rival European zoological factions.[22] It was just possible that some of this same kind of scientific diplomacy could help him influence the outcome of the pork dispute.

[18] [D.] E. Salmon, "United States Meat Inspection," *American Veterinary Review*, April-May 1895, *19*:31–42, 41. For a brief biography of Salmon, see the *Dictionary of American Biography*.

[19] Salmon's long and decisive role in American meat inspection procedures may be seen through the successive reports of the bureau. In particular, see D. E. Salmon, "Trichiniasis," pp. 475–477. See also D. E. Salmon, "The Federal Meat Inspection," *Yearbook of the U.S. Department of Agriculture, 1894* (Washington:GPO, 1895), pp. 67–80; and D. E. Salmon, *The United States Bureau of Animal Industry at the Turn of the Century* (Washington: Printed by the author, 1901), passim.

[20] D. E. Salmon, in *Annual Reports of the Department of Agriculture, for 1897–98* (Washington:GPO, 1898), pp. 190–191.

[21] These included the French Académie de Médicine (from 1896), and the Zoological Society of London (from 1899).

[22] Stiles had been found to be particularly effective in mediating between the temperamental French delegate, Raphael Blanchard, and the German, Julius V. Carus. See account in Brooks, "Stiles," pp. 167–168.

The sending of Agriculture Department employees abroad was not a new thing. The Bureau of Animal Industry had had three veterinary inspectors at British ports regularly since 1890 to examine American cattle for infectious diseases, originally pleuro-pneumonia. But these men were solely technicians.[23] Stiles' work in Germany was a good deal more, as department officials recognized. They soon found that having an "agricultural attaché" was highly useful to the department. In fact, the success of Stiles' assignment ultimately led the department to designate attachés to other countries and, in the 1930s, to operate its own Foreign Agricultural Service.[24]

Ambassador White, in contrast to the Agriculture officials, quickly dropped the word "agricultural" when referring to Stiles and consistently spoke of him as "our scientific attaché."[25] Perhaps White's terminology was not as accurate as Salmon's. Stiles did not, after all, concern himself with the whole spectrum of science in Germany. At the same time, his work involved more areas of science than those ordinarily associated with agriculture. Stiles' contacts in carrying out this mission were German pathologists, parasitologists, physicians, and sanitarians more often than veterinarians or agricultural officials. His sources of information were in the zoological, medical, public health, and vital statistics literature, more than in farm reports. Since Stiles' inquiry was related to several of the laboratory-oriented biological and medical sciences, White had good reason to use the term "scientific attaché." The nature of his work marks Stiles as a predecessor of the National Institutes of Health representatives who have served in Europe, Asia, and Latin America since the beginning of the 1960s. On the basis of Ambassador White's broader terminology, however, Stiles was also a forerunner of present-day science attachés in American embassies abroad.[26]

Designation of a science or "scientific" attaché at that particular time—1898—represented something of an innovation in America's handling of its international scientific affairs. As part of the growing maturity of United States society and the increasing vigor of its intellectual institutions, American science was rapidly expanding in size and complexity in the post-Civil War decades. Up to this point diplomatic activities which were scientific in nature had usually been taken care of by the ministers or consuls themselves, with occasional matters handled directly by Washington or by designated scientific organizations or individuals.[27] However, until late in the nine-

[23] These inspectors participated in post-mortem examinations of cattle, verified findings of any diseases reported, and forwarded positive findings to the United States to be traced to the sources. Negotiations leading to this activity had been handled in Great Britain by Minister Robert Todd Lincoln. Salmon himself went to England to help start the work. See *Sixth and Seventh Annual Reports of the Bureau of Animal Industry, for the Years 1889 and 1890* (Washington: GPO, 1891), pp. 70–71.

[24] The Foreign Agricultural Service was absorbed in 1939 in the regular Foreign Service, as was the Foreign Commerce Service, established in 1927. See William Barnes and John Heath Morgan, *The Foreign Service of the United States: Origins, Development, and Functions* (Washington: Department of State, 1961), pp. 72, 222.

[25] White to Hay, Apr. 22, 1899, Roll 88.

[26] Following Stiles' assignment, the next American use of the term "science attaché" appears to have been during World War I, when science missions with information gathering and exchange functions were sent to Paris and London. Employed and paid by the National Research Council, the professional persons attached to these missions were designated science attachés. The Paris mission, for example, included Dr. William F. Durand (pioneer in aerodynamic research), Dr. Roger Perkins (medical sciences), Mr. Warren Vinton (general science), and Dr. Carl Compton (physics). I am indebted to the late Mr. Vinton for personal recollections of this mission.

[27] No United States diplomatic posts had ambassadors before 1894, when the chiefs of mission to France, Germany, Great Britain, and Italy were elevated from ministers.

teenth century, the American government had but few substantial scientific problems or formal continuing scientific relationships with other countries. The main official science activities were those occasioned by participation in the work of *ad hoc* international commissions or by the need to have delegates at international congresses.

The lack of international scientific involvement was fortunate, for chiefs of mission who had scientific backgrounds or interests were few. Before 1800, it is true, the American diplomatic corps had included several representatives—Benjamin Franklin, Thomas Jefferson, Dr. Arthur Lee, John Quincy Adams, and Dr. Edward Stevens—who had medical competence or general interest in science. During the nineteenth century, the pioneer ecologist and etymologist George P. Marsh served some twenty-five years prior to 1881 as envoy to Turkey and Italy, while the chemist Henri Erni was United States consul in Basle during Grant's presidency, and the physician Nathaniel Niles served in various European posts from 1830 to 1850. But such men were exceptions. American diplomats then were drawn conspicuously from the ranks of poets, lawyers, educators, and historians. Andrew D. White was himself an historian and educator, but he also had a strong appreciation of the scientific ideas and forces of his time.[28] Now, as ambassador, he acted as something of a midwife in assisting one phase of America's emerging scientific internationalism.

White conceived of his new science attaché mainly as a trouble-shooter who would stand ready to rush with his microscope to any spot in Germany where cases of trichinous American pork might be reported in the Agrarian press. There Stiles would examine the charges, explode the unfounded allegations, and refer any well-grounded cases back to the United States for correction. Such action was intended to undercut Agrarian charges by reassuring the German people both of American good intentions and of the safety provided by the American system of microscopic inspection. The mission thus called for a great deal of travel around Germany.

Besides making microscopic spot-checks on these trips, Stiles found it essential to gather local published information and statistics. "How far do [the Agrarian] newspaper polemics find support in the medical statistics?" Stiles wanted to know, "or, in other words, how many cases of, and deaths from, trichinosis in Germany have been definitely traced to eating American pork?" Bound closely to these questions was the unpublicized element which Salmon had added to Stiles' mission—the evaluation of the German microscopic inspection system in action. Here Stiles needed to determine "whether it is a practical system in public hygiene, giving a protection proportionate to its cost."[29] Again, the answer was mainly to be found in the German medical literature and public health statistics. Under Germany's federal system, as in the United States, many of these kinds of sanitary data were to be found at the local level. Stiles' search for documents and data thus closely involved American consuls in the various German cities as well as the embassy staff itself.

Having obtained the science attaché he needed, White made Stiles an integral part

[28] Though White's *History of the Warfare of Science with Christendom* appeared only in 1896, he had been working on the subject since the 1870s. For a brief account of White's life, see *DAB*. See also the *Autobiography of Andrew Dickson White*, 2 vols. (New York: Century, 1905).

[29] White to Sherman, Dec. 27, 1897, Roll 83; Salmon, in *Annual Reports of the Department of Agriculture* (Washington: GPO, 1898), p. 191; and Stiles, *Trichinosis in Germany*, pp. 3, 12–13.

of his small staff and gave him his fullest backing.[30] Through the German Foreign Office, he obtained authorization for Stiles to communicate directly with officials of the Imperial Health Office, the Prussian Ministry of Agriculture, and other appropriate departments. As it became necessary, he obtained comparable authorization for Stiles to contact officials in the individual states, to visit slaughterhouses, to observe customs operations at various ports of entry, and to examine microscopic inspection procedures where they were conducted.[31] It was especially important, White wrote, that the Foreign Office "place such papers and general letters of introduction in the hands of Dr. Stiles, as will enable him to have free access to the records and material (original certificates, microscopic preparations, . . . , etc.) of slaughterhouses and custom houses, etc., in so far as will be necessary" in carrying out his investigation.[32]

The Imperial German Government was far from enthusiastic about having an American expert poking around in the German scientific and sanitary arrangements. It originally tried to discourage Stiles' appointment and, after his arrival, took its time in granting him the necessary access to officials and institutions. However, it could hardly refuse outright, since the Germans had their own scientific expert at the embassy in Washington.[33]

With Central Government policy supporting the Agrarians and public opinion opposing American intervention in Cuba, officials in the German Foreign Office continued their hard line on American pork imports for months after Stiles' arrival.[34] They sent the embassy periodic lists of trichinosis cases alleged to have come from American pork. They asserted, from their reading of American documents, that the Bureau of Animal Industry did not have enough inspectors and implied that some of those were being bribed. They even quoted from one of Stiles' own 1896 reports to try to prove that trichinosis was widespread in American hogs. As for American microscopy, it seemed to be beneath discussion. Besides, the Germans knew little about it.[35]

It took some time for Stiles to build up the information needed to reply to the German diplomatic notes.[36] Meanwhile, there was little likelihood that the United

[30] White's professional staff in 1898 consisted of the first secretary, second secretary, military attaché, naval attaché, and "scientific attaché." While Stiles enjoyed considerable independence of action, he had joint responsibility to the Ambassador and the Department of Agriculture. His letter of appointment stated: "You will perform, under the supervision of the Ambassador there, such services as may be assigned to you from time to time by the Secretary of Agriculture." Sherman to Stiles, Mar. 9, 1898, Diplomatic Instructions of the Department of State, 1801–1906, Microcopy 77 of material in National Archives, Roll 71, Germany.

[31] See, e.g., White to Secretary von Bülow, Apr. 28, 1898, U.S. Embassy, Berlin, Records, LII, Notes Feb. 26, 1898–Apr. 24, 1901, National Archives; Chargé d'affaires John B. Jackson to Privy Councillor von Deventhall, Sept. 9, 1898, *ibid.*; and Note Verbale, Embassy to Foreign Office, Feb. 4, 1899, *ibid.*

[32] White to Acting Secretary Baron von Richthofen, Jul. 18, 1898, *ibid.*

[33] Von Bülow to White, May 24, 1898 (translation), Roll 85, and White to Hay, Oct. 1, 1898, Roll 86.

[34] War was declared by Spain and the United States only about two weeks after Stiles reported for duty in Berlin. White noted, however, that throughout this period and later, "the Chancellor Prince Hohenlohe, and the Secretary of State for Foreign Affairs Mr. [later Count] von Bülow, are fair, just, and really anxious to maintain the best of relations with us." The Agrarians and allied interests were led, he wrote, by the Prussian Minister of the Interior, Baron von der Recke. White to Hay, Oct. 29, 1898, Roll 86.

[35] Von Bülow to White, May 24, 1898, Roll 85.

[36] The embassy's efforts for the American meat interests were not helped, White told Washington, by frequent "loose talk," sometimes by representatives of those interests themselves, about the presence of trichinae in American pork. White to Sherman, Apr. 7, 1898, Roll 84.

States would change its microscopic inspection system, and in any case not in time to affect German opinions of American pork in any near future. Stiles persuaded White that their overall case for American pork depended on vindicating the established American scientific rationale in the matter. This meant a defense of the cured pork that had been cleared under the present American microscopic inspection system. In order to hope to do this, the United States had first to ensure the closing of all loopholes through which uncured, uninspected, and otherwise unsafe American pork might enter Germany. This first involved making a tactical American retreat on some matters. As one step, White and Stiles suggested that American exporters completely stop their small exports of fresh uncured pork in refrigerator ships, since refrigeration was no guarantee against the later development of trichinosis. Cases of trichinosis arising from such shipments might well jeopardize the far larger market of cured pork.[37] The advice was subsequently followed.

A further desirable step, Stiles later thought, was to sacrifice part of the sausage trade, which had become the "weakest point" in their whole argument. The Germans claimed that "microscopic inspection of sausage for trichinae is unsatisfactory and even absurd." Stiles told White that this German argument "cannot be assailed." Hard sausages could be defended on scientific grounds, he noted, since research had shown that trichinae were killed in their processing; soft sausages, however, like fresh pork, might well harbor potent trichinae.[38]

An equally serious obstacle to proving America's scientific case appeared to lie in the malfunctioning of the German customs system. At some customs posts uninspected American pork had for some time been reported to be passing through with no formalities. At other places middlemen were suspected to have substituted uninspected pork for inspected American goods and passed it through under the official United States microscopic inspection stamps. Such occurrences, of course, compromised the integrity of the United States microscopic inspection system. Stiles investigated reports of all such incidents that came to his attention. This involved special trips, not only to points within Germany but to Swiss, French, Belgian, and other foreign border towns.[39] When he obtained proof of such frauds he notified White, who passed the findings on to the Foreign Office for correction.[40]

If it was important for Stiles to keep the customs officials and their practices in view, it was equally necessary for him to talk to German scientists. He knew that few of these men were swayed by the Agrarian propaganda against American pork. They were, however, capable of being persuaded by valid scientific arguments. Stiles was well aware that German investigations conducted during the past dozen years or so had

[37] White to Sherman, Apr. 18, 1898, Roll 84.

[38] White to Secretary of State William R. Day, June 24, 1898, Roll 85; and White to Hay, Jan. 25, 1899, Roll 87. The Agriculture Department did not go along with the suggestion to prohibit shipments of soft sausage. Hay to White, Jan. 24, 1899, Roll 87.

[39] Stiles' travels around Europe on a few occasions took on something of a cloak and dagger aspect. On one trip he learned about secret Russian plans to open up new lands, build new railroads, and build American-type (large-scale) slaughterhouses in order to develop a pork export industry challenging that of the United States in northern Europe. Stiles to Secretary of Agriculture, Jul. 30, 1899, typed letter, National Archives. This is apparently the only surviving Stiles manuscript among the Department of Agriculture material at the Archives in contrast with State Department material. See also White to Hay, Oct. 14, 1899, Roll 89, and Brooks, "Stiles," passim.

[40] See, e.g., Jackson to von Bülow, Dec. 20, 1898, and White to von Bülow, Feb. 4, 1899, Embassy Records, Berlin, LII, Notes.

already cleared American pork of any blame in certain of the German outbreaks of trichinosis. These included studies by such authorities as Rudolf Virchow, Carl Fraenkel, and Robert Ostertag, men whom Stiles characterized respectively as "the greatest pathologist Germany has ever produced, one of the leading sanitarians of Germany, and the recognized authority in Germany on meat inspection." Stiles exploited these authorities to the fullest in building up the embassy's case with German officialdom. White, in turn, fully recognized the need for Stiles to exchange opinions with such scientists. In at least one instance he had to ward off State Department criticism of Stiles' informal activities of this type outside the usual diplomatic channels: "Only in this way has he been able to accomplish as much as he has."[41]

Establishing the credibility of the American scientific argument with the German scientific community was essential. But ultimately, of course, the argument had to stand up in the diplomatic context. From the beginning White used Stiles' special knowledge in his own dealings with the Germans, and the embassy's notes to the Foreign Office on the pork situation came to be essentially the work of Stiles during this period. Stiles himself supplemented his own findings with additional data supplied by Salmon and others in Washington. From his various sources he worked up detailed replies to the official German arguments. By the fall of 1898 the scientific aspects of the situation were sufficiently clear to permit him to prepare a major note for White. This was sent to the Foreign Office in November.

The note was correct but forceful. With it American pork diplomacy passed from the defensive to the offensive. The note conceded not a single point to the Germans' earlier criticisms of American pork and American science; rather, with close logic and abundant facts, it established the points of error in the Germans' own case, particularly in their vaunted microscopic inspection system. "Germany's own scientific men" had failed to connect any trichinosis to United States pork. On the other hand, reports of these same scientists pointed up the many cases which had occurred in Germany during the years when American pork was excluded, 1883–1891, cases which had been traced to microscopically inspected and certified European pork.

The German microscopic system, it turned out, had been far from perfect during the previous fifteen years or so. Salmon had known since 1884 of many of its faults. These occurred partly because a substantial number (roughly three-fourths) of the inspectors were part-time and poorly trained. In the villages the inspectors often were barbers, tradesmen, or butchers themselves, had rough hands, and were generally unsuited for fine work. Official German reports through the 1880s and 1890s frequently complained of careless work done by such microscopists. Salmon had particularly noticed in these reports cases of instruments rendered useless by filthy lenses or rusty adjustment screws. Even where instruments were taken care of, he found that "the glass compressors for preparing the specimens of meat and the microscopes used in the German

[41] See, e.g., White to Hay, Jan. 25, 1899, Roll 87. Stiles also knew that such leading French scientists as Henri Bouley, Paul Brouardel, and Raphael Blanchard had found no evidence that European trichinosis cases had been caused by American pork. In fact, Stiles noted, the French restrictions of the 1880s against American pork had been made "in spite of the repeated reports of the French Academy of Medicine and the Consulting Committee of Public Hygiene, to the effect that such an interdiction was not justified by sound hygienic arguments." Ch. Wardell Stiles, "Trichinella Spiralis, Trichinosis, and Trichina-Inspection: a Zoologic Study in Public Hygiene," *Proceedings, Pathological Society of Philadelphia*, May 1901, N.S. *4*:137–153; p. 140. See also Jackson to von Richthofen, Nov. 10, 1898, Roll 86.

inspection were . . . clumsy and not adapted to accurate or rapid work." Practical Americans, already accustomed to demanding efficiency as well as organization in their own competitive society, were clearly not impressed by a scientific system that was large in size but short on quality and results. The facts in this case, they concluded, "show conclusively that the health of German citizens would be greatly improved by the complete substitution of American pork for that raised and microscopically inspected in Germany."[42]

The embassy's November note did not stop with pointing out the shortcomings of German pork microscopy; it also particularized the practical innovations and advantages of the American methods. United States pork microscopists, while relatively few in number, could do a large job effectively and rapidly. By contrast with the Germans, who were all men, American operators were young women chosen for their nimble fingers and dexterity. In addition, the Bureau of Animal Industry had had American-built microscopes especially adapted for the trichina inspection work. One innovation was the use of unusually broad vision lenses. In addition, the instruments were fitted with special compressors.

> [These] have been arranged to allow rapid work, and to permit a large surface of the specimen to be exhibited. The stage of the microscope has been grooved in such a manner that the distance between the grooves corresponds exactly with the field of vision given by the magnifying lenses, and the compressors are made to fit in these grooves, so that by advancing them one groove at a time it is impossible to miss any part of the specimen. With this apparatus an inspector can examine a given specimen in little more than half the time which the same inspector would require if he used the microscope and compressors adopted in Germany.[43]

The American note closed by asking that the Germans seal their customs loopholes against uninspected meat, cease multiple local re-inspections, and halt unwarranted aspersions on the quality of American pork. It expressed the hope that "this question of American pork may be quickly settled in accordance with the principles of practical modern sanitary science" rather than go on indefinitely with the support of spurious scientific arguments. Finally, White and Stiles urged that "if the German Government desires to protect the German sausage-makers and hog-raisers against American competition, a method may be adopted which will not be a scientifically unwarranted reflection, liable to injure American trade with other nations."[44]

Sending of the embassy's November 10 note ended the early phase of Stiles' scientific work in Germany. During the few weeks following this, as he travelled around Germany, Stiles began to notice decided changes for the better in the official and public attitudes

[42] Jackson to von Richthofen, Nov. 10, 1898, Roll 86; *Yearbook of the U.S. Department of Agriculture, 1897* (Washington: GPO, 1898), p. 248; and Salmon, "United States Meat Inspection," esp. pp. 37–41. Salmon in 1895 had taken nearly four pages of an article in a national veterinary journal to list examples in the German literature of the German microscopic shortcomings. He did this, he said, not "because it affords me an exquisite pleasure to discredit the German inspection," but to silence Germanophile critics at home who wished to replace the American microscopic system by the German (*ibid.*, p. 41).

[43] Jackson to von Richthofen, Nov. 10, 1898, Roll 86.

[44] *Ibid.* This 25-page note was prepared by Stiles and approved by White before the latter went on leave. See Jackson to Hay, Nov. 10, 1898, Roll 86. Further details on the American microscopy is found in a report by Salmon, in *Yearbook of U.S. Department of Agriculture, 1897*, p. 248.

toward Americans and American pork. No one at the embassy pretended that the American note had much if anything to do with this. Rather, they saw this as part of a wave of good feeling toward the United States which was brought about by the end of the Spanish-American War.[45] They did observe, however, that officials of the Imperial Government had apparently made a political decision "to pay less attention in the future to agrarian demands."

In Washington, even before the change in German feeling occurred, Salmon and other Agriculture Department officials were convinced that Stiles had finished what he had been sent to do. Toward the end of September 1898 they asked for his release. Greatly dismayed, White immediately sent off a telegram followed by several letters to try to head off such an order. The proposal to recall Stiles, he wrote, came just as the Reichstag had begun debates on legislation providing for uniform meat inspection throughout the German Empire. It was "absolutely necessary," White told Secretary of State John Hay, that Stiles be kept on in Berlin at this critical stage to help look out for American interests.[46]

During the previous six months White had come to rely heavily upon his science attaché.[47] He now detailed Stiles' special qualities in a letter to Hay:

> Stiles is exceedingly valuable to us. He is most thorough in his knowledge of the whole subject involved, is well abreast of the latest literature, is on the best of terms with its representatives in German science, and is alert watching the press and every point in the Empire where anything can occur likely to injure the interests entrusted to him. Moreover, he is ready and pleasing in address, speaks German and French, holds his own thoroughly well in any discussion, cannot be overawed by officialism, and yet is thoroughly a *persona grata* in social intercourse, as is also Mrs. Stiles . . . .
>
> I recently presented him at my own house to various Imperial Ministers and functionaries, as well as to sundry Ambassadors and other Diplomatic Representatives, and he was soon on the best of terms with them. His selection was most happy and he should not be allowed to return to America until the main question referred to is settled.[48]

As a result of White's urgent plea, Stiles stayed on in Germany for another year. He continued to travel extensively, and he continued to send his reports, statistical data, documents, and newspaper clippings back to the Department of Agriculture. Increasingly, however, he was drawn into behind-the-scenes discussions on the scientific aspects of the proposed German meat inspection legislation. Stiles and White had no thought that Germany would suddenly open the gates wide to American pork. They could only hope that whatever legislation was adopted, it would not discriminate against American products on unsupported scientific grounds. Their talks with German officials and scientists were aimed at making this minimum diplomatic objective known.

Stiles' continued presence was necessary to refute scientifically the renewed allegations against American pork and inspection methods. While German Central Government officials had become better disposed toward the United States, the Agrarians used the debate over the bill to air their old charges again. The Agrarian position was

[45] Jackson to Hay, Dec. 19 and Dec. 27, 1898, Roll 86, and White to Hay, Jan. 28, 1899, Roll 87.

[46] White to Hay, Oct. 1, 1898 (letter and telegram), Roll 86.

[47] *Ibid.*

[48] White to Hay, Oct. 29, 1898, Roll 86.

helped along some in mid-1899 by current German-American differences over Samoa, as well as by newspaper accounts alleging the supply of diseased meat to American troops during the Spanish-American War.[49] But protectionist sentiment was partly offset by free-trade interests. Advocates of a fair inspection bill—Central Government officials, meat importers, scientists, labor spokesmen, and others—well knew by now that Stiles had collected scientific and statistical data which was damaging to the Agrarian position. Many of them thus sought him out in order to advance the legislation. White wrote about this to Hay: "Most valuable service has been rendered by Mr. Stiles, our Scientific Attaché, who in discussing the various questions raised by such Ministers as Count Posadowski and Baron Hammerstein, as well as with various German experts and dealers, has done very much to get the Bill into good shape."[50]

In October 1899, with the bill still under discussion, American officials uncovered a confidential Prussian document which clearly aimed at drying up Stiles' local sources of information. This government order specifically directed local officials not to provide American consuls with any further information deemed "likely to work against German interests." It particularly prohibited "statistical data" pertaining to the incidence of disease, trichinae in pork, and violations of food laws.[51]

White recognized this document as a form of political appeasement of the Agrarians by the Central Government to gain support for the rest of the government's legislative program. The fundamental factor back of the order, he reported to Washington, lay in

> . . . the fact that our Consular Officers, ably assisted by Dr. C. W. Stiles . . . have been very active in combatting by means of German official figures and [scientific] authorities the unfair and unjust discriminations against American meats and fruits which have been made in the interest of the Agrarians. These German figures and authorities, by which it has been demonstrated that Trichinae are more frequently found in German than American pork, and that German methods of inspection have failed to discover trichinosis when it actually existed in German meats . . . have been a great embarrassment to the Imperial Government.

There were possibly other factors in the Prussian ban. Yet its language left no doubt that it was basically directed against American consuls "as gatherers of data of a medical, hygienic, and scientific nature, that is, against the kind of data upon which the work of the Embassy's Scientific Attaché, aided by the Consuls, has been largely founded."[52]

The Prussian ban was part of an increasing unwillingness throughout Germany to give out local sanitary and scientific information. Inquiries on these subjects now were to be referred to the generalized reports of the Imperial Health Office. This policy, of course, was hardly novel. It only reflected a return to the traditional European attitude of secrecy with respect to official statistical data. As far as Stiles was concerned, the

[49] White to Hay, Apr. 22, 1899, Roll 88, and White to Hay, May 10, 1899, Roll 88.

[50] White to Hay, Apr. 22, 1899, Roll 88.

[51] White to Hay, Oct. 13, 1899 (telegram) and Oct. 14, 1899, Roll 89.

[52] White to Hay, Oct. 14, 1899. White listed three other lesser causes which he felt might have contributed to the extraordinary ban on U.S. consular activity. These included (1) fear of possible American legislation against imports of German toys containing poisonous paints; (2) injudicious activity of U.S. Treasury agents in Germany in gathering information on German exporters and manufacturers; and (3) discovery by Germans of circular State Department letters to American consuls directing them to report the secret processes of German manufacturers.

new restriction was of little consequence, for he already had orders to terminate his mission. In early December 1899 he left Berlin and sailed for home.

The German meat inspection bill had not yet had its second reading in the Reichstag when Stiles left, and it was May before final action came in that body. While it was by no means everything that American diplomats had wished, White considered that the bill was one with which the United States and its pork interests could live.[53] The legislation was frankly protectionist and resulted in an overall decline in America's pork market in Germany. Still, it had the merit of not resorting to the old sanitary rationale to defend discriminations against American pork.

Back in Washington Stiles pulled together all of his material for a long final report. As issued, it was principally a statistical review which exonerated American pork from blame for most if not all European trichinosis. It also included a final damaging critique of the enormous German microscope bureaucracy. The 100,000 or so microscopists who manned the system around the Empire cost, Stiles estimated, close to $3,275,000 per year. This was well over the then-current appropriation of the entire Department of Agriculture.[54] Superficially, this was a remarkable testimony to the German dedication to science. The only trouble was that it was not worth all the cost or effort. Recapitulating from the Germans' data, Stiles emphasized once more the incidence of trichinosis in Germany since about 1880: over 53 per cent of the 6,300 odd cases and 41 per cent of the deaths, he noted, "appear to have been due to faults of the German inspection." In the United States, on the other hand, where there was no microscopic inspection of pork for domestic consumption, only about 900 cases had been recorded during twice as long a period. Stiles' conclusions in this matter were fairly easily arrived at. "It can not be admitted," he wrote, "that the German microscopic inspection system is a hygienic measure which it would be well to introduce into this country."[55]

Stiles' report on trichinosis in Germany contributed an important element of rationalism to subsequent discussions leading to the 1906 reform of American meat inspection. In particular, it played a considerable role in persuading American authorities not to rely on microscopic inspection of pork for domestic consumption.[56] Despite the current prestige of German science, the United States in this case opted against following the German example, and instead rejected the cumbersome, expensive, and useless microscope bureaucracy. The work of Stiles and Salmon in discrediting

[53] White to Hay, May 22, 1900, Roll 91. The Bundesrath approved the bill after the Reichstag action, and the Emperor signed it in July.

[54] Stiles, *Trichinosis in Germany*, pp. 28–29. The part-time inspectors who formed a large part of this total, Stiles reported, still included not only physicians, pharmacists, and chemists, but teachers, blacksmiths, and even butchers themselves who wanted extra income. American costs of microscopic inspection were about $180,000 in 1892, its first year, and varied substantially from year to year depending on the level of trade. See *Twenty-First Annual Report of the Bureau of Animal Industry, for the Year 1904* (Washington: GPO, 1905), p. 16. In addition to his official report, Stiles summarized his German conclusions in his shorter paper for the scientific community: "Trichinella Spiralis, Trichinosis, and Trichina-Inspection: a Zoologic Study in Public Hygiene," *loc. cit.*

[55] Stiles, *Trichinosis in Germany*, p. 37.

[56] Benjamin Schwartz, "Trichinosis in Swine and its Relationship to Public Health," *Annual Report of the Board of Regents of the Smithsonian Institution for the Year Ending June 30, 1939* (Washington: GPO, 1940), p. 419.

The history of microscopy in the United States remains a fruitful area for further study. One topic, related to the present, which might be of interest is that of the role of the microscope industry, German and American, throughout the years of the pork inspection dispute.

this system deserves to rank with Charles V. Chapin's contemporary (1900–1906) rejection of terminal disinfection after infectious diseases as a classic reevaluation of an outmoded sanitary fetish.[57]

Stiles, meanwhile, resigned from the Bureau of Animal Industry in 1902 to take a position as medical zoologist in the newly enlarged and strengthened Hygienic Laboratory. There he continued his work on zoological classification and nomenclature, and soon made a name as a bibliographer with his monumental *Index Catalogue of Medical and Veterinary Zoology*. But the largest part of his time for the next fifteen years or so was taken up by his involvement with hookworm disease and the campaign for its eradication. In this campaign he made a major contribution to the health of the entire southern region of the United States. Unlike trichinosis, however, hookworm disease afforded Stiles abundant justification for urging the spread of microscopy and laboratory techniques through the United States.

[57] James H. Cassedy, *Charles V. Chapin and the Public Health Movement* (Cambridge, Mass.: Harvard Univ. Press, 1962), pp. 114–117.

The American Scientist as Social Activist

Franz Boas, Burt G. Wilder, and the Cause of Racial Justice, 1900–1915

*By Edward H. Beardsley**

ACCORDING TO TRADITIONAL INTERPRETATION American scientists who joined in the public debate over the Negro question in the early twentieth century entered the dispute on the side of those favoring greater discrimination and repression. To be sure, some scientists opposed racist ideas and practices, but supposedly they did so in passive and private fashion, leaving public involvement to their more bigoted colleagues. Eventually, of course, American science became an active force for racial egalitarianism, but allegedly the shift began only in the late 1920s, reaching its peak in the 1930s, when Nazi brutalities against European Jewry made the inherent dangers of racism more clear. In sum, American scientists were Johnny-come-latelies in advocating racial justice for Negroes.[1]

Like many of the time-honored historical generalizations concerning the Negro, however, this one also suffers from some distortion. Far from being bigots, a number of influential scientists of the early twentieth century strongly supported the concept of racial justice by their research and in more visible ways. Preliminary investigation of the racial attitudes of American scientists indicates that among that group were anthropologist Franz Boas, psychologists E. B. Titchener and Livingston Farrand,

* Department of History, University of South Carolina, Columbia, South Carolina 29208. I am grateful to the National Endowment for the Humanities for a 1969 summer grant which supported the research for this article.

[1] For examples of what I have termed the traditional view see Oscar Handlin, *Race and Nationality in American Life* (Garden City, N.Y.: Doubleday, 1957), pp. 136, 141, 145, 149; Arnold Rose, *The Negro in America* (Boston: Beacon Press, 1962), p. 34; I. A. Newby, *The Development of Segregationist Thought* (Homewood, Ill.: Dorsey Press, 1968), p. 4; I. A. Newby, *Jim Crow's Defense. Anti-Negro Thought in America, 1900–1930* (Baton Rouge, La.: LSU Press, 1965), pp. xi, 20, 50. August Meier, in *Negro Thought in America, 1880–1915* (Ann Arbor: University of Michigan Press, 1963), pp. 162, 183, does identify Boas as actively involved in the early civil rights movement but does not mention Wilder. T. F. Gossett in *Race: The History of an Idea in America* (Dallas: SMU Press, 1963), pp. 81–83, 94, 424, noted defections from scientific racism as early as the late nineteenth century, but he did not find scientists taking a public stand against racism until the 1920s, One important exception to the traditional viewpoint is George Stocking's *Race, Culture, and Evolution* (New York: The Free Press, 1968), which argues (pp. 261–267) that the behavioral sciences were in a transitional stage in the early twentieth century, characterized by a shift away from a biological explanation of race and toward a cultural one. By about 1917 the shift was virtually completed (at the level of theory) for anthropology and sociology, Focusing on the development of ideas, Stocking's book does not, however, deal with scientists' involvement with public aspects of race. The implication is that this came later.

anatomists Burt Green Wilder and F. P. Mall, and sociologists Frances Kellor and R. E. Park.[2]

Although further study is needed to establish the depth of involvement of most of those men, at least two—Boas and Wilder—were deeply and publicly committed to the ideal of justice for black Americans. While Boas' and Wilder's activism was not necessarily typical of that of other scientists sympathetic to the Negro cause, the vigor of their involvement, along with the fact that theirs was a shared outlook, suggests a need for revising the traditional racist image of American scientists of the early twentieth century.

Boas and Wilder, in many ways very dissimilar men, shared two important characteristics. Each was solidly grounded in his particular science and each exhibited an independence of mind that kept him in perpetual tension in relation to surrounding society. Wilder, the elder of the two, was born in Boston in 1841. As a student at the Lawrence Scientific School, on the eve of the Civil War, he studied under Louis Agassiz and Jeffries Wyman and proved himself an exceptionally able pupil. Agassiz, in fact, considered Wilder his best student in the ante-bellum period. While Wilder reciprocated the feelings of admiration and was greatly influenced by his teacher, he did not share Agassiz's scientific outlook on race. On that question the more egalitarian views of anatomist Wyman had far greater impact on his thinking.[3]

When the war broke out, Wilder enlisted as a medical cadet and later served as a surgeon in the all-Negro Massachusetts Fifty-fifth Infantry Volunteers. Afterward, following the completion of medical studies at Harvard, he joined the faculty of the newly opened Cornell University as professor of comparative anatomy and zoology, a position he held until his retirement in 1910. At Cornell he soon established himself as a leading scholar and something of a character besides. A perpetual and energetic crusader, he was a mainstay in the national campaign against smoking, an ardent enthusiast for temperance, and along with Theodore Roosevelt gave lifelong allegiance to a more simplified spelling. Most undergraduates remembered him, however, for his annual lectures on hygiene and sex, presentations so graphic that coeds were known to faint from shock. A thorough-going moralist, Wilder exhorted freshmen to

[2] Farrand was active in the formation of the N.A.A.C.P. See *The New York Times*, May 31, 1909, p. 3. On Titchener's views see E. B. Titchener to B. G. Wilder, May 22, 1915, B. G. Wilder Papers, Regional History and University Archives, Cornell University, Ithaca, N.Y.; an example of Kellor's outlook is Francis Kellor, "The Criminal Negro," *The Arena*, 1901, *26*. For Park's view see R. E. Park, "Negro Home Life and Standards of Living," *Annals of the American Academy of Political and Social Science*, 1913, *49*: 147–163. Mall, in addition to Boas and Wilder, is discussed below. To say that these men were concerned with what we would today term "racial justice" does not mean that they had what we would call a modern outlook. All were, to varying degrees, influenced by the intellectual climate of their time. Park, for example, apparently came to his egalitarianism by way of a Lamarckian analysis of racial development (see Stocking, *Race, Culture, and Evolution*, p. 247). For all his commitment to equality, Boas believed it possible that the Negro race, on the average, was slightly (but inherently) inferior to the white in mental ability. See, e.g., Franz Boas, "The Negro and the Demands of Modern Life," *Charities and the Commons*, 1905, *15*: 85–86.

[3] Morris Bishop, *A History of Cornell* (Ithaca, N.Y.: Cornell University Press, 1962), pp. 110–111. Wilder later commented on his two teachers and their racial views. See his article on Wyman in David Starr Jordan, ed., *Leading American Men of Science* (New York: Henry Holt, 1910), pp. 190–192, and "What We Owe to Agassiz," *Popular Science Monthly*, July 1907, *71*: 10. While Wilder conceded Agassiz's scientific racism, he insisted that Agassiz never wavered from his support of civil rights for Negroes.

shun all evil, whether in the form of hazing parties, cigarettes, coffee, narcotics, alcohol, or women of pleasure.

But his scholarship attracted as much attention as his causes, with the result that his laboratory drew many of the ablest students at Cornell. One of them was Theobald Smith, a future leader in bacteriology, who later told Wilder: "'It was a fortunate thing for us that your laboratories were so small and crowded, because all of your work was done in the presence of your pupils, and we could not very well escape the infection of your enthusiasm'"[4] Wilder first earned his reputation in the field of zoology, but toward the end of his career he shifted his research interest to neurology and became a specialist on the morphology of the brain.[5]

Boas came from a different background. Born in Germany in 1858, the son of a prosperous and politically conscious middle-class family, he was also a product of the German university tradition. Following work at Heidelberg and Bonn he took his Ph.D. at Kiel University in physics and mathematics. But even as a graduate student his interests began to veer in other directions. Shifting first to geography, by 1890 he was firmly committed to cultural anthropology as a result of opportunities he had had to study North American Indians. While he was defining his scholarly interests, he was also laying down roots in the United States. In 1899, after two museum appointments, a period at Clark University, and a stint as an editor of *Science*, Boas accepted a permanent position as head of anthropology at Columbia University, a post he would hold until 1937.[6]

As an anthropologist his chief contribution to American scholarship, in addition to his elaboration of the idea of culture as independent of race, was the introduction of the rigorous methods of the physical sciences into a field previously characterized by bias and unsupported generalization. Boas' approach to anthropology, one student recalled, was one of "'icy enthusiasm,'" which meant an "ardent, unsparing drive for understanding, completely controlled by every critical check."[7] As a teacher he was brusque, demanding, and supremely self-confident. He inspired awe from colleagues and fear from most students, although those who could work independently found him a superb mentor. And yet he had great compassion for those students with severe personal problems. One student remembered that he worried over them "like an anxious father or grandfather."[8]

What catalyzed the social activism of Boas and Wilder, however, was not their education or their independent-mindedness. The key to understanding their involvement with the Negro question lay in certain experiences which each had as mature men. The pivotal influence on Wilder was his war service. He accepted the assignment as a regimental surgeon to black troops largely in hopes of furthering his scientific study of Negro anatomy, an interest which he inherited from his teacher, Wyman. Ultimately that interest won official sponsorship when B. A. Gould of the United

[4] Bishop, *Cornell*, pp. 110–111; see also *Who Was Who in America*, Vol. I (Chicago: Marquis, 1950), p. 1345. The lecture remarks are from B. G. Wilder, "Hygiene and Morality," Miscellaneous Pamphlets (1899–1901), Wilder Papers.

[5] J. H. Comstock, "Burt Green Wilder," *Science*, May 22, 1925, *61*:533.

[6] A. L. Kroeber, "Franz Boas: The Man," *Memoirs of the American Anthropological Association*, 1943, *61*:5–13.

[7] *Ibid.*, pp. 22, 26.

[8] Margaret Mead, *An Anthropologist at Work. The Writings of Ruth Benedict* (Boston: Houghton Mifflin, 1959), p. 346.

States Sanitary Commission put him in charge of gathering physical data on Negro soldiers and supplied him with instruments to do the work.[9] That project, plus his duties as surgeon, brought him into intimate contact with the troops of the Massachusetts Fifty-fifth. As a result the young scientist soon formed a number of strong friendships with his men, several of whom he taught to read and write.[10]

Besides friends, his military experience also gave him a new view of the race. One thing that impressed Wilder was the Negro's devotion to principle. When Massachusetts authorized the Fifty-fifth the state set the Negroes' pay at the level of laborers ($10 per month) rather than soldiers ($13). The black troops responded by refusing to accept the discriminatory wage, even though the state declared that that was all they would get. In all, the Fifty-fifth served for over a year without pay before they finally won their point. More remarkable to Wilder was the fighting spirit of the regiment during that difficult period. In June 1864 the Fifty-fifth played a crucial role in capturing a Confederate fortification on James Island, South Carolina. The impressive thing about the Negroes' participation was its apparent spontaneity. From what Wilder could learn no white officer ordered the blacks into battle; they simply took themselves in when they saw the attack floundering.[11]

Wilder's Civil War experiences were so central to his later involvement in racial matters that even if his research had not pointed up the essential equality of the Negro, he would perhaps have still played an activist role. As he remarked years afterward, his "personal observation" had convinced him "that the [Negroes'] title to . . . rights and opportunities was earned during the Civil War by the general conduct of soldiers of African descent, by their valor, by their initiative, and by their deliberate self-sacrifice for the sake of a principle."[12] In responding as actively as he did to the growing pattern of discrimination in American life after the 1890s Wilder was merely honoring a basic rule of friendship which said that when comrades were unfairly attacked one came to their aid.

If Wilder's commitment had its basis in personal experience, Boas was an activist for what were essentially professional reasons. A cautious and skillful investigator who deplored hasty and unfounded generalization, Boas took strong exception to American scientists' readiness to embrace what he regarded as unsupported speculation about the Negro race.

When Boas assumed his Columbia University post in the late 1890s the concept of evolutionism was still the fashion in anthropology. A doctrine that arranged all the races of man in a sort of evolutionary line-up from the savage at the rear to the most civilized races up front, this theory provided not only a system for rating achievements of contemporary cultures but also a description of the pathway the higher races had

[9] Jeffries Wyman to Wilder, July 25, 1864, box 1, Wilder Papers; also see Wyman to Wilder, Oct. 12, 1864, box 1, and Mary A. Q. Gould to Wilder, Apr. 14, 1868, box 1, Wilder Papers.

[10] Wilder Diary, Apr. 6, 1865, p. 215, box 13, Wilder Papers.

[11] B. G. Wilder, "Two Examples of the Negro's Courage, Physical and Moral," *Alexander's Magazine*, Jan. 1906, *1*: 22–25.

[12] B. G. Wilder, "The Brain of the American Negro," *Proceedings of the First National Negro Conference* (n.d.), p. 23. Certainly it would be wrong to claim for Wilder the outlook of the modern-day civil rights activist. Like most racial liberals of his time, Wilder was to some degree a captive of prevailing racial ideoogy, although he struggled against it. In one publication he noted that it was common knowledge that the white race had, on the whole, evolved to a higher position than had the black. See Wilder in Jordan, ed., *Leading American Men of Science*, p. 192.

followed in getting to the top. But the way up was clearly not open to all. To the evolutionist the term "civilized race" carried the additional connotations of superior and white (more precisely, Anglo-Saxon), while "savage" and "barbarian" implied inferior and non-white. Furthermore, certain of the non-white peoples assigned to the lower orders of being (the Negroes, for example) were destined to remain there. The belief was that evolution had simply stopped working for them.[13]

By the mid-1890s Boas strongly rejected such thinking, largely because anthropologists had never put their evolutionary model to the test, had never paused in their zeal for sweeping generalization to make those detailed analyses of specific cultures upon which generalization had to rest. Boas in that period was as interested as the evoluionist to find laws of cultural development, but he knew that scientists must first "understand the process by which the individual culture grew before we can undertake to lay down the laws by which the culture of all mankind grew."[14]

But if evolutionism with its belief in superior races was yet unproven, available evidence, Boas believed, also suggested that it was erroneous. A common misconception about primitive people was that they had a lower order of intelligence than white men, rendering them fickle, excitable, and unable to concentrate at length. As one contemporary scientific account put it, even short conversations wearied the native, "'particularly if questions are asked that require efforts of thought or memory. . . .'"[15] Boas' research led him to quite different conclusions. Willing to study primitive peoples on their own cultural terms, he realized early that the alleged weakness of the savage mind was merely a mirage which owed its existence to anthropologists' insistence on viewing native mores and actions from the vantage of white cultural values. If natives lost interest in conversing with an investigator, it was usually because his questions seemed trifling and were often posed in the white man's language besides. "I can assure you," Boas reported to a group of colleagues, "that the interest of . . . natives can easily be raised to a high pitch and that I have often been the one who was wearied out first."[16]

Fundamentally, then, Boas' public involvement on the issue of race sprang from a scientist's desire to rid his discipline of prevailing amateurism and to substitute the

[13] On the connection between evolutionism and the idea of racial inferiority see George Stocking, "American Social Scientists and Race Theory, 1890–1915" (unpublished Ph.D. dissertation, University of Pennsylvania, 1960), pp. 466–467. The theory that Negroes lay outside the evolutionary process is treated in John Haller, *Outcasts from Evolution. Scientific Attitudes of Racial Inferiority, 1859–1900* (Urbana: University of Illinois Press, 1971), pp. ix, 100, 198–199.

[14] Franz Boas, "The Growth of Indian Mythologies," *Journal of American Folklore*, 1896, *9*:11. On Boas' rejection of evolutionism (by the 1890s) see Stocking, "American Social Scientists and Race Theory," pp. 522–523, 526–527. Boas was not the first anthropologist to question linear evolution or to emphasize the empirical approach to research. The German anthropologist Rudolph Virchow argued the inherent equality of the races, and his influence on Boas is believed to have been substantial. Also, American anthropologist Horatio Hale, who supervised Boas' British Columbia research, stressed not only the need for an empirical approach but also skepticism of the ideas of the evolutionists. (On Virchow see Clyde Kluckhohn and Olaf Prufer, "Influences During the Formative Years," *The American Anthropologist*, 1959, *61* (No. 89): 21–22. On Hale, see Jacob Gruber, "Horatio Hale and the Development of American Anthropology," *Proceedings of the American Philosophical Society*, 1967, *111*:18–19, 25–33. Perhaps Hale's influence was more a reinforcement than a catalyst to Boas' thinking, for Gruber himself notes that Boas had already begun to question prevailing ideas before he met up with Hale.)

[15] Quoted by Boas in "Human Faculty as Determined by Race," *Proceedings of the American Association for the Advancement of Science*, 1894, *43*:320–321.

[16] *Ibid.*

rigorous methods of science for facile generalization. The broad issue at hand was the matter of sovereignty in science; and those professionals, like Boas, who regarded their approach as the more scientific were vitally concerned to educate not only colleagues but also the more enlightened part of the public to distinguish between kings and pretenders. When Boas began work at Columbia, anthropology was still largely the preserve of those he regarded as amateurs in their approach. As an aggressive professional, holding strong views about the need for his discipline to become more truly scientific, he felt obliged to oppose that amateurism wherever he found it. In segregationist America there was no lack of targets.[17]

Although Boas and Wilder had differing perspectives on the race issue, their response to problems of discrimination and oppression was similar. Each played essentially an educative role, whose goal was to advance more rational thinking on race among both whites and Negroes. With whites—both scientists and laymen—their purpose was to liberalize opinion. With blacks, the aim was to dissolve inbred feelings of inferiority and encourage greater self-esteem. Both men proved equally skillful in cutting through the inconsistencies and strained reasoning of the segregationist defense.

In dealing with racism among whites Wilder's approach was more personal than Boas' and had more of the earmarks of a moral crusade. While not a belligerent man, he was argumentative and often blunt. His targets tended to be individuals rather than groups or institutions, and usually people with some influence on public opinion. One of those targets was the novelist Owen Wister, a Harvard-trained Philadelphia lawyer, who first achieved popularity with *The Virginian* (1902). In 1905 the *Saturday Evening Post* began serializing a new book, *Lady Baltimore*. The story of a young Northerner on a genealogical mission in contemporary South Carolina, the novel was in large part a literary endorsement of both the Southern view of Reconstruction and the national decision to return the management of Negroes to the former master class. Annoyed by Wister's patrician racism, Wilder was particularly angered by the author's attempt to make science serve his white supremacist notions. One passage in the book had the main character comparing skulls of a gorilla, an Aryan, and a South Carolina Negro. The Negro and Caucasian skulls were easily distinguishable, but between the ape's and the black man's the hero could see only a "'kinship which stares you in the face.'"[18]

[17] Certainly there was more impelling Boas on the question of Negro inferiority than just his concern for scientific professional standards. His own students and colleagues testify to this. In describing the motive forces in his life, they cite not only his dedication to facts and the scientific method but also his liberal outlook on social and political questions, his great affection for the primitives he studied, his sense of an obligation to use his research in the interest of freedom, and his passionate opposition to anti-Semitism. While all were factors in his public involvement on the Negro question, his commitment to scientific objectivity and reliability still seems the most basic and fundamental explanation. His involvement on the anti-Semitism issue supports this. Although from a Jewish background and a foe of anti-Semitism since his youth, Boas did not become actively and publicly involved on that issue until the 1920s, when Nazi racists made a major effort to enlist science in support of their views. Boas also never involved himself with the Indian's plight as he did with the Negro's or Jew's. And it is suggestive, I think, to note here that the idea of Indian inferiority was never a major tenet of scientific racism. See Melville Herskovits, *Franz Boas. The Science of Man in the Making* (New York: Charles Scribner's, 1953), pp. 113–114. Also see the comments of other students and colleagues in *Amer. Anthropol.*, 1943, *45* (memoir 61): 8, 14, 23, and 1959, *61* (memoir 89): vi, 2, 6, 21–22, 30.

[18] Wilder, "The Brain of the American Negro," p. 26; for a sketch of Wister see *Who Was Who in America*, Vol. I, p. 1370; also see Owen Wister, *Lady Baltimore* (New York: The MacMillan Co., 1906), pp. 289–300, *passim*.

Frankly astonished that a "liberally educated writer who had every opportunity for ascertaining the facts" could harbor such ideas, Wilder was even more disturbed by the probable influence of Wister's story on its readers. So he wrote the novelist, pointing out his errors and inquiring about the source of his racial notions. Through his secretary Wister replied crisply that his anatomical views merely reflected common knowledge, that kind "'found in any museum of anatomy or academy of natural sciences.'"[19] With that, Wilder dropped his cover of amiability. "Pardon my persistence," he wrote, "but there is more to be said." Reputable scientific opinion had long held—and Wilder spoke, he said, from the vantage of forty-five years in science—that the differences between Negro and white crania were insignificant compared to the vast gap between Negroes and the higher apes. Even a child could distinguish skulls of various human races from those of primates. What Wister needed to do was to re-visit his local museum for a closer look. But more than that, he should immediately and publicly retract his inaccuracies. If he did not, Wilder assured him that "unwelcome as the task may be" he would have to try to "arrest further diffusion of the scientific error and the political venom that characterize the passages in question."[20]

Ultimately the dispute ended in a standoff. Wister promised to "'set the matter right'" if further inquiry proved his statements incorrect, but when the book version appeared, the passage still left the impression of an affinity between apes and Negroes. Meantime, Wilder published his own correction in *Alexander's Magazine*, for he had decided that even full atonement by Wister would be inadequate. The novelist never offered that, for he regarded Wilder's demand as unreasonable. Wilder, of course, disagreed: from his standpoint a man who should have known better had involved himself in the sorry business of spreading racism and should be called to account.[21]

That rule also applied to women. In 1915 Margaret Deland, a well-born Easterner who had attained a fair measure of popularity and wealth from her writing, published a story entitled "A Black Drop." In it she argued that racial intermarriage was not only harmful to the Caucasian race but instinctively and inherently repulsive to whites. If a white man did not feel repulsed, "'there is something wrong with him,'" the authoress insisted.[22] What Wilder basically objected to was Deland's claim that racial aversion was instinctive. He was certain that it was merely the result of prejudice or social convention. "No natural instinct," he wrote her, "has prevented the existence of millions of mulattoes."[23] Southern whites might proclaim their inherent repulsion in public, but that merely reflected their hypocrisy; in private they obviously observed a different set of ethics. Indeed, if there were only some way to identify kinship through the blood, Wilder felt certain that "very few of the proud Southern families would fail of representation in the veins of their despised poor relations."[24]

Deland was unconvinced. There were undeniably millions of mulattoes, but that was no argument against her theory; it merely showed that "there is something wrong with

[19] Wilder, "The Brain of the American Negro," p. 26.

[20] *Ibid.*, pp. 26–29.

[21] *Ibid.*, pp. 29–30; also see Wister, *Lady Baltimore*, p. 171, for the author's revision of offensive passages. Wilder's attack clearly put Wister on the defensive, for in the prefatory remarks in his book he insisted that he bore no malice toward the colored race, and he expressed surprise that anyone could have thought otherwise.

[22] Quoted in Wilder to Margaret Deland, May 13, 1915, box 2, Wilder Papers; on Deland see *Who Was Who*, Vol. II (1950), p. 151.

[23] Wilder to Deland, May 13, 1915, box 2, Wilder Papers.

[24] Wilder notation, "Negro Ravia" file, n.d., box 13, Wilder Papers.

a vast number of white men!"[25] But Wilder should not think her prejudiced: Booker T. Washington, she enthused, was "one of the most splendid people in the World" and superior to a large proportion of whites, even though, racially considered, he was "nearer the primitive man than is the Anglo-Saxon."[26]

Such encounters were surely frustrating. But Wilder, an old hand at the aggravating business of reform, was undeterred. In fact, at the same time he was battling literary figures he was also keeping watch on the scientific community. One colleague whom he sought to expose was the Virginia anatomist Robert B. Bean. In 1906 Bean published three articles on the Negro brain. One, a highly technical and fairly cautious treatment, appeared in a professional journal of anatomy and was the product of his doctoral research under the anatomist Franklin P. Mall. Its central point was that the Negro brain was smaller than the white in the region controlling the higher intellectual faculties. The other two articles, published in a popular periodical and less restricted by the demands of experimental evidence, emphasized what Bean saw as the social implications of his research. The first argued that Negroes, because of the size and shape of their brain, not only lacked the white man's intellectual capacity but were also creatures of uncontrollable passion. The second concluded that since the Negro mind was essentially unredeemable, schoolmen should discourage college education for blacks and train them instead for work more in line with their ability, such as farming and the manual trades.[27]

Wilder, who regarded even the technical article as a challenge to his standing as a neurologist, did not discover the two popularized treatments (and the extent of Bean's departure from his evidence) until 1909. When he did, he quickly decided that he must do more than merely answer Bean in a scientific journal; he must expose him publicly.[28] And he had best do as thorough a job as possible, for it appeared that Bean's ideas were already having an impact. One man who read the articles was the editor of *American Medicine*, who shortly afterward urged disfranchisement of all Negroes on the ground that they were "'an electorate without brains.'"[29] To lay a firm basis for his rejoinder Wilder redid much of Bean's research himself, using his own sizeable collection of Negro and Caucasian brains. By spring 1909 he had his results in hand. He also had a forum from which to present them: a national civil rights group, planning a conference in New York, invited Wilder to address them on the Negro brain. In that speech he devoted major attention to Robert Bean.[30]

The basic flaw in Bean's claim about Negro intellectual potential, Wilder argued, was that it rested on too small a sampling of Negro brains. Indeed, at one point Bean even tried to generalize from "*a* Negro and *a* Caucasian [brain] . . . as if they represented a constant racial difference. . . ."[31] Invidious comparisons, Wilder objected,

[25] Deland to Wilder, May 19, 1915, box 2, Wilder Papers.

[26] *Ibid.*

[27] On Robert B. Bean see *Who Was Who*, Vol. II, p. 51; the technical article appeared in *American Journal of Anatomy*, 1906, *6*; the two popularized treatments are Robert B. Bean, "The Negro Brain," *The Century Magazine*, Sept. 1906, *50*, and "The Training of the Negro," *ibid.*, Oct. 1906, *50*.

[28] Wilder had read the *Journal of Anatomy* and was preparing to comment on it in a paper to be delivered to a meeting of the American Philosophical Society in April 1909. He had to ask Bean for the references to his *Century* articles, however. See Wilder to Robert B. Bean, Feb. 22, 1909, box 2, Wilder Papers.

[29] Wilder commented on this in "Brain of the Negro," p. 36.

[30] Wilder's brain collection is noted in J. H, Comstock, "Burt Green Wilder," *Science*, May 22, 1925, *41*:532–533; also see Wilder, "Brain of the Negro," pp. 22, 35–39.

[31] "Brain of the Negro," p. 38.

were not science; they were merely tricks which anyone could play. Wilder had in his collection a large brain of a onetime murderer and a smaller one once belonging to an esteemed judge. Surely he was not to conclude that the criminal class was superior to "those who pass on their misdeeds."[32] But more was involved than faulty methodology, Wilder suspected. Bean's findings also reflected his own personal bias against Negroes. One man who shared his suspicions, Wilder noted, was Bean's teacher, Mall, who had recently accused Bean of shaping his evidence to support his anti-Negro views.[33]

Although the speech later appeared in print as part of the published proceedings of the conference, Wilder knew that such a document would give his views on the Negro brain and on Bean only limited exposure among those white readers he wanted to reach. So he decided to circulate additional copies of the article himself. By summer 1910 he had spent $500 for printing and distribution and had put his article into the hands of a thousand additional readers.[34]

In Bean's case Wilder apparently did not know the man he attacked, which made it easier to say unpleasant things. But even when personal acquaintance was involved, Wilder did not back away, as was clear in his encounter with anatomist Robert W. Shufeldt, a former student who had spent a number of years in scientific work in Washington, D.C. In 1915 Shufeldt published a book titled *America's Greatest Problem: The Negro*, which represented a high point in a long career of Negro-baiting. Wilder was thoroughly offended by it and wrote a lengthy and devastating review which he sent to the editors of *Science*. Just afterward he told a friend that "it has been no easy or pleasant task, and will probably cost me his regard; he was an old pupil, but I think we never have met since and I have not been attracted by a certain flavor in his writings which also characterizes the present work; some of it is really disgusting."[35] Ultimately Wilder had to settle for publication of a much shorter review than he wanted. *Science* editor James McKeen Cattell not only found the article too long but also thought it improper to "take up in *Science* the question of the social relations, etc. of the Negro."[36] Still, even the abbreviated review was stinging: in it Wilder ridiculed Shufeldt's obsession with the supposed sexuality of Negroes.[37]

Franz Boas also tried to check the more unfounded assertions of his colleagues. His approach, however, was less rancorous, largely because he addressed his criticisms not to individuals but to all his colleagues. One of his platforms was the annual meeting of the American Association for the Advancement of Science, whose subsection of anthropology provided the first national home for American workers in the field. In 1894 and again in 1908 Boas served the subsection as its presiding officer and thus had the responsibility for delivering a state-of-the-profession address. Both times he used that speech to call for more rigorous standards in the study of the Negro race.

In 1894 his subject was the racial bias of many African studies. Too often, he said, such investigations began with the assumptions that the white race represented the highest perfection in man and that deviation from whiteness was a measure of race in-

[32] *Ibid.*

[33] *Ibid.*, p. 39.

[34] Wilder to G. Spiller, Aug. 15, 1910, box 2, Wilder Papers.

[35] Wilder to Dr. Lamb, Aug. 8, 1915, "Shufeldt" file, box 13, Wilder Papers; see Shufeldt, Biographical Sketch, Apr. 1902, Collection 239, Academy of Natural Sciences, Philadelphia; Shufeldt's racist attitudes are well expressed in Robert W. Shufeldt to Wilder, Aug. 17, 1915, box 13, Wilder Papers.

[36] James McKeen Cattell to Wilder, Aug. 31, 1915, box 13, Wilder Papers.

[37] *Science*, Nov. 26, 1915, N.S. *42*:768.

feriority. Such ideas were patent myths, Boas insisted, which owed their existence partly to the anthropologists' tendency to confuse actual cultural achievement with aptitude for achievement. Because black Africans did not then exhibit high culture, anthropologists concluded that they lacked capacity for it. That view was not warranted, for in earlier times certain African regions which had encountered Mohammedan influence had developed sophisticated civilizations.[38]

A second fundamental error was the belief that achievements of various races were the result of biology. Actually, history, far more than inheritance, caused one race to dominate over others. For example, Boas noted, many favorable conditions quite apart from race aided the rapid growth of civilization in Europe. Later, when European culture began to spread to other continents the peoples there, again for nonracial reasons, "were not equally favorably situated. . . ." As a result, European culture had smothered "all [the] promising beginnings" which had appeared elsewhere.[39] But that was the story only for the modern era. Once, history stood on the side of certain colored races: then they played the lead roles, while white men were mere primitives.[40]

A decade and a half later, when Boas made his second address, racial segregation had entered its most intense phase. Those clamoring for still greater repression were raising the specter of racial interbreeding and arguing that the Negro posed a threat not only to the biological integrity of the white race but to civilization itself. In Boas' mind that situation made any racial bias of scientists more regrettable than ever, for racial fear-mongers were then looking to science to substantiate their claims.

Matters were grave enough, he felt, to call for a moratorium on unfounded scientific statements on race mixing. In his 1908 address Boas issued that call: "I feel that it behooves us to be most cautious and particularly to refrain from all sensational formulations of the problem, that are liable to add to the prevalent lack of calmness . . .; the more so since the answer to these questions concerns the welfare of millions of people."[41] But scientists could do more, Boas suggested, than merely keep quiet. Since there was a lack of even the "most elementary facts" about interbreeding, one important service they could perform was to investigate thoroughly the whole question of the fitness of mulattoes.[42] Boas did not know what conclusions might result, but he suspected they would remove much of the fear of amalgamation. His own studies had convinced him that all so-called pure races—even the Anglo-Saxon—were products of considerable mingling of diverse people, and such mixing had not proved a serious barrier to high cultural attainment.[43]

Focusing on the scientific community might do much to alter the racial views of his colleagues, but it was a poor way, Boas recognized, to change popular thinking. Greatly concerned to widen the influence of his views, in the early twentieth century Boas began an assault on middle-class white racism that was extraordinary for a man so heavily engaged in strictly professional tasks. His ideas were presented in various forms, in newspaper and periodical articles, book reviews, and a major book of his own. The periodical pieces appeared or were noted in such journals as *Everybody's Magazine*, *Van Norden's Magazine*, *Charities and the Commons*, and *The Century*, and

[38] Boas, "Human Faculty," pp. 301–306.

[39] *Ibid.*, pp. 307–308.

[40] *Ibid.*, p. 303.

[41] Franz Boas, "Race Problems in America," *Science*, May 28, 1909, *29*:845–846.

[42] *Ibid.*

[43] *Ibid.*, pp. 848–849, 840–841.

his articles in *The New York Sunday Times* extended his range even further. Altogether he argued his case at least a dozen times in the decade before World War I.[44]

His emphasis was chiefly the African origins of the Negro. As he explained to Booker T. Washington, "I am particularly anxious to bring home to the American people the fact that the African race in its own continent has achieved advancements which have been of importance in the development of civilization. . . ."[45] In fact, he told a Columbia colleague, the evidence from Africa "is such that we must class the Negro among the most highly endowed human races."[46] Awareness of that, he reasoned, would force fair-minded men at least to give up the notion that licentiousness, shiftlessness, and criminality were racial characteristics of black men.[47]

But winning wide acceptance—even consideration—of such heretical ideas was clearly beyond the power of a single individual. If prevailing racist attitudes were to be substantially changed, a large-scale, collective assault was needed. Accordingly in 1906 he threw his energy behind the establishment of an African museum, an institution which would combine public exhibits with facilities for extensive scholarly research on the Negro. In its main aim of substituting truth for error in the public mind, the proposed museum illustrated well the almost boundless faith in reason of that generation of scientists. Downplaying the force of irrationality, Boas seemed convinced that people instinctively opted for truth when presented with a choice. As he told one colleague, the African museum would "serve eminently well to counteract the prejudices which hinder the advancement of the Negro race."[48] He even foresaw basic alterations in national policy toward Negroes as a result of the institution's work.[49]

As Boas conceived it, the museum would have two primary functions. On the one hand it would attempt to persuade the public by means of exhibitions and publications that Africa had once been a seat of highly developed civilizations. Its second function, and the one with greatest potential for influencing government policy, would be a comprehensive investigation of Negro anatomy. Scientists had previously written a great deal on such questions as the Negro brain, the deleterious effects of race mixing, and the arrested development of Negro children. But all those accounts were written "entirely from a partisan point of view."[50] Objective investigation not only promised to elevate white concepts of the Negro but also would enable government to utilize "unprejudiced scientific investigation" in deciding "what policy should be pursued."[51]

The major obstacle before the project, not surprisingly, was cost. Physical facilities alone would require about $500,000, and operating expenses demanded an additional $670,000 endowment. In the first decade of the twentieth century such funding was almost unavailable. Certainly the federal government offered little avenue of hope. Washington might spend millions of dollars annually to combat livestock and plant diseases, but support of social research lay far in the future. Private sources were like-

[44] For a full listing of Boas' publications in this period see "Bibliography of Franz Boas," in *Amer. Anthropol.*, 1943, *45* (No. 3): 82–92; also see Wilder to Boas, May 11, 1909, Boas Correspondence, American Philosophical Library, Philadelphia; for the *Times* articles see *N. Y. Times*, Sept. 24 and Oct. 1, 1911.

[45] Boas to Booker T. Washington, Nov. 8, 1908, Boas Correspondence.

[46] Boas to Felix Adler, Oct. 30, 1906, Boas Correspondence.

[47] Boas, "The Negro and the Demands of Modern Life," pp. 86–87.

[48] Boas to Adler, Oct. 30, 1906, Boas Correspondence.

[49] *Ibid.*

[50] Boas to Lucius P. Brown, June 5, 1909, Boas Correspondence.

[51] Boas to Starr Murphy, Nov. 23, 1906, Boas Correspondence.

wise unpromising. Large scale philanthropy was of recent origin and up to then had gone mostly to noncontroversial projects in such fields as public health, medicine, and graduate education.

But Boas refused to concede the odds and confidently undertook to enlist the country's major givers behind his effort. Included in his solicitation were John D. Rockefeller and Andrew Carnegie, who had recently endowed institutes bearing their names, and two supporters of Southern Negro education, Robert C. Ogden and George F. Peabody. In each appeal Boas stressed the irrationality and social danger of racism and the value of the African museum idea as a vehicle for dissolving such tension. The worth he himself placed on the project was evident in his appeal to the head of the Carnegie Institution: "You are aware that I am exceedingly anxious to interest you in other kinds of anthropological research. However, if I were given the choice, I do not know whether I should not think the Negro work more important on account of its practical bearings at the present time."[52]

If Boas seriously hoped for support, he was disappointed. None of the wealthy givers whom he solicited regarded the project as sufficiently important to merit a contribution. Carnegie merely referred Boas to the director of his Institution, who in turn referred him to the federal government. Apparently all found the proposal too radical and needlessly upsetting to white sensibilities. Peabody for one preferred a more gradual approach to race problems: "What you have in mind will be a most desireable thing to do, but there are special developments now underway in the South . . . which will make it perhaps more advantageous to have some delay in the carrying out of any such program. . . ."[53]

Rebuffed by private sources Boas turned to the one public agency which might aid him, the Smithsonian Institution's Bureau of Ethnology. Seeking to capitalize on growing national interest in the "new" immigration, he redesigned his project and in 1909 proposed an investigation of the relative influence of heredity and environment which would focus on Negroes, mulattoes, and ethnics. The investigation was tailor-made for the Bureau, he argued, and certain to strengthen its relationship with Congress, which was then vitally interested in the so-called European races.[54] But the Bureau would have none of it. Such research would only arouse the "race feeling" of Congress and jeopardize Bureau appropriations.[55] Incensed by such timidity Boas complained to a colleague that the policy of the Smithsonian "has been for years one of hesitation when scientific questions of considerable importance have come up. . . ."[56]

In the absence of substantial public and private support Boas saw no recourse but to try to create a program of Negro studies on his own at Columbia. His ability to coax modest sums from his many institutional and individual contacts (Peabody ultimately

[52] Boas to R. S. Woodward, Dec. 4, 1906; also Boas to Andrew Carnegie, Dec. 8, 1906; Boas to Starr Murphy, Nov. 23, 1906; Boas to Robert C. Ogden, Feb. 11, 1907; Boas to George Peabody, Feb. 11, 1907, all in Boas Correspondence. Ogden and Peabody were deeply involved in efforts to provide technical and vocational education for Negroes. Both served as trustees for Hampton and Tuskegee institutes and both were officers of the Southern Educational Board. See *Dictionary of American Biography*, Vol. VIII, pp. 641–642 (Ogden) and Supplement, Vol. II, pp. 520–526 (Peabody).

[53] Peabody to Boas, Feb. 13, 1907, Boas Correspondence.

[54] Boas to Charles D. Walcott, May 4, 1909, Boas Correspondence.

[55] J. W. Hodge to Boas, Mar. 18, 1910, Boas Correspondence.

[56] Boas to J. W. Jenks, Mar. 21, 1910, Boas Correspondence.

became one of his staunchest supporters) and his success in attracting a number of capable students, several of them Negroes, combined to produce at least a scale-model version of the museum idea. By the mid-1920s he could report a variety of projects: he was sending one student to Africa, and several others were studying the American Negro. One was gathering folk tales in the Caribbean, another was measuring Harlem blacks, and a third was studying the structure of a Negro community in Virginia.[57]

What made Boas' and Wilder's involvement particularly exceptional for that day, however, was not that they did something *for* Negroes, but that they did something *with* Negroes—especially with a certain group of them. The first decade of the century witnessed the rise of a vigorous civil rights movement in America. Led by outspoken individuals of both races who accepted no compromise with white supremacy, it demanded immediate justice for at least the talented minority of blacks. Whereas the vast majority of whites sympathetic to the Negro's plight recoiled from the new militance, Boas and Wilder did not hesitate to stand with it.

One reflection of their commitment was their many appearances before civil rights conferences in that period. Wilder's appeal stemmed only in part from his special knowledge of anatomy; he was also what that generation called an inspirational speaker. His zeal for the Negro's cause and his ability to puncture white pretensions were amply illustrated by a 1905 address at the Boston A.M.E. Zion Church on the occasion of the William Lloyd Garrison Centennial. His major theme that day was the bravery of the black Massachusetts soldiers during the Civil War. By all measurements, Wilder noted, the Negroes had served as ably as any troop of whites. But where was the justice, he asked, of always imposing the white race as the standard? Certainly history gave no endorsement of white supremacy. It sometimes argued the opposite: "Who are the leaders of Tammany Hall?" Wilder asked. "Who are the mismanagers of insurance companies? Who compose the lynching mobs and the gangs of college hazers? Whites." But the so-called superior race had not just ridiculed the law. It had also evaded basic responsibilities: "Who were the importers of the first Negro slaves, and who were the progenitors of multitudes of their descendents? White men again."[58]

Less a spellbinder than Wilder, Boas nonetheless had something important to contribute to the budding civil rights movement. One Negro who recognized that was W. E. B. DuBois, then teaching at Atlanta University, Southern center of the new black activism. In 1906 he invited Boas to deliver the university's commencement address and take part in its annual Negro conference.[59] Boas' message to Atlanta Negroes was the one he characteristically gave to black people: despite insistent claims of whites, there was no proof of the racial inferiority of Negroes. To the contrary, the significant achievements of Negroes in pre-modern Africa pointed to the essential equality of the black race and justified its playing as active a role in American life as any other people. Of course, white-dominated society, however unfairly, would stiffly resist that par-

[57] For a review of projects for 1926 and 1927 see Boas to G. F. Peabody, June 25, 1926; also see Boas to Zora Hurston, May 3, 1937; Melville Herskovits, Memorandum to Boas, "End of December, 1927" file, all in Boas Correspondence. On Boas' relationship with Peabody see Boas to Peabody, June 25, 1926, and Aug. 4, 1926, and Peabody to Boas, Aug. 5, 1926.

[58] B. G. Wilder, "Two Examples of the Negro's Courage," *Alexander's Mag.*, Jan. 1906, *1*:25; see also pp. 22–25; on Wilder's oratorical skills see Bishop, *Cornell*, p. 111.

[59] W. E. B. DuBois to Boas, Mar. 31, 1906, Boas Correspondence.

ticipation, which meant that Negroes must develop both patience and perseverence in their search for fulfillment. But if Boas warned against expecting the impossible and against unruly agitation, he did not advise letting white society set the timetable for change. Instead, he urged the Atlanta students to press continuously for "full opportunities for your powers."[60]

The most notable landmark in Boas' and Wilder's emergence as social activists was their participation in the 1909 and 1910 conferences of the National Negro Committee. Convened by a biracial group who saw need for an immediate and forthright attack on Southern segregation, the conferences proved to be staging sessions for the creation in 1910 of the National Association for the Advancement of Colored People. In supporting those meetings Boas and Wilder were out of step with the great majority of whites claiming to be liberals on race. Most reacted as did William James and Andrew Carnegie, who boycotted the New York City sessions on the ground that they would do the Negro no good and only fan the animosity of white racists.[61]

Wilder, one of about a dozen whites on the 1909 program, spoke on his speciality, the Negro brain. His opening remarks revealed again that his was not merely an academic interest. "The American Negro is on trial, not for his life but for the recognition of his status, his rights, and his opportunities. At this, as at most other trials, experts disagree." But Wilder claimed to be one witness who could testify impartially, even though he was a white man. For one thing he believed in evolution and therefore minimized not just differences between men but even those between man and the apes. His own past was his second qualification: "during both my army and university experiences, there have been occasions when I was tempted to exclaim, 'Yes, a white man is as worthy as a colored man—provided he behaves himself as well.'"[62] After roasting again his old adversary Wister, and after dealing with the scientific missteps of Bean, Wilder then carefully weighed the evidence for and against the Negro in the matter of innate mental capacity. He concluded that "as yet there has been found no constant feature by which the Negro brain may be certainly distinguished from that of a Caucasian."[63]

The next year Boas appeared before the conference. His topic, "The Anthropological Position of the Negro," promised little more than a review of past research. What he actually delivered, however, was a veiled endorsement of racial intermarriage as the best long-run solution to problems of race.[64] Most white Americans, Boas knew, recoiled from such a proposition, for they viewed Negroes as inferiors, and mulattoes—who supposedly inherited the worst traits of both parent races—as even worse. Yet available evidence belied both beliefs. Historians and anthropologists had found, for example, that race mixing not only had been occurring since prehistoric times but also had produced many peoples of marked capability and achievement.[65]

[60] Franz Boas, "Commencement Address at Atlanta University," in *Race and Democratic Society* (New York: J. J. Augustin, 1945), p. 69; also see pp. 63–66.

[61] For a discussion of the origins of the N.A.A.C.P. and the reactions to the Negro conferences see Elliott M. Rudwick, *W. E. B. DuBois: Propagandist of the Negro Protest* (New York: Atheneum, 1968), pp. 120–131.

[62] Wilder, "The Brain of the Negro," p. 22.

[63] *Ibid.*

[64] On arrangements for Boas' speech, see Oswald Garrison Villard to Boas, Mar. 16, 1910; and Boas to Villard, Mar. 17, 1910, both in Boas Correspondence. Boas' speech was reprinted in *The Crisis*, Dec. 1910, *1*:22–25.

[65] *Crisis*, pp. 22–24.

Although Boas did not say it explicitly, his meaning was clear: white Americans had no reason to fear large-scale intermarriage between the races. At any rate, such mixing was inevitable, he proclaimed. White and black lived side by side in America, and outside additions to the population were almost exclusively European. Those two facts "must necessarily lead to the result that the relative number of pure Negroes will become less and less. . . . The gradual process of elimination may be retarded by legislation, but it cannot possibly be avoided."[66] Of course such a development lay far in the future, but once extreme physical differences were eliminated, existing race tensions would appreciably lessen.[67] Boas' solution, while surely repugnant to the vast majority of Americans, left no doubt about his allegiance to the principle of equality.

A fundamental reason why Boas and Wilder made themselves available to the emerging civil rights movement was their conviction that Negroes, as much as whites, needed to learn about the potential of the black race. Both understood, as so many of the white "friends" of the Negro did not, that slavery and segregation had convinced even Negroes of their inferiority. Even among the "best classes of the Negro," Boas had noted a strong "feeling of despondency. . . ."[68] Until the black man gained confidence in himself he could never play the role in national affairs that he ought. One way to foster race pride, both scientists were convinced, was to convey to the Negro a sense of the achievements of his forebears.[69]

Corroboration of that reasoning and a tribute to the work of Boas and Wilder came in 1911 from a young New York Negro. Willis Huggins was an industrial arts student at Columbia University. After reading two articles by Boas on native African culture, he wrote the scientist to thank him for the "timely information you gave out" on the achievements of the Negro race. Huggins had re-read the pieces to a meeting of young blacks at a local YMCA, and "to a man those present voiced appreciation of your careful research and . . . fair play." Indeed, Huggins concluded, the whole "Army of Negro youth" who were seeking to advance their race had "much to be grateful for in the products of your research."[70]

Like the student Willis Huggins, later historians would generously applaud the humanism of Franz Boas, but unlike the New York Negro most would recall only his efforts in the period after World War I. Burt Green Wilder they would neglect completely.

Overlooked by scholars, the activism of Boas and Wilder in the early twentieth century deserved much better. For one thing, their influence within their respective disciplines surely created a pressure on colleagues to proceed with caution in applying their research to contemporary racial problems. Reputable scientists from other disciplines who also opposed such distortions in that early period likely exerted a similar influence in their fields. To the degree that professional scientists were responsive to that kind of pressure—or example—from their leaders (and the authoritarian nature of scientific organizations would seem to make this likely), then the meaning of Boas'

[66] *Ibid.*, p. 25.

[67] *Ibid.* Radical as these ideas were for the time, Boas' solution would be regarded by some today as a racist scheme, for it would solve the race issue by a gradual purge of an identifiable black race and culture.

[68] Boas to Starr Murphy, Nov. 23, 1906, Boas Correspondence.

[69] *Ibid.*

[70] Willis N. Huggins to Boas, Oct. 16, 1911, Boas Correspondence.

and Wilder's involvement is that early-twentieth-century American science was less racist than commonly portrayed and actually featured a vigorous egalitarian tradition.[71]

In the realm of contemporary popular attitudes Boas' and Wilder's impact was much more limited. But they did exert some influence. Along with a small band of courageous men and women of both races, they at least succeeded in keeping a surging American racism off balance during those bleak decades after 1890. Outside the South, racist doctrine never gained full ascendancy, partly because of the kind of resistance they posed. Historian Oscar Handlin once commented on the value of such holding actions: "Always the 'glittering generalities' of the Revolutionary period have been remembered by some, and the respect due them remained an impediment in the way of those who wished to deny the equality of man—not enough of an impediment to prevent the development of racism, but enough to keep alive a sense of discordance with tradition."[72]

If they failed to win many converts from racism in that early day, Boas' and Wilder's influence on later Americans was far greater. One product of their labors (and that of colleagues sharing their views) was a countervailing ideology of brotherhood, which a recent generation of activists has used with telling effect aginst white supremacist ideas. Indeed, many of the key arguments of the modern civil rights movement were elaborated by Boas and Wilder in the early twentieth century. When later spokesmen talked of the impact on the Negro of a hostile social environment, his need for race pride, the contributions of Africa to culture and civilization, the essential equality of the races, and the insignificance of some racial differences in the face of larger individual variations, they were merely voicing notions advanced a half-century before by scholars like Wilder and Boas.

[71] See the beginning of this article for a listing of other scientists who apparently shared Boas' and Wilder's outlook. To be sure, there were reputable scholars in the period who were committed to the concept of inferior races. E. A Ross in sociology and C. B. Davenport from genetics are notable examples, although the historian of the eugenics movement describes Davenport as a careless and manipulative scientist, whose methods were often faulty. See Mark Heller, *Eugenics. Hereditarian Attitudes in American Thought* (New Brunswick: Rutgers University Press, 1963), pp. 67–68. An alternative way of viewing the racial attitudes of American scientists would be to focus on academic disciplines rather than on individuals. Doing this reveals what appears to be a common pattern in the development of many of the behavioral and biological sciences. Over time, an increased commitment to inductive scientific procedures came to characterize these disciplines. As that happened, an earlier attachment to sweeping generalization gave way to a more critical approach emphasizing empirical data and de-emphasizing grand theory. One result of the shift was a reconsideration of the presumed inferiority of the Negro. These developments occurred at different times for various disciplines, but it seems apparent that anatomy had adopted a more cautious, empirical approach by the end of the nineteenth century, anthropology had reached that stage before World War I, genetics and sociology shifted over by the 1920s, and psychology was reconsidering some of its earlier racial theories by the 1930s. Discussions of this phenomenon for particular disciplines are found in Heller, *Eugenics*, pp. 66–67, 163, 167; Roscoe C. and Gisela J. Hinkle, *The Development of Modern Sociology* (Garden City, N.Y.: Doubleday, 1943), p. 22; and Ruth Benedict, *Race, Science and Politics* (New York: Viking Press, 1968), pp. 71–78, which looks at changing racial views in one area of psychology. Stocking considers this process of professionalization and its impact on race thinking in a different way. Focusing on the behavioral sciences, Stocking argues that what caused the decline of the idea of racial determinism within those disciplines was the adoption of the culture concept from anthropology. Although assimilation proceeded at varying rates for the several disciplines, by the 1930s the idea of a cultural explanation for human behavior, he says, had achieved the status of a paradigm throughout the behavioral sciences. See *Race, Culture, and Evolution*, pp. 303–305.

[72] Handlin, *Race and Nationality in American Life*, p. 146.

Finally, their efforts also had significance for the social history of science in the United States in that their fight for racial justice set one of the historic precedents for the recent and widespread social activism of American scientists. Following World War II scholars in a variety of fields came to regard such involvement as a normal professional role: nuclear physicists compaigned publicly for disarmament and control of atomic weapons; biologists fought against pollution and population explosion; psychiatrists and biologists moved in the vanguard of the civil rights movement. Not since the period of the American Revolution, when "philosophers" such as Thomas Jefferson, David Rittenhouse, Benjamin Rush, and Benjamin Franklin assumed leading roles in the political and social movements of their day, had American scientists so eagerly sought to relate their studies to the reform needs of their time. Franz Boas and Burt Green Wilder, as much as any other scholars of their day, helped revive the idea that a scientist owed more to society than mere pursuit of knowledge and its economic application.

The Introduction of *Drosophila* into the Study of Heredity and Evolution: 1900–1910

*By Garland E. Allen**

IN THE HISTORY OF SCIENCE advances in theory often depend upon the introduction of new methods, techniques, and "favorable" material. In biology, the latter often means the right organism for investigating a particular phenomenon. All too frequently the introduction of techniques or organisms into research is regarded as the inspiration of a single genius, yet research into the history of such developments generally shows that techniques and materials do not crop up suddenly, in one stroke, within the scientific community. Rather, they are often widespread among a number of workers just prior to or contemporaneous with the individual or group who is frequently credited with "discovering" them. No case illustrates this general principle more clearly than the introduction of the small fly *Drosophila melanogaster* into the study of heredity during the first decade of the twentieth century.[1]

The "discovery" of *Drosophila* as a favorable organism for experiments in heredity is usually credited to Thomas Hunt Morgan (1866–1945).[2] As this paper will attempt to show, Morgan did not "discover" *Drosophila* in 1910. The organism had been used by a number of workers in the decade before Morgan accidently hit upon it for his studies on evolution. In so doing, however, he and his co-workers revolutionized the science of genetics and propelled it into the central position it has occupied ever since in the history of twentieth-century biology.

*Department of Biology, Washington University, St. Louis, Missouri 63130.

[1] In the early years of the century *Drosophila* went under a variety of names. The common variety, later called *Drosophila melanogaster*, was prior to 1910 referred to as *Drosophila ampelophila*, or *D. amplophora*. In common usage *Drosophila* was often referred to as the pomace fly, vinegar fly, fruit fly, or banana fly. Throughout this paper I will use the terms that have been in general use since 1910: *Drosophila melanogaster* for the scientific name, and fruit fly for the common name.

[2] E.g., in Bernard Jaffe's *Men of Science in America* (New York: Simon and Schuster, 1946), pp. 387–388:

> The evening primrose did not satisfy Morgan. He wanted a short-lived organism and one that could be easily bred in the laboratory under changing conditions. He tried the mouse, the rat, the pigeon, and even undertook some painstaking experiments on the intricate life cycle of a plant louse, until one day he heard of another insect, which W. E. Castle of Harvard had been using in connection with certain investigations in inbreeding. . . . *Drosophila*, said one wit, must have been created by God expressly for Morgan.

Between 1910 and 1925 Morgan and his group at Columbia University were able to show that Mendelian "factors," or "genes" as they came to be known after 1909, could be viewed as discrete entities arranged linearly on the chromosomes in the cell nucleus. This work went far toward establishing the material basis for the Mendelian theory and suggested the precise mechanism by which the process of hereditary transmission—its regularities and irregularities—took place. Morgan and his co-workers not only provided the intellectual satisfaction inherent in showing that a theoretical scheme (such as Mendel's) has a basis in material reality, they also were able to make concrete gains in elucidating many otherwise inexplicable phenomena associated with the Mendelian theory. For instance, with the chromosome theory, they explained why certain traits seemed to be inherited together (linkage) and others not; why linkage groups occasionally appeared to dissociate (through crossing-over and recombination); how the expression of some traits is affected by specific alterations in the chromosomes (such as deletion, the loss of a chromosomal part); and why certain traits appear much more frequently in one sex than in the other (what came to be called sex linkage). They even provided maps of the positions of specific genes relative to each other on the chromosomes and eventually related these to actual chromosomal structure. For this elegant work Morgan received the Nobel Prize in Physiology and Medicine in 1933.

How did Morgan become interested in heredity, and how, of all the organisms available to use for breeding studies, did he hit upon the fortuitous choice of *Drosophila*?

Morgan was originally trained as an embryologist under W. K. Brooks (1848–1908) at Johns Hopkins University (Ph.D. 1891). Brooks introduced Morgan, like all of his Hopkins students, to embryology in its broadest form—as general morphology. The purpose of morphology was ultimately to determine evolutionary relationships by using a variety of methods: taxonomy, comparative anatomy, embryology, physiology, and ecology. Embryology was a cornerstone of the morphological method, for it was thought that by comparing the patterns of embryonic growth phylogenetic (evolutionary) relationships could be more accurately determined. One aspect of the evolutionary process which particularly fascinated Brooks was heredity: the origin and transmission of variations. He passed this interest on to his students as one of the compelling problems of modern biology.[3] William Bateson, who worked under Brooks' guidance at the seashore laboratory in Beaufort, North Carolina, during the summers of 1883 and 1884, wrote: "For me this whole province [heredity] was new. Variation and heredity with us had stood as axioms. For Brooks they were problems. As he talked of them the insistence of these problems became imminent and oppressive."[4]

Thus, from the beginning, Morgan was primed to see the importance of

[3] Dennis M. McCullough, "W. K. Brooks' Role in the History of American Biology," *Journal of the History of Biology*, 1969, *2*:411–438, esp. p. 433.

[4] William Bateson in "Wm. Keith Brooks: A Sketch of His Life by Some of His Former Pupils and Associates," *Journal of Experimental Zoölogy*, 1910, *9*:1–52; see esp. pp. 5–8 (quotation on p. 7).

heredity to all biological problems—in evolution, embryology, and physiology. However, his own early work (between 1891 and 1910) was largely in embryology, especially in experimental embryology. Morgan was an enthusiastic proponent of the new experimental school of embryology known as *Entwicklungsmechanik,* established in the late 1880s by Wilhelm Roux (1850–1924) and supported by a large number of younger European biologists such as Hans Driesch (1867–1941). Given his background with Brooks, Morgan's work in embryology gradually led him closer to the interrelated problems of heredity and evolution.

Through studies on regeneration, Morgan came to question the Darwinian theory of natural selection as a means of explaining how evolution occurs.[5] How, he asked, could the selection of minute, randomly occurring variations account for something so complex and adaptive as the power to regenerate? What causes variations? How is it that new variations are not swamped as soon as they occur? And, how can a small variation, even though moving toward an ultimately adaptive function, be useful in its incomplete, early stages?[6]

To Morgan, Darwin's theory was weakest in its insistence on minute, randomly occurring variations as the raw material on which selection acts. As a result, Morgan became especially interested in the "mutation theory" published between 1901 and 1903 by the Dutch plant physiologist and breeder Hugo De Vries (1848–1935). De Vries had attempted to provide an alternative to Darwinian theory—an alternative which did not depend upon selection acting on minute, individual variations. De Vries based his theory largely on observations of a single species, the evening primrose *Oenothera lamarkiana,* claiming that in this form he had observed variations sufficiently large to produce a new species in one generation.[7] Darwin had referred to such large-scale variations as "sports" or "monstrosities" and accorded them little importance in the general theory of evolution. (Today biologists know of no cases where a new species arises in a single generation by such large-scale variations.) To Morgan, like many of his contemporaries, De Vries' theory seemed a very attractive alternative to orthodox Darwinism as they understood it.

Through another aspect of his embryological work prior to 1910—the study of sex determination—Morgan again came up against the problem of heredity and variation. As an embryologist he had become interested, around 1900, in the problem of how sex is determined: is it largely hereditary (determined at the moment of fertilization by factors inherent in the egg, sperm, or both), or is it developmental (determined by an interaction between the cells of the embryo and external factors such as temperature, nutrition, ions in solution)? For several decades biologists had been divided on this issue, attempting with

[5] Garland E. Allen, "T. H. Morgan and the Problem of Natural Selection," *J. Hist. Biol.*, 1968, *1*:113–139.

[6] Morgan's most thoroughgoing criticism of the Darwinian theory is found in *Evolution and Adaptation* (New York: Macmillan, 1903).

[7] Hugo De Vries, *The Mutation Theory,* trans. J. C. Farmer and A. P. Darbyshire (Chicago: Open Court, 1910), Vol. I, p. viii. De Vries' work had originally been published in German as *Die Mutationstheorie,* 2 vols. (Leipzig: Von Veit, 1901–1903). For a further treatment of De Vries' theory and its reception, see Garland E. Allen, "Hugo De Vries and the Reception of the Mutation Theory," *J. Hist. Biol.*, 1969, *2*:55–87.

many observations and experiments to prove one case or the other.[8] As an embryologist Morgan inclined toward the view that sex was a developmental phenomenon, though he admitted that the experiments attempting to show that sex ratios could be influenced by environmental factors were inconclusive.[9]

Morgan's growing interest in the problem of heredity was stimulated even more after he moved (from Bryn Mawr) to Columbia University in 1904, where his friend (and department chairman) Edmund Beecher Wilson (1856–1938) was currently engaged in cytological studies on the chromosomal nature of sex determination. Though he admired Wilson's cytological work in general, Morgan was not disposed to see a complex adult trait like sexual differentiation "explained" by reference to discrete cell structures such as chromosomes.[10] More and more in the period after 1905, Morgan found his own laboratory work turning explicitly toward problems of heredity as the newly discovered (1900) Mendelian theory began to spread. He carried out breeding experiments with mice, rats, pigeons, and chickens to test the validity of the Mendelian theory of which he was at that time highly skeptical. He made a detailed analysis of the breeding experiments of the French biologist and Mendelian Lucien Cuénot on yellow mice.[11] He tried to study the effects of temperature

[8] For a summary of some of these experiments and theories, see Mark Green, "The Development of a Chromosomal Theory of Sex Determination" (Cambridge, Mass., unpublished A.B. thesis, Harvard College, 1966).

[9] Garland E. Allen, "Thomas Hunt Morgan and the Problem of Sex Determination," *Proceedings of the American Philosophical Society*, 1966, *110*:48–57.

[10] In a letter to Hans Driesch in 1905, Morgan wrote:

> As to chromosomes, I am in the thick of it here. Wilson's recent discovery that in certain bugs the spermatozoon that has the extra chromosomes makes the *female every time,* and the one without it makes the male (exactly the reverse of McClung's supposition) makes it look at first sight as though the chromosomes were *the thing.* But work it through as he has done and you will find that it lands you in an absurdity for that same chromosome will be a male determining one in the following generation. Wilson is going to indulge in generalities about it. I will confess that when he first showed me his results I was somewhat staggered, especially as I had just sent a little paper to the press in which I had attacked the well-known assumption that the nucleus must be the bearer of the hereditary qualities of the male. . . . Now however when it is evident that the chromosome theory will not even explain the case of sex determination I feel assured of my position once more.

(Letter from T. H. Morgan to Hans Driesch, Oct. 23, 1905, p. 2. Microfilm supplied through the courtesy of Professor Dr. Reinhard Mocek, University of Halle, who discovered these letters while preparing a study of Driesch's philosophy and was kind enough to make them available to me. All further citations from these letters are from the same source.)

Morgan's statement about the logical absurdity of Wilson's conclusion is based on the following argument: the single X chromosome (called the accessory chromosome throughout the period prior to 1908) carried by the male inevitably determines a female offspring in the next generation (since eggs contain only X chromosomes). Thus, an X-bearing sperm fertilizing any egg produces a chromosome complement of XX, or female. However, that female will pass on the X from her father in one-half the eggs she produces. If any of those eggs are fertilized by a sperm bearing a Y chromosome, or in some cases no complement to the accessory X, the resulting offspring is male. To Morgan it was absurd to think that a single chromosome could have any direct determining effect on sexual differentiation when the same chromosome could be found in either a male or female. In Morgan's view, to conclude that the X chromosome *determines* femaleness, it would be necessary to show that the X appears always, and only, in the female sex.

[11] Lucien Cuénot, "Les races pures et leur combinaisons chez les souris," *Archives de Zoologie Expérimentaie et Générale*, 1905, *3*: notes cxxiii–cxxxii. See also T. H. Morgan, "Some Experiments in Heredity in Mice (Abstract)," *Science*, 1908, *27*:493; and "Recent Experiments on Inheritance

and salt solutions on altering sex ratios, and he carried out a detailed cytological analysis of the chromosomes in several species of naturally parthenogenetic forms (aphids and rotifers).[12] He had also begun to breed various species in an attempt to find De Vriesian "mutations" in animals. It was this latter issue which led Morgan directly to the breeding of *Drosophila.*

In 1904 De Vries gave the inaugural address at the opening of the Station for Experimental Evolution, founded by C. B. Davenport with the financial backing of the Carnegie empire, at Cold Spring Harbor, New York. In that address De Vries suggested that mutations might be induced by exposing animals and plants to the recently discovered roentgen and curie rays produced by the "decay" of radium. Perhaps directly inspired by this suggestion, Morgan had apparently begun to expose a number of species of animals, especially *Drosophila,* to radium and other substances such as salts, sugars, acids, alkalis, and proteins (injected principally into insects in the region of the reproductive tissue) with the hope of obtaining some new species. As he wrote to the botanist A. F. Blakeslee many years later:

> I did quite a lot of work by treating the flies with radium and, as a matter of fact, some of the descendants produced mutants of the type we are now familiar with [1935]. But since I did not get them in the immediate offspring of the treated flies, I thought the results not worth publishing, and made only a brief statement with regard to the facts in the case.[13]

Similar experiments were carried out about the same time by Morgan's friend Jacques Loeb (1858–1924) and a colleague, Wilder Dwight Bancroft (1867–1953).[14] Neither the experiments of Morgan nor those of Loeb and Bancroft produced any certain or conclusive results. Although Loeb and Bancroft claimed that they obtained some definite, though small (non–De Vriesian) mutations,

---

of Coat Color in Mice," *American Naturalist,* 1909, *43*:494–510, esp. pp. 503–504. That Morgan was carrying out these experiments as early as 1905 is confirmed in a letter to Driesch, Aug. 15, 1906, where Morgan describes the work which he and his students were carrying out at the Marine Biological Laboratory, Woods Hole, Massachusetts, during the summer of 1906 (p. 3). Cuénot had obtained unexpected (non-Mendelian) results (i.e., ratios among offspring in a cross between two yellow parents) which he had explained by reference to an hypothesis Morgan disliked (selective fertilization—the idea that sperm carrying certain hereditary traits will only fertilize eggs carrying those same or some other specific trait). Today Cuénot's results are explained by reference to "lethal" factors, genes which if present in the homozygous (double-dose) condition, cause the embryo to die; this, of course, alters the ratios which would otherwise be expected among the offspring.

[12] T. H. Morgan, "Sex Determining Factors in Animals," *Science,* 1907, *25*:382–384; "Sex Determination and Parthenogenesis in Phyllozerans and Aphids," *Science,* 1909, *29*:234–237; "Chromosomes and Heredity," *Amer. Natur.,* 1910, *44*:449–496.

[13] T. H. Morgan to A. F. Blakeslee, May 27, 1935. From the Morgan Papers, California Institute of Technology Archives, Box 1 (hereafter referred to simply as Morgan Papers).

[14] Jacques Loeb and W. D. Bancroft, "Some Experiments on the Production of Mutants in *Drosophila,*" *Science,* 1911, *33*:781–783; that Loeb and Morgan worked independently on this topic for a period between 1909 and 1910 is indicated in a letter from Loeb to Morgan, May 17, 1911 (Loeb Collection, Library of Congress, Box 9, Folder 1).

Morgan doubted that their results were any more certain than his.[15] Nonetheless, it appears that Morgan began the serious and systematic breeding of *Drosophila* in the laboratory primarily for attempted studies of "mutation." Two questions immediately come to the fore: when did Morgan first begin to breed *Drosophila* in the laboratory, and how did he come to use this fruit fly?

To answer the first question, we can say with some certainty that *Drosophila* was being used in Morgan's lab at least by 1907; furthermore, he probably had direct contact with laboratory breeding of the insect as early as 1906. In 1908 Morgan wrote that one of his students was centrifuging many kinds of eggs, a common procedure among experimental embryologists under the influence of the *Entwicklungsmechanik* school in the early 1900s. One of these students had worked with the eggs of the fruit fly, and by centrifuging the eggs had produced an "abnormal development which has been inherited in the next generation."[16]

To answer the second question (what led Morgan directly to the idea of using *Drosophila* in his own experiments on mutation?) requires some historical detective work. It is impossible to date precisely when Morgan first learned that this was an easy organism to raise in the laboratory or when it first occurred to him to use it in his own work. Some attempt to pinpoint the date, however, gives an indication of the wide use to which *Drosophila* was being put in the first decade of the twentieth century.

It is possible that Morgan first explicitly encountered the breeding of *Drosophila* in the laboratory as early as 1905. For a number of years, one of Morgan's former graduate students from Bryn Mawr, Nettie M. Stevens, had been carrying out cytological studies on chromosomes in male and female insects. One of the insects she had worked with was *Drosophila,* which she had bred in the laboratory at Bryn Mawr as early as 1905 (Morgan had left Bryn Mawr in 1904).[17] Morgan had always been particularly interested in Stevens' work and thus must have known about the feasibility of breeding *Drosophila* in the laboratory through contact with her.

More evidence suggests, however, that *Drosophila* become known to Morgan through the work of W. E. Castle (1867–1962) and his associates at Harvard. Castle, like Morgan, was originally trained as an embryologist but had much earlier given up embryology for studies in evolution and heredity. In the early 1900s, partly motivated by the popular eugenics movement, Castle had become interested in the problem of inbreeding (brother-sister matings within a single

[15] Morgan wrote to Blakeslee on May 27, 1935: "Even after x-raying the number of mutants is not very large, and there are many kinds of mutants. . . . In regard to the experiments of Loeb and Bancroft, I am quite sure that what they got were mutants that were already present in the stock supplied to them" (Morgan Papers, Box 1).

[16] Morgan to Driesch, Jan. 20, 1908, p. 1.

[17] N. M. Stevens, "A Study of the Germ Cells of Certain Diptera, with Reference to the Heterochromosomes and the Phenomena of Synapsis," *J. Exp. Zool.,* 1907, *5*:359–374. As she claims here (p. 359) that the germ cells "were examined in the autumn of 1906," and that her studies on *Drosophila* "extended over more than a year" (p. 365), we can conclude that she must have been working with the organism in the laboratory at least by the summer or fall of 1905.

line over a number of generations).[18] According to J. Walter Wilson (1896–), whose teacher at Brown University, H. E. Walter, had been one of Castle's students (Ph.D. 1906), Castle had planned to conduct experiments on inbreeding with rabbits; but there were certain difficulties with what he proposed. As Wilson reported:

> Questions were raised, by Morgan among others, as to the interpretation of the results in the light of the deleterious effects of inbreeding. As a result of these questions Castle took as a rapidly breeding animal Drosophila and inbred it for 20 generations without any detectable loss of vigor.[19]

It is obvious that *Drosophila* would fill the bill for any selection or heredity experiments because of its short generation time of three weeks. How did Castle come to use *Drosophila,* and how did Morgan learn of Castle's work?

Drosophila is an extremely common insect which has been known since ancient times. A graduate student in Castle's laboratory around 1900, C. W. Woodworth, began breeding *Drosophila* for the purposes of embryological investigation. Woodworth "was growing *Drosophila* on concord grapes and suggested the animals as a suitable organism in Castle's works on the effects of inbreeding. Later when the grape crop gave out, Castle substituted bananas which seemed to be the technique for many years."[20]

At least one other person in Castle's laboratory also began working with *Drosophila* about the same time, F. W. Carpenter, who studied behavioral responses in the insect.[21] The first hereditary studies published from Castle's laboratory appeared in 1906 and included the results of the inbreeding experiments.[22] Castle even saw the inbreeding work in a Mendelian light, designating specific traits—for example, high or low productiveness—as a Mendelian dominant and recessive, respectively. Castle's reports represent the first published work on heredity in *Drosophila.*

Drosophila appears to have been "in the air" (no pun intended) by 1906 or 1907. At least two other persons took up breeding experiments with this insect, stimulated directly by Castle's work. One was W. J. Moenkhaus at Indiana University, who had worked with Castle for two years at Harvard (1896–1898), prior, however, to Castle's earliest known work with *Drosophila.* In the early 1900s Moenkhaus had taken up the question of heredity and sex determination, and had bred *Drosophila* for this purpose.[23] The other was Franke E. Lutz,

[18] See, e.g., W. E. Castle, "The Laws of Heredity of Galton & Mendel, and Some Laws Governing Race Improvement by Selection," *Proceedings of the American Academy of Arts and Sciences,* 1903, *39*:223–242. Also J. Walter Wilson to A. H. Sturtevant, Sept. 20, 1965 (Sturtevant Papers, California Institute of Technology Archives; this source will be cited simply as Sturtevant Papers).

[19] Wilson to Sturtevant, *op. cit.*

[20] Blakeslee to Morgan, May 22, 1935 (Morgan Papers, Box 1).

[21] Carpenter's paper on *Drosophila* behavior was chronologically the first published report on laboratory studies with *Drosophila;* see *Amer. Natur.,* 1905, *39*:157–171.

[22] W. E. Castle, "Inbreeding, Cross-breeding and Sterility in *Drosophila,*" *Science,* 1906, *23*:153.

[23] W. J. Moenkhaus, "Report of Dr. W. J. Moenkhaus" in *Carnegie Institution Yearbook No. 5, 1906, Annual Report of Department of Experimental Evolution* (New York, 1907), p. 105; also "The Effects of Inbreeding and Selection on the Fertility, Vigor and Sex Ratio of Drósophila Ampelophila," *Journal of Morphology,* 1911, *22*:123–154.

working at the Station for Experimental Evolution, Cold Spring Harbor, from 1904 to 1909 (and after 1909 at the American Museum of Natural History). Lutz had become interested in heredity and evolution in insects, especially the problems of use, disuse, and degeneration, sexual selection, the inheritance of abnormal features (wing venation), and inbreeding. Lutz had published four papers on heredity and selection in *Drosophila* prior to 1909. According to J. Walter Wilson, Lutz had come to work with *Drosophila* out of an interest in Castle's inbreeding experiments. Castle had inbred *Drosophila* for twenty generations, and Lutz had obtained some of Castle's stock in order to continue the experiment (particularly, testing the effects of inbreeding on fertility).[24]

Morgan undoubtedly knew of most of this work. He was an avid reader of the professional literature, which in the years between 1906 and 1909 contained over a dozen references to specific studies on *Drosophila* (by Carpenter, Castle, Moenkhaus, Lutz, and Stevens). Furthermore, two papers, by Lutz and Stevens, had been read at the Seventh International Congress of Zoology (Boston, 1907) and later published in their proceedings.[25] Morgan had been present at that congress as chairman of the section on experimental zoology,[26] and it is most likely that he heard both Lutz's and Stevens' papers.

But it was apparently the experimental breeding of *Drosophila,* initiated by Castle in particular, which most specifically attracted Morgan's attention. Morgan, in fact, acknowledges his direct indebtedness to Castle's work for calling his attention to *Drosophila.* Three years before his death, Morgan wrote in a short article titled "Genesis of the White-Eyed Mutant": "I was, of course, familiar with the important paper of Castle, Carpenter, Clarke, Mast and Barrows and used it in my lectures on experimental zoology [at Columbia]. It seemed to me that *Drosophila ampelophila* might furnish excellent material for genetic work since it might be easily and cheaply bred in the laboratory."[27] And in a letter to Blakeslee in 1935, Morgan wrote:

> It is quite correct that Woodworth, working in Castle's laboratory, was the first to show the availability of Drosophila for experimental purposes. I knew of this work, of course, and while looking for available material for work which could be carried out in New York, and specifically at Columbia where there were no funds for raising larger animals, I got some material . . . from outside to see whether I could find characters suitable for genetic work, which turned out as you know to be the case.[28]

As important as the increasing references to *Drosophila* may have been in

[24] J. Walter Wilson to O. P. Jones, Oct. 8, 1965 (Sturtevant Papers).

[25] F. E. Lutz, "Inheritance of Abnormal Wing Venation in *Drosophila.* Read before the 7th International Congress of Zoology," and N. M. Stevens, "The Chromosomes in *Drosophila ampelophila.* Read before the 7th International Zoological Congress, Boston (1907)," *Proceedings of the 7th International Zoological Congress,* 1912, pp. 411–419; 380–381.

[26] Morgan to Driesch, Sept. 15, 1907.

[27] T. H. Morgan, "Genesis of the White-Eyed Mutant," *Journal of Heredity,* 1942, *33*:91–92 (p. 92). The paper that Morgan refers to is W. E. Castle, F. W. Carpenter, A. H. Clarke, S. O. Mast, and W. M. Barrows, "The Effects of Inbreeding, Cross-breeding, and Selection upon the Fertility and Variability of Drosophila," *Proc. Amer. Acad. Arts Sci.,* 1906, *41*:729–786.

[28] Morgan to Blakeslee, May 27, 1935 (Morgan Papers, Box 1).

stimulating Morgan's actual use of the organism, there was another equally if not more powerful stimulus acting on him directly. Fernandus Payne, formerly an undergraduate at Indiana University, was a graduate student at Columbia from 1907 to 1909 (Ph.D. 1909). While at Indiana, Payne had worked with Moenkhaus investigating blindness in cave fish as one example of the current theory of the inheritance of acquired characters. Payne knew of Moenkhaus' work with *Drosophila*, and of the ease with which the insect could be cultured and bred in the laboratory.[29]

When Payne entered Columbia in the fall of 1907, he enrolled in Morgan's course in experimental zoology, where Morgan discussed the breeding experiments on *Drosophila* by Castle and his associates. As a part of this course each student carried out a laboratory research problem, and in discussing prospective topics, Payne and Morgan came to the question of the role of Lamarckian principles in evolution. Morgan knew of Payne's earlier interest in blind fish, and between them they questioned whether the influence of darkness on inheritance of eyesight might be tested in the laboratory. It may have been Payne who suggested using *Drosophila* for this purpose. The project which Payne designed with Morgan was straightforward: "We planned the simple experiment of breeding *Drosophila* in the dark to see whether we could change the reactions of this fly to light."[30] Over the next several years Payne carried out his experiments with a total of forty-nine generations, in which no decrease in visual function was noted.[31] This was in the fall of 1907; Payne began his breeding experiments in October. He recounts that Morgan specifically urged him to collect his own initial flies from the wild rather than getting them from ongoing cultures. Payne wrote: "I used the simple procedure of laying some ripe bananas on the window sills; the flies thus caught were the start of my experimental work."[32] Payne went on, under Morgan's supervision, to try to induce mutations in *Drosophila* by a variety of factors: heat, cold, centrifuging the eggs, and even the effects of X rays.

Thus, by the time Morgan had decided to try to find (or induce) De Vriesian mutations in animals (around 1908, as best we can surmise), he was well aware of the general use to which *Drosophila* had been put in at least three other laboratories (Castle's, Stevens', and Moenkhaus') and the specific procedure and utility of *Drosophila* from firsthand acquaintance with Payne's experiments in his own laboratory.

Three stories have circulated among older geneticists and biologists about where Morgan obtained his initial culture of *Drosophila:* (1) that he obtained flies directly from Castle, who had been using them for the inbreeding studies; (2) that he simply collected his own as he had told Payne to do (or that he used some of Payne's ongoing stock); and (3) that he obtained a pure-line culture from F. E. Lutz (see above). The first story was originated by H. E.

[29] Fernandus Payne to A. H. Sturtevant, Oct. 16, 1947 (Sturtevant Papers).
[30] *Ibid.*
[31] F. Payne, "Forty-Nine Generations in the Dark," *Biological Bulletin*, 1910, *18*:188.
[32] Payne to Sturtevant, *op. cit.*

Walter, of Brown University, and relayed to A. H. Sturtevant by J. Walter Wilson.[33] Since there is, however, no independent corroboration of this story, it must be considered an unlikely possibility. The second is much more likely, though Payne himself does not recall that Morgan actually collected any flies of his own or used any from Payne's stock.[34] The third story is the one which Morgan himself told, though not until many years after the fact (1942):

> It seemed to me that *Drosophila ampelophila* might furnish excellent material for genetic work since it could be easily and cheaply bred in the laboratory. I knew that Lutz was breeding these flies and asked him to give me a culture, which he kindly did.[35]

Although Payne specifically disputes Morgan's claim,[36] it has been further verified by Sturtevant, quoting Morgan's long-time assistant, Edith Wallace. She remembered Lutz actually bringing the flies to the Columbia laboratory, these being the only cultures she worked with consistently.[37] Since Lutz moved even closer to Morgan in 1909 (from Cold Spring Harbor, Long Island, to the American Museum, just downtown from Columbia) it would have been more obvious for Morgan to approach Lutz (rather than, for example, Castle) for *Drosophila* stocks.

How can we explain what appears to be a contradiction between the procedures Morgan used to obtain his own stocks of *Drosophila* and the procedures he advocated to Lutz? Given Morgan's usual approach to problems, it is likely that he started his search for mutations with stocks collected himself but switched somewhere around 1910 to cultures obtained from Lutz. Morgan often began lines of research on a hunch, working with whatever forms were available. But by 1909 he may well have become aware that a largely wild stock of flies would have disadvantages for the detection of mutations. Genetic heterogeneity within a stock adds many variables to the study of heredity. If the original stocks were highly variable, then the appearance of any new variation could not be ascribed necessarily to mutation but could be simply the reappearance of a latent or recessive trait. Thus, there was considerable advantage, from the experimental point of view, to working with an inbred and thus genetically pure stock in searching for newly arisen mutations.

Why, then, did Morgan insist that Payne collect his own flies for the experiments on the inheritance of acquired characters? It is obvious that any studies of the effects of selection on evolution should begin with as heteroge-

[33] Wilson to Sturtevant, Sept. 20, 1965 (Sturtevant Papers).

[34] Payne to Sturtevant, *op. cit.*; Sturtevant to J. Walter Wilson, Sept. 22, 1965; O. P. Jones to J. Walter Wilson, Oct. 5, 1965 (all letters in Sturtevant Papers).

[35] Morgan, "Genesis of the White-Eyed Mutant," p. 92.

[36] In a letter to Sturtevant on Oct. 16, 1947, Payne wrote: "I am certain he [Morgan] did not get [a stock] from Lutz" (Sturtevant Papers).

[37] Sturtevant wrote in the letter to Wilson on Sept. 22, 1965: "I can add that Miss E. M. Wallace, who was Morgan's assistant, told me (about 1960) that she remembered that Lutz did bring Drosophila to Columbia, and that it was with these cultures that her own experience with the animal began."

neous, or natural, a population as possible. In other words, the value of selection experiments lies in being able to relate the effects of selection populations found in the wild, which normally contain a fairly wide range of variability. For this purpose flies collected on the window sill would be highly appropriate.

* * *

It is clear from the foregoing that Morgan was by no means the first to conceive the idea, or carry out the practice, of breeding *Drosophila* in the laboratory. The organism had been exploited by several researchers prior to 1908 when Morgan appears to have taken up his first breeding experiments. Morgan's interest in evolution, and particularly in De Vries' mutation theory, stimulated his search for the means of producing mutations in animals. After trying some mammalian forms, he appears to have turned to *Drosophila* largely because, through Payne's studies on selection, he saw the practical advantages of the organism in the laboratory.

To say that Morgan's mind was "prepared" to see the values of *Drosophila* as a favorable organism for the study of heredity is to beg the most important question. He was "prepared" largely because of a series of developments affecting many others as well: (1) the intense questioning of Darwinian theory which led Morgan, De Vries, and others (for example the neo-Lamarckians), to seek organisms which could be experimentally studied in the laboratory; (2) the belief that a precise knowledge of patterns of inheritance was essential for any full-scale understanding of the evolutionary process; and (3) the increasing use of *Drosophila* itself as a favorable laboratory organism by a number of workers between 1900 and 1910.

What Morgan brought uniquely to the situation was the ability to move quickly and incisively once he discovered the crucial but accidental appearance of the white-eyed male fly in 1910. Lutz claimed that he observed what he thought was the dead body of a white-eyed fly in the original culture he gave to Morgan, but he had assumed it was a result of the not-uncommon practice in which larvae eat the eyes out of dead adults.[38] More important, he had not examined it closely and, by his own admission, did not recognize its importance. Had he seen the value such a mutant could have, Lutz wrote, he would not have given it away so readily. However, he admitted, "it fell into good hands."[39] There is, of course, a real question whether Lutz had ever seen a bona fide white-eyed mutant. Whatever exactly Lutz had observed, it did not catch his attention in the way the white-eyed mutant caught Morgan's

[38] Morgan, "Genesis of the White-Eyed Mutant," p. 92. Even if this had been a genuine white-eyed mutant, however, Morgan felt it could not have been ancestral to his white-eyed mutant. If the fly had mated before dying, then white-eyed males would have turned up within the next several generations (just as they turned up in Morgan's F generation), and this did not happen. Thus Morgan was reasonably satisfied that the white-eyed mutation which he had observed in 1910 had originated spontaneously in the culture in his possession and was not present in the initial flies he received from Lutz.

[39] As quoted by Morgan (*ibid.*).

a year later. Once Morgan saw the use to which the mutant could be put, he acted quickly (as was his way) to design some sort of experiment around it.

Contrary to popular view, once Morgan had discovered the white-eyed fly in 1910, he did not begin to breed it to test the applicability of the Mendelian scheme. In fact he remained skeptical of the basic tenets of Mendelism even after the mutant fly appeared. Morgan bred the mutant to see if it would remain true, like a De Vriesian mutation, or whether it would be swamped, like Darwin's small-scale variations.[40] However, when the white-eyed trait disappeared in the F_1 generation but reappeared in the F_2 in a roughly 3:1 ratio, Morgan recognized the similarity to Mendelian predictions. Because he could change his mind when confronted with the evidence, Morgan was able to exploit the white-eyed mutant, and *Drosophila* in general, to elucidate the very theory he had once so strongly opposed. The development of *Drosophila* studies from there on has been detailed elsewhere.[41]

None of this narrative is meant to detract from Morgan as a person or from the very real contributions which he made to the study of heredity. It is meant, rather, to emphasize the collective and social nature of the scientific enterprise. The work of Morgan and his group with *Drosophila* in the years after 1910 was made possible because many other people had contributed techniques, knowledge, and enthusiasm about the use of this insect in the laboratory. By emphasizing only Morgan's genius, as significant as it was, the myth is perpetuated that science is advanced primarily by great individuals (the Newtons, Darwins, and Mendels). A more valid view of the history of science lies in seeing the vital, even determining, role which a particular community of workers has on the development of new theories. The danger lies not so much in overestimating the genius of one or two but more in underestimating the genius of the many.

[40] See, e.g., T. H. Morgan, "Hybridization in a Mutating Period in *Drosophila*," *Proceedings of the Society of Experimental Biology*, 1910, 7:160–161. The term "mutating period" is directly from De Vries, who thought that mutations occurred in large numbers at certain times, falling off in frequency at others.

[41] See Garland E. Allen, Introduction to T. H. Morgan, A. H. Sturtevant, C. B. Bridges, and H. J. Muller, *The Mechanism of Mendelian Inheritance* (New York: Johnson Reprint, 1972). Also, A. H. Sturtevant, *A History of Genetics* (New York: Harper & Row, 1965), and E. A. Carlson, *The Gene: A Critical History* (Philadelphia: W. B. Saunders, 1966).

George Ellery Hale, the First World War, and the Advancement of Science in America

*By Daniel J. Kevles**

EUROPE WAS AT WAR, the *Lusitania* had been sunk, and early in July 1915 Secretary of the Navy Josephus Daniels asked Thomas A. Edison to gather together the nation's "keenest and most inventive minds" to find a defense against the submarine.[1] By October the venerable inventor had formed the Naval Consulting Board, a coalition, save for two mathematicians, of representatives from America's major engineering societies. The National Academy of Sciences, the government's official scientific adviser, had been omitted.

George Ellery Hale considered the omission "depressing."[2] A distinguished astronomer, editor of the *Astrophysical Journal,* Director of the Mt. Wilson Observatory, and Foreign Secretary of the National Academy, Hale was an advocate of the NAS. Its members, he was sure, ought to have as much a role in national defense as the Edisons. Besides, Hale had ambitions for the NAS that went beyond any martial purpose to the peacetime advancement of science. If only, he dreamed, its reputation could be raised "to a point where it may penetrate to the sanctum of the Secretary of the Navy."[3] During the war Hale did quite a bit to try to enhance the effectiveness of the National Academy of Sciences.

Hale combined a promoter's talents with a cultural and professional commitment to science. His father was a highly successful entrepreneur who made a great deal of money in the elevator business; his mother, a genteel lady, impressed upon her son the importance of cultivation. Hale had no taste for gilt and ostentation. His ample resources were better spent for books and fitting out an upstairs laboratory, for a telescope on the spacious grounds of the family's South Side Chicago home, and for travel to meet the leading scientists of Europe. Catholic in his interests—they ranged from Egyptology to English novels—Hale was equally at home poring over early editions of Newton and Galileo or reading Keats aloud to pass away long nights at the telescope.[4]

* California Institute of Technology.

[1] Daniels to Edison, 7 July 1915, Josephus Daniels MSS, Library of Congress, Washington, D.C., Box 39 (hereafter, Daniels MSS).

[2] Hale to Edwin Grant Conklin, 12 Oct. 1915, George Ellery Hale MSS, Mt. Wilson and Palomar Observatories, Pasadena, Calif., Box 11 (hereafter, Hale MSS).

[3] *Ibid.*

[4] For the general biographical details of

For all his gentility he had the skyscraper drive of his father. He walked with a quick, nervous step; nothing but frequent illnesses ever dissipated his energetic ambition. Before his graduation from M.I.T. he invented the spectroheliograph, an ingenious photographic device for analyzing the elements in the prominences of the sun (the invention won him election to the Royal Astronomical Society at the age of twenty-two). Hale was no less energetic—and successful—at the promotion of astronomical research. The businessman's son got along well with men of wealth and was naturally at ease marshalling arguments for science to a wide variety of powerful people. Corporate and foundation executives found him a slim bundle of animated enthusiasm, difficult to refuse. Hale himself had launched the *Astrophysical Journal,* convinced the traction magnate Charles Tyson Yerkes to finance the Yerkes Observatory, and persuaded the trustees of the Carnegie Institution to inaugurate and sustain Mt. Wilson.[5]

After his election to the National Academy in 1902 Hale came increasingly to ponder the general advancement of science in the United States. In the first fifteen years of the twentieth century American research knew a quickening pace. The scientific community was swiftly expanding (between 1900 and 1914 membership in the American Association for the Advancement of Science jumped from 1,925 to 8,325, and the special societies showed the same kind of growth). The journals were bulging ever more thickly with research papers. In the universities scientists were finding more sympathetic administrative support—and more money—for their work. The budgets of the federal scientific agencies, like the newly created National Bureau of Standards, were climbing.[6]

Science was also acquiring a new and powerful patron—American business. Through the nineteenth century manufacturers had paid almost no attention to basic research. Now some, like American Telephone and Telegraph and General Electric, were opening industrial research laboratories. Sympathetic men of wealth were contributing money to university laboratories, and John D. Rockefeller and Andrew Carnegie went so far as to endow independent research institutions. Business was coming to realize that an investment in science could pay dividends.[7]

For all its growth Hale considered American science, particularly physical science, underdeveloped. It still suffered from limited funds and inadequate public appreciation. Moreover, despite the increasing number of papers published, a good deal of American research was insignificant. Hale thought he knew the source of the trouble. It is "well-known . . . that few broad generalizations or fundamental discoveries . . . , are to be found in the history of this country. We seem to be too much interested in the details and the merely technical elements of investigation, and therefore do not see the woods for the trees."[8]

Over the years Hale groped toward developing a way of improving things. The National Academy, in whose affairs he was increasingly involved, figured prominently in his thinking, and in 1915 he published a slim, programmatic book, *National Academies and the Progress of Research.*

Hale's proposals flowed directly from his own outlook and experience. Hale the

---

Hale's life, see Helen Wright, *Explorer of the Universe: A Biography of George Ellery Hale* (New York: Dutton, 1966).

[5] *Ibid.*

[6] Daniel J. Kevles, "The Study of Physics in America" (unpublished Ph.D. dissertation, Department of History, Princeton University, 1964), pp. 165–302.

[7] *Ibid.*

[8] Hale to Simon Newcomb, 21 Mar. 1906, Simon Newcomb MSS, Library of Congress, Washington, D. C., Box 25 (hereafter, Newcomb MSS).

cultivated gentleman thought the public had to be better educated to the intellectual adventure of science. Hale the persuader of businessmen shrewdly recognized the potential for financial support in the growing interest of manufacturers in research. More of them had to be convinced of what the managers of A.T.&T. and G.E. already realized. If scientists would only emphasize how Faraday's work had laid the foundation of electrical engineering, Hale stressed, they could "multiply the friends of pure science and receive new and larger endowments."[9]

Hale the astrophysicist was accustomed to working with physics, chemistry, and astronomy, and an interdisciplinary awareness, he believed, would turn American scientists away from mere details toward the "large relationships." For Hale the relative insignificance of the country's research was a product of too much narrow specialization. While he recognized the inevitability of specialization, he thought that American scientists ought to pay more attention to fields other than their own, even to cooperate with workers in other disciplines. Then, Hale was sure, they would be more likely to produce meaningful results.[10]

Hale was also convinced that American scientists would benefit from cooperation within their own specialties. Group efforts stood a much better chance of financial support than the isolated appeals of individuals. More important, group efforts tended to avoid wasteful duplication and to encourage a fruitful division of work (many astronomers had to cooperate to map the heavens). Hale, who was a principal founder of the International Union of Solar Research, put particular faith in the benefits of international cooperation. If American scientists were to work more closely with Europeans in the same specialty, their professional efforts were bound to take on more significant dimensions.[11]

In Hale's opinion the most appropriate organization to put these ideas into practice was the National Academy. The NAS included scientists from most disciplines. By emphasizing the "unity of knowledge"—Hale assumed the existence of such a unity—it could "do much to broaden and stimulate its members." Institutionally the Academy's "national and representative character," its "unique position among American societies," gave it a most favorable opportunity to promote both foreign and domestic cooperation.[12]

The NAS, Hale argued, had to be properly equipped to fulfill its potential. The Academy's funds were limited; it neither had a building of its own nor did it publish a journal of research. It ought to issue a Proceedings, Hale urged, to keep American scientists abreast of progress outside their own specialties. Moreover, with its own home the Academy could sponsor lectures and exhibits to call public attention to the wide import of science (Hale considered the daily press notices of scientific progress "synonomous with rank sensationalism"). Properly fitted out, the Academy could uphold the dignity of science and help persuade the public of its intellectual and practical benefits.[13]

But Hale was certain that the NAS needed more than what money could buy. He had studied Europe's great scientific academies, those bodies born of monarchical beneficence: all of them enjoyed the prestige of government patronage. Hale believed

[9] Hale, *National Academies and the Progress of Research* (Lancaster, Pa.: New Era Printing Co., 1915), pp. 130–131.

[10] *Ibid.*, pp. 100–101; for more explicit expressions of this view, see Hale to Newcomb, 5 Jan. 1899, 10 Mar. 1906, Newcomb MSS, Box 25.

[11] Hale, *National Academies*, pp. 102–104; see also the letters cited in n. 10.

[12] Hale, *National Academies*, pp. 100–101, 89, 93.

[13] *Ibid.*, pp. 110–111, 102–106, 117, 53; the quotation is from p. 115.

that to accomplish "great results" any academy "must enjoy the active cooperation of the leaders of the state."[14] The NAS, a child of Congress, might have been created in 1863 to provide scientific advice upon the request of any federal agency, but by 1915 the government's relations with its official scientific counselor were hardly active.

The growth of the government's own scientific bureaus had made federal requests for the Academy's help increasingly redundant. In more than fifty years the NAS's services had been called for on only fifty-one occasions; only four of them had been in the twentieth century, the last in 1908.[15] The NAS has had "no influence upon legislation," one member sighed; for the most part its advice had been "ignored." In 1915 Congress expressed its opinion of the Academy's usefulness by asking for a report on the feasibility of a universal alphabet.[16]

The Academy's desultory relations with the government were the result of more than superfluousness. Many of its members stressed the organization's exclusive and honorific side. Only 16 of its 133 members were associated with federal scientific agencies; most did not take an interest in the problems of federal science. Some critics charged that smug satisfaction made NAS scientists content to enjoy the esteem of membership in preference to advising the government. Arthur Schuster, Foreign Secretary of the Royal Society, knew the ways of academies intimately, and, as a friendly guest at the NAS's semi-centennial celebrations, he warned against mistaking "the signs of gray hairs for the stamp of an enviable dignity."[17]

Self-satisfaction aside, many members of the NAS were simply worried about maintaining the Academy's independence from political machinations. The Academy considered itself a disinterested source of scientific advice. If it were to thrust itself upon the government, it might get deleteriously enmeshed in interest-group politics; political compromises would stain its authority. Better to remain passive and pristine than to become active and tainted. Nothing in the Academy's charter prohibited it from offering advice to the government; nothing had led the NAS to aggression on the battlegrounds of Washington politics. The Princeton biologist Edwin Grant Conklin deplored the Academy's increasingly do-nothing record: the NAS seemed to have no other purpose than to act as a "blue ribbon society."[18]

Hale, an outright activist in general Academy affairs, agreed with the conservative members on the issue of the government. The NAS might recommend appointees for federal scientific posts. It might advise on subjects that cut across the responsibilities of established federal agencies. Beyond that Hale was reluctant to tighten the Academy's bonds with the government until the NAS had more fully developed its "standing and prestige."[19] Otherwise, in its weak condition, it would become a political supplicant. Mere supplicants, Hale was sure, were too insecure to provide the kind of disinterested advice upon which the Academy prided itself.

14 *Ibid.*, p. 53.

15 See the list of the Academy's services to the government, *ibid.*, pp. 78–81. A French observer noted that the Academy's relationship with the government was *"assez theorique."* Maurice Caullery, *Les universités et la vie scientifique aux États-Unis* (Paris: A. Colin, 1917), p. 249.

16 Charles R. Van Hise to Arthur L. Day, 30 Dec. 1913, Hale MSS, Box 53; *Report of the National Academy of Sciences, 1916,* p. 20.

17 Quoted in *Report of the National Academy of Sciences, 1913,* p. 86.

18 Conklin to Hale, 28 Mar. 1913, Hale MSS, Box 11. Othniel C. Marsh said while president of the Academy in 1889 that the NAS "would lose both influence and dignity by offering its advice unasked." Quoted in Charles Schuchert, "Othniel C. Marsh," *National Academy of Sciences Biographical Memoirs,* Vol. XX (1939), p. 30.

19 Hale, *National Academies,* pp. 83, 164–165.

But Hale was staunchly pro-Ally, and surely in a national emergency the Academy could approach the government without fear of political taint. In the middle of the *Lusitania* crisis Hale suggested privately that the NAS offer its services to President Woodrow Wilson. The fading of the crisis made the proposal untimely.[20] After the sinking of the *Sussex* in March 1916 Hale moved again. On Wednesday, 19 April, the day after Wilson delivered an ultimatum to Germany, the Academy, spurred by Hale—and the crisis—to unprecedented initiative, resolved at its annual meeting to offer the President its help. One week later a distinguished delegation from the NAS presented the resolution to Wilson at the White House.

The President's acceptance led in June 1916 to the formation of a new agency of the Academy, the National Research Council. At the end of July the President agreed to apppoint representatives of the government to the Council upon nomination by the Academy.[21] National preparedness had set the NAS on its way to obtaining that item crucial to Hale's program—the active cooperation of the leaders of the state.

Between mid-1916 and early 1918 the Council developed around the ideas expressed in Hale's book. Hale had argued specifically for the Academy's bringing together the country's major research institutions: the NRC was a federation of representatives from scientific societies, industrial and government laboratories, and universities. He had argued for demonstrating the utilitarian value of science: the NRC had an Engineering Division and the support of leading industrial figures. He had suggested international cooperation: the Council had scientific attachés at the American embassies in London and Paris. The official status of the Council's foreign representatives was just one among many examples of federal support. Many powerful military and executive officials sat on the NRC's multifarious committees. In addition the Council enjoyed the prestige of serving as the Department of Science and Research for the Council of National Defense, the Cabinet group authorized by law to mobilize the nation's resources.[22]

As a federation of research interests the NRC was proving a useful instrument for the mobilization of science. The Council had persuaded many government agencies, traditionally jealous rivals, to cooperate. University and industrial laboratories, historically remote from the government's needs, were working on military problems. The bureaus of the Army and Navy, normally hostile to advice from nonmilitary sources, were now paying attention to the ideas of academic and industrial scientists. The NRC's results reinforced the value of cooperation. The Council's scientists were developing myriad weapons, devices, and techniques for the military. Physicists could point to instruments for the detection of submarines; chemists to apparatus for gas warfare; psychologists to methods for the efficient assignment of personnel. Scientists from many disciplines were making their contributions.[23] The Research Council, in short, had brought about an unprecedented and fruitful collaboration of university and industrial scientists with the military.

Through it all Hale never forgot his fundamental goal—the advancement of research in the United States.[24] An early report of the NRC had argued that "true

[20] Hale to Conklin, 10 June 1915 and 13 June 1915, Hale MSS, Box 11.

[21] Robert A. Millikan, *The Autobiography of Robert A. Millikan* (New York: Prentice-Hall, 1950), p. 124; Wright, *Hale*, pp. 287–288; Wilson to Hale, telegram, 24 July 1916, Hale MSS, Box 60.

[22] For the organization and work of the NRC, see Robert M. Yerkes, ed., *The New World of Science: Its Development During the War* (New York: Century, 1920), *passim.*

[23] *Ibid.*

[24] Years later Hale wrote: "While most of my own experience as Chairman was during the war, the plans I always had in mind looked

preparedness" included the encouragement of pure science. After the diplomatic break the NRC had focused on military problems, and so long as the war continued it would concentrate on defense. But the Research Council's "close contact" with military and civil authorities and with industrial and engineering organizations, Hale believed, had created "many new friends for pure science."[25] The Rockefeller Foundation had approached the Council to develop a plan for the endowment of research in physics and chemistry. The Carnegie Corporation had expressed an interest in funding the construction of an Academy building.[26] Corporate leaders like John J. Carty of A.T.&T. had spoken for industrial support of basic research. This "coalition of widely different interests . . . ," Hale could say of the NRC, "is certain to be beneficial to research in all of its aspects."[27]

In March 1918 Hale set out to make the coalition permanent. The NRC, officially created in response to the President's emergency request, had no guarantee of government cooperation beyond the war. The Academy might legally maintain the NRC of its own accord; Wilson might not continue to appoint government representatives to the Council. Whatever Wilson's peacetime policy, future presidents would not be bound by it. Hale, eager to insure the continuing cooperation of the government, wanted the NRC endowed with an enduring legal foundation.

He had a choice of two procedures. The Academy could go to the Congress for a revision of its charter, or, on the basis of its existing charter, it could seek an Executive Order from the President. The choice was not difficult for Hale to make. In March 1918 the Congress was Democratic. Hale, the businessman's son and a starched-collar Republican, considered the party of William Jennings Bryan a string-tie crowd dominated by red-neck anti-intellectuals. Democratic senators had been known to speak "eloquently against all . . . organizations for research and advanced study"; they preferred to concentrate on "the needs of 'the Little Red School House on the hill,' standing for light and leading to the lowly of the land!"[28] Were the Academy, an aristocrat of the higher learning, to approach the Congress, surely it would place itself in jeopardy. Hale chose to seek an Executive Order from the Democratic President, who, after all, had been a leading academic.

Hale drafted an Order and on 27 March 1918 sent it to the President through the helpful Colonel Edward M. House. In a covering letter he came down hard on the national importance of science. America could not "compete successfully with Germany, in war or peace, unless we utilize science to the full for military and industrial purposes." Besides, Germany was not the United States' only competitor ("the movement to establish councils for the promotion of scientific and industrial research," Hale exaggerated, "has swept over the whole world since the outbreak of the war"). As a federation of the nation's major research interests the NRC had effectively mobilized science. It was clearly capable of "unlimited" peacetime development.[29]

---

forward to work under peace conditions." Hale to Isaiah Bowman, 2 May 1933, Hale MSS, Box 4.

[25] National Academy of Sciences, *Proceedings*, 1916, *2*:507–508; Hale to Franz Boas, 14 Mar. 1918, National Research Council MSS, National Academy of Sciences-National Research Council, Washington, D. C., Institutions file (hereafter, NRC MSS); Hale to Arthur Schuster, 18 Apr. 1918, Hale MSS, Box 47.

[26] Millikan, *Autobiography*, p. 180; Hale to Schuster, 10 Dec. 1917, Hale MSS, Box 37.

[27] Carty, "Science and the Industries," *Reprint and Circular Series of the National Research Council*, 1920, No. 8; Hale to Schuster, 18 Apr. 1918, Hale MSS, Box. 47.

[28] Hale to H. H. Turner, 6 Mar. 1916, Hale MSS, Box 41.

[29] Hale to the President, 26 March 1918, Woodrow Wilson MSS, Library of Congress, Washington, D. C., File VI, Case 206 (hereafter, Wilson MSS).

The proposed Executive Order was designed to give the presidential seal to the burgeoning activities of the wartime NRC. Besides providing for appointments of government representatives, it charged the Council itself with eight different duties: stimulating pure and applied research for the national welfare; surveying the "larger possibilities of science"; promoting cooperation at home and abroad; "correlating and centralizing . . . the research work of the government"; calling attention to wartime scientific needs; helping industry to focus on critical research problems; furthering technical liaison; spurring the establishment of research fellowships. The duties added up to a mandate of sweeping scope.[30] Only three of them explicitly mentioned the military or the national defense. The document clearly aimed at glossing the Council's postwar possibilities with the prestige of a government attachment.

The President thought that the Order would simply codify what the Research Council was already doing. But, sensitive to the niceties of power, he sent it to the Council of National Defense. Would the CND ascertain whether "any embarrassments or conflicts of authority will be created by acting on Dr. Hale's suggestion?"[31] To the CND Dr. Hale's suggestion hardly seemed an innocent codification. It decided, in the bloodless phrases of its minutes, that the "functions described in this order duplicated the functions of many of the executive departments . . . [and would] result in confusion."[32]

No member of the CND was more opposed to the Order than Secretary of Agriculture David Houston. A political scientist by training and a university president by experience, the Secretary was acutely sensitive to matters of administrative jurisdiction. The author of a study of the nullification crisis under President Andrew Jackson, he was particularly alive to questions of constitutional authority.[33] The National Research Council was a private agency, and Houston did not think a private agency ought to correlate and coordinate the work of the government's own departments. It would be "scarcely appropriate," he argued, for the President officially to suggest that a "semi-governmental agency" encourage the establishment of research fellowships. Legalities aside, Houston had political objections. The President might find it ultimately "embarrassing" to approve, even tacitly, industrial research activities "by an agency over which the Government has very slight control. Such investigations might take a course quite at variance with the policy of the administration."[34]

Sympathetic to the NRC, the President asked Hale for a clarification. Did the Research Council want "authority" to coordinate the scientific work of the government? Wilson doubted his "right to give any outside body such authority" and doubted "the practicability and wisdom of doing so."[35]

Hale hurried to see House in New York. Encouraged by the Colonel, who obligingly said he would push the Order through, he was determined not to let the

[30] Copy of the proposed order is with David Houston to Wilson, 30 Apr. 1918, Wilson MSS, File VI, Case 206.

[31] Wilson to Josephus Daniels, 12 Apr. 1918, Daniels MSS, Box 14.

[32] Minutes of Meeting of the Council of National Defense, 15 Apr. 1918, Daniels MSS, Box 451.

[33] *Dictionary of American Biography*, Vol. XXII (Supplement Two) (New York: Scribner's, 1958), pp. 321–322; E. David Cronon, ed., *The Cabinet Diaries of Josephus Daniels, 1913–1921* (Lincoln, Neb.: Univ. Nebraska Press, 1963), p. 300.

[34] See Houston to Wilson, 30 Apr. 1918, Wilson MSS, File VI, Case 206.

[35] Wilson to Hale, 19 Apr. 1918, Wilson MSS, File VI, Case 206.

matter drop because of opposition in the Cabinet.[36] But House or no House, Hale still had to deal with the issue raised by the President.

Hale did not really want authority for the Council. Whatever his desire for federal patronage, he still feared too close a connection with the government. The NRC clearly could not expect to wear the mantle of authority without inviting some federal control.

Throughout the war the fundamental issue of the Council's independence had never left Hale's mind. The NRC might serve as the scientific department of the Council of National Defense; Hale took "great precautions" to have the NRC "*act* as a department of the Council . . . , without becoming a part of it and without surrendering any authority to it." The government might have representatives on the Research Council; nominations to the Council lay in the hands of the Academy, which made it "possible to eliminate all question of political preferment." Not in any matter would the NRC surrender authority to the government. The Council had the "advantage of Government co-operation," Hale noted. But it retained "the freedom of action which Government Departments could not enjoy."[37]

The Research Council had "no desire for *authority* to coordinate the . . . scientific agencies of the Government . . . ," Hale replied to Wilson. It merely wished "such official recognition as will give it the influence which it needs to secure co-operation" both at home and abroad. Now Hale waited, afraid that if the Order went back to the Cabinet it would "surely be dished."[38] It did go back to Houston. But House apparently had intervened, and the President simply asked his Secretary of Agriculture to revise the text to express Hale's "real purpose."[39]

Houston revised it—with the clear aim of maintaining the political and Constitutional prerogatives of the federal government. He denied the NRC the power not only to correlate the scientific work of the government, but even to promote cooperation among federal agencies. He also refused the Council any explicit responsibility for assisting industry and establishing research fellowships. All of his revisions, from the finest changes in language to the slashing away of complete paragraphs, bespoke his determination not to yield any of the government's official functions to the NRC.[40]

The revised Order, with Wilson's signature, arrived from the White House on 10 May 1918, and Hale considered it "entirely satisfactory."[41] He might have been eager to strengthen the Council's ability to help the government, but he was basically concerned with keeping the government connection so as to strengthen the Council. Sure of his political purpose, he had guided the NRC through the maneuver-filled enterprise of wartime Washington with its administrative independence intact. Now he had an Executive Order that assured both the permanence of the NRC and federal representation without federal control. Moreover, the basic purposes of the Council—the stimulation of pure and applied research, the promotion of cooperation at home and abroad—had the permanent imprimatur of the presidential office. The NRC's ambitious architect could say: "We now have precisely the connection with

[36] Hale to Evelina Hale, 22 Apr. 1918, Hale MSS, Box 80.

[37] Hale to James R. Angell, 13 Aug. 1919, Hale MSS, Box 3; Hale to Schuster, 18 Apr. 1918, Hale MSS, Box 47.

[38] Hale to the President, 22 Apr. 1918, Wilson MSS, File VI, Case 206; Hale to Evelina Hale, 26 Apr. 1918, Hale MSS, Box 80.

[39] Wilson to Houston, 24 Apr. 1918, Wilson MSS, File VI, Case 206.

[40] See Houston to Wilson, 30 Apr. 1918, and accompanying draft of the order with penciled changes, Wilson MSS, File VI, Case 206.

[41] Hale to Evelina Hale, 10 May 1918, Hale MSS, Box 80.

the government that we need and . . . it will be our own fault if we do not make good use of it."[42] Making good use of it meant going ahead on the basis of the ideas outlined in *National Academies and the Progress of Research*: specifically, funding the Council and setting it up so as to encourage cooperation.

Hale argued the economic importance of science to appeal to corporate and foundation America. In the spring of 1918 the NRC created an Industrial Advisory Committee of "strong men," in the proud words of a Council report, "with the imagination to foresee the general benefits" of science. The group, drawn from the board rooms of the nation's major industries, included Cleveland H. Dodge, George Eastman, Andrew W. Mellon, Pierre S. DuPont, Elihu Root, Ambrose Swasey, and Edwin W. Rice, Jr.[43] The members were not to assume any administrative responsibilities but were simply to lend their prestige to what Hale frankly called a propaganda program for science. Most of them obliged with weighty statements like Root's: without pure science, the "whole system" of industrial progress would "dry up."[44]

Within a year of the Executive Order, the NRC had its principal financial gifts in hand. The Carnegie Corporation (Root was a powerful trustee) donated $5,000,000, partly for endowment, partly for the construction of a permanent building for the NAS and NRC. The Carnegie gift was contingent upon the Academy's raising $150,000 for the site, and a frenetic campaign brought the funds from many members of the Industrial Advisory Committee as well as figures like Henry Ford. A.T.&T. had promised the Council $25,000 annually; General Electric was considering a matching grant.[45] From the Rockefeller Foundation came a grant of $500,000 for postdoctoral fellowships in physics and chemistry.[46]

In 1919, while the fund-raising was in progress, the Council was organized to encourage cooperation. The NRC's Division of General Relations included slots for federal representatives appointed by the President under the terms of the Executive Order. Its divisions of Science and Technology brought together scientists from a wide variety of disciplines, both as individuals and as representatives of the country's numerous scientific and technical societies. In addition, the NRC belonged to the International Research Council, which was another product of Hale's wartime efforts. The IRC was designed to include constituent unions in such disciplines as astronomy, geodesy, physics, and chemistry. The predecessor of today's International Council of Scientific Unions, the International Research Council was Hale's vehicle for achieving cooperation with foreign scientists.[47]

All the while that Hale urged cooperation under the NRC he emphasized that the Council would not be dictatorial. Scientists, he knew, were exceedingly jealous of their prerogatives and disliked being told what kind of research to do or how to go about it. Hale was sensitive enough to this feeling to write an assurance into the Executive Order. In all "co-operative undertakings" the NRC was "to give encouragement to individual initiative, as fundamentally important to the advance-

[42] Hale to James R. Garfield, 16 May 1918, NRC MSS, Hale file.

[43] Council of National Defense, *Second Annual Report, 1918* (Washington, D. C.: G.P.O., 1918), p. 63.

[44] Copies of the statements are in Hale MSS, Box 52, Industrial Research file; Root, "Industrial Research and National Welfare," *Science*, 29 Nov. 1918, p. 533; Hale to Willis R. Whitney, 19 June 1918, Hale MSS, Box 43.

[45] Wright, *Hale*, p. 312; Millikan, *Autobiography*, pp. 188–189; Hale to William W. Campbell, 14 May 1919, Hale MSS, Box 9.

[46] Hale to Campbell, 28 June 1933, Hale MSS, Box 9; Minutes of the Meeting of the NRC Fellowship Board, 16 Apr. 1919, Hale MSS, Box 59, NRC Fellowship Board file.

[47] See the NRC's statement of organization, *Science*, 16 May 1919, p. 459; Wright, *Hale*, pp. 299–304.

ment of science."[48] Just as he was determined to keep the Research Council free of government control, Hale wished to avoid the NRC's appearing to control American research. Instead of exercising authority within the scientific community, the NRC had to bring about cooperation by "moral pressure."[49]

But moral pressure by itself could not easily overcome the facts of scientific life in the United States. American scientists might gladly cooperate under the NRC in a war emergency, but the guns had hardly fallen silent before those who had come to Washington were hurrying home. Once back on the campuses they were spread over so vast a geographical area as to make active cooperation mechanically difficult. In any case the vast majority of scientists, wherever they had been during the war, were naturally returning to their own special research interests. Whatever Hale's belief in the value of interdisciplinary cooperation, many disciplines did not readily lend themselves to cross-fertilization from other fields. Moreover, cooperation within a discipline like physics was much less natural than in one like astronomy. The evidence suggests that during the 1920's and 1930's the NRC managed to achieve very little active cooperation of any sort.[50]

One item in the Research Council's program did affect the development of American science significantly. The fellowships funded by the Rockefeller Foundation gave promising young men a special opportunity: a postdoctoral year or two free of teaching duties during which they could work with leading scientists in their specialties. Moreover, in 1923 John D. Rockefeller, Jr. created the International Education Board to provide, among other things, postdoctoral fellowships for study in other countries.[51] The IEB program, which was administered with advice from the Research Council, gave young and able Ph.D.'s the same kind of special opportunity abroad that the NRC fellows enjoyed at home, and Europe had many more first-rate scientists. Many of the IEB and NRC fellows won distinction in their fields.[52] The total number of them was limited, but they seeded a generation of students with a taste for the kind of high-quality research that was much more significant than mere details and technicalities.

In the end the NRC's peacetime effectiveness depended less on moral pressure than on money, and its fiscal resources were limited and private. Had the Council come out of World War I able to write checks against the federal treasury it might have been able to make much more of Hale's program workable. With ample funds it could have supported cooperative projects, furthered interdisciplinary communication, helped specialists buy research assistance, equipment, and time away from teaching. But the only federal money the NRC got was for its annual dues to the International Research Council.

Just after the armistice, when the NRC's influence was at its height, federal funds for science were difficult to get. Americans, turning to isolationism, were not particularly interested in maintaining a competitive military capacity; in any case, few people outside the Army and Navy—and not all that many within the services—

[48] See copy of the Order in Houston to Wilson, 30 Apr. 1918, Wilson MSS, File VI, Case 206.

[49] Hale to John C. Merriam, 5 Feb. 1918, NRC MSS, Hale file.

[50] See the *Annual Reports* of the National Research Council, 1919–1939.

[51] George W. Gray, *Education on An International Scale: A History of the International Education Board 1923–1938* (New York: Harcourt, Brace, 1941), pp. 3–11, 16–23. The IEB fellowships, which also went to Europeans, did a good deal to facilitate the migration of young scholars in the 1920's; twenty-five IEB fellows worked under Niels Bohr in Copenhagen (*ibid.*, p. 102).

[52] Raymond B. Fosdick, *The Story of the Rockefeller Foundation* (New York: Harper, 1952), pp. 146, 149.

identified research with national defense. Congress was in a mood for economy and for the most part saw no reason to appropriate money for scientific activities outside the government's own bureaus.[53]

The political and military facts of World War I may have made such appropriations improbable; Hale himself made them even more unlikely. His fear of political interference with American science was virtually consuming, and federal aid might open the door to federal control. The war had been a major opportunity for the ambitious author of *National Academies and the Progress of Research.* With the Executive Order in hand he was thoroughly satisfied, as he told a colleague convinced of the need for federal support, to advance science in the United States by depending on "private funds for a long period in the future."[54]

[53] Clarence G. Lasby, "Science and the Military," in David D. Van Tassel and Michael G. Hall, eds., *Science and Society in the United States* (Homewood, Ill.: Dorsey Press, 1966), pp. 262–263.

[54] Hale to Willis R. Whitney, 19 June 1918, Hale MSS, Box 43. It is interesting to note that Vannevar Bush, the head of the Office of Scientific Research and Development in World War II, was also quite eager to avoid political interference with science. Bush was largely responsible for the original National Science Foundation bill, which would have so insulated the NSF from federal supervision that President Harry S. Truman vetoed it. See Truman's veto message in James L. Penick *et al.*, eds., *The Politics of American Science, 1939 to the Present* (Chicago: Rand, McNally, 1965), pp. 86–89.

"Into Hostile Political Camps": The Reorganization of International Science in World War I

*By Daniel J. Kevles**

IN 1946 HUGO R. KRUYT, the president of the International Council of Scientific Unions, warned the delegates at the opening of the organization's first postwar meeting that they would do well to "keep politics as far from science as possible." For as Kruyt reminded his colleagues, the development of international cooperation in research had been paralyzed for over a decade after Versailles, largely because in World War I his predecessors had reconstructed the major scientific organizations of the world on a political basis.[1] Apart from the importance of its aftereffects, the reorganization during the Great War deserves historical attention in its own right. And now that the papers of the chief figure in the story, George Ellery Hale, have become available to scholars, the episode can be told in revealing detail.

Fifty-six years old when the war broke out, Hale was a distinguished astrophysicist, the director of the Mt. Wilson Observatory in Pasadena, California, and the Foreign Secretary of the National Academy of Sciences. The son of a well-to-do Chicago businessman, he was also a highly energetic and successful promoter of research, having founded the *Astrophysical Journal*, the Yerkes Observatory, and the facility atop Mt. Wilson itself. Over the years Hale's entrepreneurial vision had extended beyond his own discipline to the entire realm of science. As an astrophysicist accustomed to working with chemistry, physics, and astronomy, he had a special interest in furthering interdisciplinary cooperation. To his mind, group efforts stood a much better chance of financial support than the isolated appeals of individuals; equally important, they tended to avoid wasteful duplication and lent themselves to a fruitful division of work. The principal founder of the International Union of Solar Research, Hale had long cherished the dream of promoting interdisciplinary cooperation in other sciences on an international scale.[2]

As an internationalist, Hale expressed a proud tradition in the world of learning.

*Division of the Humanities and Social Sciences, California Institute of Technology, Pasadena, California 91109.

[1] *The Fourth General Assembly of the International Council of Scientific Unions, Held at London, July 22nd to 24th, 1946, Reports of Proceedings* (Cambridge, 1946), p. 3. For the history of the new international scientific organizations from a special point of view, see Brigitte Schröder-Gudehus, *Deutsche Wissenschaft und internationale Zusammenarbeit, 1914–1928; ein Beitrag zum Studium kultureller Beziehungen in politischen Krisenzeiten* (Geneva: Imprimerie Dumaret & Golay, 1966).

[2] Daniel J. Kevles, "George Ellery Hale, the First World War, and the Advancement of Science in America," *Isis*, 1968, *59*:427–429. For the full details of Hale's life, see Helen Wright, *Explorer of the Universe: A Biography of George Ellery Hale* (New York: Dutton, 1966).

Truth knew no national boundaries, those who pursued it liked to say, and before 1914 scientists everywhere considered themselves members of an international community whose activities were unaffected by the divagations of politics. But once the war came, the tradition was quickly strained to the breaking point. In October 1914 ninety-three German professors, including thirteen scientists of eminent repute, issued a manifesto "to protest before the whole civilized world against the calumnies and lies with which our enemies are striving to besmirch Germany's undefiled cause. . . . "[3] In response, angry fellows of the Royal Society in England demanded the removal of all Germans and Austrians from the list of foreign members, and the French Academy of Sciences actually dropped the signers of the manifesto. In 1915, after the Germans had introduced gas and submarine warfare, the distinguished British physician Sir William Osler predicted that, for his generation at least, the war would spell "the death of international science."[4]

The disruption in the world of science distressed Hale. But while he was a staunch pro-Allied interventionist, even the sinking of the *Lusitania* left his sense of scientific fraternity intact. If Germany was guilty of "abominable military practices," Hale absolved her scientists of responsibility for the actions of their government. The scholars who had signed the manifesto had simply "lost their heads," and it was "very unfortunate" that they had been dropped from the Académie des Sciences. Even were the United States to enter the war, Hale's personal regard for scientists everywhere would continue undiminished. He simply failed to see why "political considerations" should dilute "our interest in the progress of international science."[5]

Hale the promoter of interdisciplinary cooperation remained particularly interested in the progress of the International Association of Academies. Founded at the turn of the century amid the proliferation of scholarly enterprises, the IAA was in theory to coordinate the increasingly diverse activities of international learning. In practice, the association had done little more than host banquets and excursions at its triennial assemblies. Still, by 1913 it had a membership of twenty-one academies drawn from fourteen European nations and the United States.[6] In Hale's ambitious belief an institution so representative and prestigious might well be developed into "a really effective nucleus" for international scientific cooperation.[7]

If the bitterness of the war dampened Hale's hopes of invigorating the Association, he could at least try to save it from complete collapse. Fourteen of the member academies were located in belligerent nations and were unable to participate in the IAA's activities. More important, since Germany had four academies—more than any other nation—the "leading," or executive, academy was at Berlin. Hale proposed that a neutral academy assume that position for the duration. The British were willing; so were the Germans (Max Planck, a signer of the manifesto, was now counseling discreet

[3] The full text of the manifesto, with the list of signers, is in Georg F. Nicolai, *The Biology of War* (New York:Century, 1918), pp. xi–xiii.

[4] William Osler, *Science and War* (Oxford: Clarendon Press, 1915), pp. 36–37.

[5] George Ellery Hale to H. H. Turner, June 15 and Sept. 16, 1915; to Arthur Schuster, June 9, 1915; to Henricus Bakhuysen, Sept. 16, 1915, George Ellery Hale MSS, Millikan Library, California Institute of Technology, Pasadena, California (hereafter, Hale MSS), Box 41, Turner file; Box 37, Schuster file; Box 3, Bakhuyzen file.

[6] Brigitte Schröder, "Caractéristiques des relations scientifiques internationales, 1870–1914," *Cahiers d'Histoire Mondiale*, 1966, *13*: 170–174.

[7] Hale to Gaston Lacroix, May 2, 1917, Hale MSS, Box 47.

silence on political matters so as to preserve as much of international science as possible).[8] But the French refused to acquiesce in the plan unless the Berlin academy would relinquish its leading status in perpetuity. Hale fretted through a brief flurry of negotiations in 1915, then, finding the scientists of Paris implacable, let the issue drop.

The affairs of the association lapsed into a state of suspended animation. Hale himself, a staunch advocate of preparedness, turned to the mobilization of science at home. Drawing upon his special brand of promotional thinking, he persuaded the National Academy to create a National Research Council, a joint organization of experts from the universities, industry, and the government, including the military. Designed to promote cooperative efforts in science, the NRC's immediate purpose was to attack the technical problems of defense, and by the summer of 1917 it had inaugurated several projects, notably in chemical warfare and the detection of submarines.[9] Through it all Hale had paid the IAA no more than occasional and subdued attention. But that summer it was a matter of bitter concern to Professor Émile Picard of the University of Paris, an eminent mathematician whose son had recently been killed in action. A former president of the French Academy of Sciences and now the new permanent secretary of its section for mathematics and astronomy, Picard wrote Hale a strong letter about the future of international science.

In Picard's blunt report French scientists would no longer even sit down at the same table with their colleagues across the Rhine. "Personal" relations of any kind would be "impossible" with men whose government had committed such atrocities and who themselves had "dishonored" science by exploiting it for criminal ends. For the sake of justice alone, the Frenchman insisted, Germany's scientists had to be ostracized from the structure of international research. His specific recommendations: dissolve the International Association of Academies and replace it with a new organization—without the neutrals at least for the duration, without the Central Powers indefinitely.[10]

The Hale of mid-1917 was decidedly receptive to Picard's ideas. If two years earlier he was generously exonerating German scientists of all malevolence, he was now pointing accusingly to the manifesto of the enemy professors, seething over the report that 150 of them had petitioned for the resumption of unrestricted submarine warfare, and above all, counting them culpable for the acts of their government. No more willing than Picard to conjoin with such men, Hale was equally willing "to cut loose from them altogether" by leaving the IAA.[11] Besides, since the association had been a gross disappointment, it might just as well be scrapped. In the wake of its demise, Hale would recognize international science completely, both to exclude the Central Powers and to achieve his special aim of a robust multinational cooperation.[12]

While Hale saw glowing possibilities in the situation, words of caution arrived from Arthur Schuster, Secretary of the Royal Society of London. A mathematical physicist

[8] Arthur L. Day to Hale, Sept. 1, 1915, Hale MSS, Box 13, Day file.

[9] For the creation and wartime development of the National Research Council, see Kevles, "George Ellery Hale, the First World War, and the Advancement of Science in America," pp. 427–437.

[10] Émile Picard to Hale, July 22, 1917, Hale MSS, Box 47. For the general biographical details of Picard's life, see J. Hadamard, "Émile Picard," *Obituary Notices of Fellows of the Royal Society*, 1942–1944, *4*:129–150; Paul Montel, "La vie et l'oeuvre d'Émile Picard," *Bulletin des Sciences et Mathématiques*, 2nd Ser., 1942, *66*:3–17.

[11] Hale to Edwin Grant Conklin, Oct. 8, 1917; to H. H. Turner, July 25, 1917, Hale MSS Box 47; Box 41, Turner file.

[12] Hale to Picard, Sept. 17, 1917, Hale MSS, Box 47.

with a strong interest in astronomy and seismology, Schuster had been active in the International Association of Academies since its founding. Like Hale, he was dissatisfied with the organization and, his nephew dead at the front, he could no more conceive of postwar meetings with enemy scientists than could Picard. But for all his bitterness against the Germans, Schuster was not so impassioned as to forget that legally the dissolution of the IAA required a two-thirds majority of the member academies. If, as was likely, two neutral academies were to add their votes to the six of Austria and Germany, the action would be blocked. If the academies of the Entente were simply to withdraw from the association, the neutrals might choose to stay in it with the Central Powers; then it would not be Germany's scientists who were ostracized but those of her enemies. Before opening the door to so disastrous a denouement, Schuster thought that the Allied and American academies had better deliberate upon the difficulty. The Royal Society, he told Hale, had tentatively scheduled a conference for that purpose, to be held in London in the spring of 1918.[13]

Hale was all for the conference, but he could not act without the support of his own National Academy. In September 1917 he put the case to the leadership—the "time was never so opportune" for the reorganization of international science under the auspices of the Allies and the United States—and he asked for opinions.[14] Some of the academy's hierarchy responded with bitter attacks against the scientists of the Central Powers. William Wallace Campbell, the director of the Lick Observatory, brushed aside the apolitical tradition of international science: "We have learned that blood is thicker than water. Can we learn also that science is thicker than blood?" Neither Campbell nor anyone else in this group could conceive of meeting with the Germans for a long period after the war. Yet even Campbell was wary of throwing the neutrals into the arms of the Central Powers, and a sizable fraction of the exclusionists doubted the wisdom of reorganization before the conclusion of the war.[15]

Another—and much smaller—group among the academy's leaders dissented from wartime reorganization on principle, none with a more penetrating argument than Arthur L. Day, the Home Secretary of the National Academy. A geophysicist, Day relied on the cooperation of colleagues around the world. Since his wife was the daughter of the German physicist Friedrich Kohlrausch, he had special reasons for moderation toward the enemy. Hale might indict German scientists for the acts of their government, but Day refused "behind an impregnable censorship to condemn an entire nation unheard." Moreover, simple prudence dictated against formally ostracizing the enemy. How embarrassing if the original constitution of the National

[13] Schuster to Hale, Aug. 27, 1917, Hale MSS, Box 47. For biographical material, see Arthur Schuster, *Biographical Fragments* (London: Macmillan, 1932); G. C. Simpson, "Sir Arthur Schuster, 1851–1934," *Obituary Notices of Fellows of the Royal Society*, 1932–1935, *1*:409–423.

[14] Hale, "Memorandum on the Organization of International Science," attached to: Hale to the members of the Council of the National Academy of Sciences, Sept. 18 and 22, 1917, Hale MSS, Box 47.

[15] William Wallace Campbell, "International Co-operation in Science," Sept. 15, 1917, and replies from the council members, Sept. and Oct. 1917, Hale MSS, Box 47. It is difficult to pin down just how adamantly opposed was each member of this group to future international co-operation with the Germans; their apparent emotions ranged from sad to venomous, and all took the position that collaborating with Central Powers scientists would be impossible for a long time after the war ended. The members of this group included Albert A. Michelson, Robert A. Millikan, Whitman Cross, Russell Chittenden, W. H. Howell, Michael I. Pupin, W. W. Keen, and William Bowie.

Academy, which was created in 1863, had outlawed intercourse with scientists south of the Mason-Dixon line. Most important, while reorganization might have its scientific merits, Hale was calling for it presumably to wage a war in science after the war on the battlefield. Day drove to the heart of the matter: reorganization for that purpose had "little to do with either science or its advancement; the appeal is rather to the sentiments and antipathies aroused by the war." Hale would be building on "quicksand."[16]

Hale had to tell Schuster in late October 1917 that "rather an unexpectedly strong sentiment" had developed against fixing international scientific policy while the war raged.[17] Hale himself remained commmitted to the wartime reconstruction of international science. But he had to find his way out of some difficult dilemmas. How to reorganize research under the nations of the Entente without losing the neutrals to the Germans? How to persuade the more cautious and moderate members of his own academy to back a reorganization in the present which would exclude the Central Powers in the future?

Hale also faced an equally important dilemma of a different kind. It was his belief that no scientific academy could produce "great results" without the "active cooperation of the leaders of the state."[18] He was sure that any foreign venture of his own academy, particularly one of the magnitude he had in mind, would require the support of President Woodrow Wilson. Yet the reorganization which Hale was promoting contradicted a basic tenet of the President's foreign policy: Wilson's insistence that America's quarrel was with the rulers, not the people, of Germany. Hale worried that the administration might refuse to bestow its blessing upon a new international scheme which excluded enemy scientists principally on grounds of war guilt.[19]

However dissimilar each of the dilemmas, all proceeded from Day's acute charge: Hale's argument for reorganization depended entirely upon an appeal to the sentiments and antipathies aroused by the war. But if in the fall of 1917 Hale had to concede the issue to the Days, the National Research Council was soon engaged in a trans-Atlantic activity that suggested a way out of all the troublesome dilemmas.

Six months after the United States entered the war the handling of technical information had become chaotic. At home, no administrative system existed to insure that the military and civilian bureaus of the government received reports germane to their work. Abroad, numerous American agencies were independently dunning the British and French for identical bits of information. Annoyed, Allied officials were becoming uncooperative. At one point, a French administrator who had supplied seven different people with the same report politely refused the eighth request.[20] The story of the eighth request made its way back to Washington; so, too, did a British plea for some coordination of technical liaison, and both provided the Research Council with an

[16] Day to Hale, Oct. 1, 1917, Hale MSS, Box 47. Of all the scientists whom Hale asked for an opinion, the only prominent ones who seemed to side with Day were Edwin Grant Conklin and Edward C. Pickering. For biographical information on Day, see Robert B. Sosman, "Memorial to Arthur Louis Day, 1869–1960," *The Geological Society of America Bulletin*, Oct.–Dec. 1964, *75*: P147–P155.

[17] Hale to Schuster, Oct. 29, 1917, Hale MSS, Box 47.

[18] George Ellery Hale, *National Academies and the Progress of Research* (Lancaster, Pa.:New Era Printing Co., 1915), p. 53.

[19] Hale to Schuster, Sept. 22, 1917, Hale MSS, Box 47.

[20] Extract from letter of Lt. Col. E. S. Gorrell, AEF, Oct. 29, 1917, in Records of the Army Air Force, National Archives, Washington, D.C. (hereafter, AAF MSS), Record Group 18. Ser. 22, File 402. 1, Technical Information.

opportunity. Months before, the NRC had urged the appointment of scientific attachés at the embassies, but the suggestion had died somewhere in the Cabinet. Now, with the advantage of such attachés more evident, the council managed to win official approval of a Research Information Committee. A foreign arm of the NRC itself, the committee was to coordinate technical liaison with the Allies and see to it that the appropriate bureaus at home got the appropriate reports.[21]

In the opening months of 1918 the committee was fitted out for duty. The President, drawing upon his special emergency fund, provided some $38,000 for expenses. The State Department granted two members of the NRC the status of scientific attaché: Henry Bumstead, assigned to London, and William F. Durand, a Stanford engineer who left the chairmanship of the National Advisory Committee for Aeronautics to take the post in Paris. Once abroad, Bumstead and Durand found Vice-Admiral Sims particularly helpful. Sims, irritated by his liaison problems, considered the committee "a most excellent idea," and besides, he and Durand, an Annapolis graduate, were old shipmates. The admiral provided the two attachés space and furniture at Naval headquarters in London and Paris. More important, he ordered his entire European command to cooperate fully in the enterprise, and Major John Biddle, chief of the American Army in England, soon followed suit.[22]

Bumstead, who was widely acquainted with British scientists, and Durand, who spoke fluent French, had no trouble ferreting out information from their colleagues abroad. Moreover, both men commended themselves to the Allies by distributing technical literature which grew out of the multitudinous research activities going on in the United States. By mid-spring 1918 the two attachés were funneling home a large volume of reports, and military officers were frequenting the committee's Washington headquarters in increasing numbers to run over the contents of the latest diplomatic pouch. Impressed, the Italian government asked the committee in April to open a third office in Rome.[23]

For Hale the work of the attachés added up to a viable kind of international scientific cooperation, a subject which, in April, was much on his mind. The Royal Society had scheduled the London conference of academies for May 8, and in anticipation of the meeting Hale was once again considering the question of reorganization. His exclusionary purposes remained, but now he forged an argument for reconstruction which turned on something quite different from the animosities of the conflict. Its pivotal point: just as the Information Committee had been created to help prosecute the war, so, for the same end, might a new agency of international science.

In the specific proposal that Hale drew up for the London conference, each of Germany's enemies was to form its own NRC, all of which were to federate in an Inter-Allied Research Council. As a pool of NRCs, the organization would incorporate the constituencies of academic, industrial, and governmental science. In the opinion of Hale the belligerent, so diverse an agency would materially aid the Allies in the war.

[21] Howard E. Coffin to the Secretary of War, Nov. 9, 1917, AAF MSS, Ser. 87, Aircraft Board File, folder 245; minutes of meeting of the Council of National Defense, Dec. 27, 1917, Josephus Daniels MSS, Library of Congress, Washington D.C., Box 451.

[22] Adm. William S. Sims to Capt. Roser Welles, n.d., quoted in "Report of Research Information Committee," May 1, 1918, National Research Council MSS, National Academy of Sciences–National Research Council, Washington, D C. (hereafter, NRC MSS), Research Information file; National Academy of Sciences, *Annual Report, 1918*, p. 49.

[23] Hale to Secretary of War, May 1, 1918, NRC MSS, Research Information file.

In the vision of Hale the promoter, so grand a cooperative coalition could advantageously replace the International Association of Academies in the peace.[24]

The scheme neatly cut through the political difficulties blocking wartime reorganization. By creating the new agency for war purposes, the Allies could naturally omit the Central Powers without openly committing themselves to a recriminatory policy. As a result, the exclusionary innovation would neither alienate the neutrals, who could be invited to join later, nor would it give the moderates in Hale's own academy any grounds for opposition. And Hale supposed that President Wilson would surely approve an agency which would at once strengthen the war effort—and leave the issue of German scientists to the future.

By the end of April the leadership of the National Academy had united behind the scheme. Now, with the London conference imminent, Hale went to the administration by writing to the President and seeing the secretaries of State and War. To the secretaries Hale no doubt made the same case as he made to Wilson. Surely the President would wish "to pass upon any international arrangements" to be proposed by "representatives of the United States." Hale emphasized how his plan would aid the war effort and sidestep the bitter split in the camp of science. He hoped that the President would "urge" Allied scientists to coalesce in this potentially effective organization.[25] The administration, for its own unexplained reasons, dodged and delayed. The Secretary of State deferred to the Secretary of War, who in turn shied from taking the initiative.[26] The President asked Hale for a clarification—did not the United States already have scientists in contact with the Allies abroad? A series of letters followed, and by mid-May Wilson had entered no objection to the plan, but he had neither urged it upon Allied scientists nor bestowed upon it a single word of positive approval.[27]

Hale had failed to win Presidential endorsement by the scheduled date of the London conference, but with the Germans once again threatening the Marne the French delegation was unwilling to leave Paris anyway; the meeting was put off for the moment. For Hale the delay meant more time to extract explicit approval for his plan from the White House. Moreover, the London conference was quickly rescheduled for October 9, 1918, a convenient time for Hale himself to go abroad and argue for his Inter-Allied Research Council in person.

Hale was eager to do just that. In his view the scheme remained the only reasonably safe exit from the political dilemmas of reorganization. Equally important, as spring turned into summer, the plan acquired increasing appeal as an instrument with which to help prosecute the war, mainly because of the development of the Research Information Committee. In mid-May some fifty to seventy physicists, engineers, and military officers began convening every Thursday evening in the council's headquarters to discuss the week's communications from London and Paris. Soon, centralizing technical liaison made such good sense to the Army's General Staff that on July 2,

[24] Hale to Schuster, Apr. 18, 1918; Hale, "Suggestions for the International Organization of Science and Research," Hale MSS, Box 47.

[25] Hale to the President, Apr. 29, 1918, Woodrow Wilson MSS, Library of Congress, Washington, D.C. (hereafter, Wilson MSS), File VI, Case 206.

[26] See memoranda together with Hale to Secretary of State, May 1, 1918, Records of the Department of State, National Archives, Washington, D.C. (hereafter, State Dept. MSS), Record Group 59, File 763. 72/9887; Hale to Evelina Hale, May 6, 1918, Hale MSS, Box 80.

[27] Woodrow Wilson to Hale, May 8, 1918; Hale to Wilson, May 10, 1918; Wilson to Hale May 13, 1918, Wilson MSS, File VI, Case 206a.

1918, the War Department strengthened the Research Information Service, as the original committee was now renamed, by ordering all its bureaus to utilize the agency.[28] Hale could now see his proposed Inter-Allied Research Council sponsoring regular meetings in London or Paris to which the Allied and American governments would send army and navy officers, scientists and industrial engineers.[29]

However appealing the Inter-Allied scheme now appeared to Hale, over in England Arthur Schuster—ill, weary, his son at the front—gloomily doubted its viability. Many British scientists were chafing at the bureaucratic controls of wartime, and Schuster suspected that they would dislike having to cope with still another organization. But Hale's plan had won a lot of sympathy in England. While a sizable number of Schuster's colleagues could not countenance meeting with the Germans soon after the war, few wanted to promote what a report of the Royal Society called "active enmity" against the enemy's men of research. Bumstead, acting as lobbyist for the Inter-Allied Council in England, found an influential cohort in the widely respected mathematical physicist Sir James Jeans. Near the end of July Bumstead and Jeans persuaded Schuster that Hale ought at least to have a chance to argue for his plan at the coming London conference. And in the considered opinion of all three men, Hale might very well win the general support of Britain's scientists if his scheme were crowned with the approval of the highly esteemed American President.[30]

Hale, who already knew of Schuster's doubts, learned of the Presidential point by cable, and in early August he once again sought Wilson's endorsement. Since the Information Service was now a going concern, Hale shifted the argument from the problems of postwar relations to the "pressing need" for the international coordination of wartime research. Would the President express his "hope," in writing, that the London conference would lead to the establishment of the Inter-Allied Council? Would he also award the National Academy's delegation official status and, in particular, appoint to it Major Biddle and Vice-Admiral Sims? While Hale worried that his request would run into opposition in the cabinet, Colonel Edward M. House had promised his personal help. Hale waited, musing: "The interesting thing now is what the President will do."[31]

The colonel put in his good word and the President sympathized with Hale's cooperative aims. But Wilson the diplomat had to tell Hale, "frankly . . . I do not think it would be wise for me to approve the Inter-Allied conference before it has taken place." He might very well sanctify its recommendations after he knew what they were. For the moment, the conference had better remain unofficial and without the services of Biddle and Sims.[32]

Hale, while disappointed, refused to concede. If the London conference were to resolve in favor of the Inter-Allied Council, perhaps, as Wilson's reply suggested, he

[28] Robert A. Millikan to William F. Durand, June 6, 1918, NRC MSS, Research Information file; copy of the War Department order is in AAF MSS, Ser. 81, Bureau of Aircraft Production, Science and Research Department file.

[29] Hale to Picard, Aug. 14, 1918, Hale MSS, Box 47.

[30] Henry Bumstead to Hale, July 19, 23, and 25, 1918; Schuster to Hale, July 27, 1918, Hale MSS, Box 47; Royal Society, "Memorandum of Committee on International Scientific Organizations," p. 4, Hale MSS, Box 48.

[31] Hale to Wilson, Aug. 3, 1918; Edward M. House to Hale, Aug. 9, 1918, Hale MSS, Box 47; Hale to Arthur A. Noyes, Aug. 14, 1918, Hale MSS, Box 32, Noyes file.

[32] House to Wilson, Aug. 12, 1918, Wilson MSS, Ser. 2, Box 181; Wilson to Hale, Aug. 14, 1918, Hale MSS, Box 47.

could then get the coveted Presidential approval. But the White House would scarcely bless any action out of line with Wilson's foreign policy, and, as Hale knew by late August, a growing number of Allied scientists were sounding decidedly un-Wilsonian. In Hale's worried prognosis the Allies would surely want the London conference to issue sweeping denunciations of Germany. If Wilson could only be made to understand how, more than ever, creating a council for war purposes was the only way out of the dilemmas of reorganization. Hale decided to make one last effort to win the President's support, and on September 10, his departure for London imminent, he went to the White House for a face-to-face talk.

Wilson understood what might happen in London only too well. The President, as Hale set down the gist of their conversation, was "*very emphatically opposed* to any resolution directed against German men of science." Such recriminations would be the "best possible way to play Germany's game," since they would add credibility to "the claim that Germany is surrounded by vindictive enemies." Hale countered with how organizing for war would sidestep just that pitfall, and Wilson then made explicit the policy which had doubtless governed his responses to each of Hale's previous entreaties: the President objected to joining with the Allies in any more organizations of a formal kind. As Hale recorded Wilson's reasoning, "France and other European nations have felt the war much more than we have, and are therefore likely to take drastic action that might bind us to do things contrary to our natural intent."

Hale jockeyed. On the point of a new formal organization Wilson (apparently not the Wilson who allegedly went to Versailles entirely naïve about the British and French) refused to budge. He would acquiesce only to inter-Allied scientific conferences of a strictly informal nature.[33] With that miniscule concession the interview ended, and six days later Hale sailed for England.

London in early October was cold and damp but, with the prospect of the pending victory, by no means dreary in spirit. To Hale's ambitious mind the imminence of peace pointed to a way of slipping around Wilson's proscription. As Hale now apparently chose to read the meaning of his White House interview, the President had objected only to joining formally with the Allies while the war lasted. But given the evident march of international events, the proposed organization was unlikely to take real shape until some time after the surrender of the Central Powers. Under the circumstances, at least so Hale believed, the Inter-Allied Council could be created without violating Wilson's instructions.[34] He might have been paying more attention to the President's words than to his intent; at times the promoter in Hale could dominate the responsible citizen. Whatever the intricacies of his reasoning, Hale aimed to request adoption at the conference of his entire scheme, even to ask for the President's approval after the fact.

But to win the President's ultimate sanction, Hale would have to keep the outcome of the conference in harmony with Wilson's foreign policy, and the task appeared difficult. The French and Belgians, determined more than ever to read the Germans out of international science indefinitely, had come to London armed with a program of sweeping revanche. The British remained more moderate, but they had prepared a

[33] "Memorandum of Interview with President, September 10, 1918," Hale MSS, Box 60, Wilson file.

[34] See Hale to Wilson, Oct. 15, 1918, Wilson MSS, File VI, Case 206c.

resolution for the conference that was not quite Wilsonian. The document rejected the resumption of personal relations with enemy scientists even after the resumption of diplomatic contacts. It also indicted "the Central Powers—and more particularly Germany . . . " for breaking the "ordinances of civilization."[35]

Hale and his fellow American delegates—Bumstead, Durand, and three other members of the National Research Council—hammered away at the British. Evidently over the weekend before the conference, the leadership of the Royal Society agreed to bring their resolution more in line with the American position. Wilson had said that a reformed Germany could enter the League of Nations; the revised version allowed the resumption of personal relations in science when "the Central Powers can be readmitted to the concert of civilized nations." Wilson had warned against specially recriminatory denunciations; the British agreed to drop the phrase "and more particularly Germany" from the charge against the Central Powers.[36]

The conference opened on schedule in Burlington House, the headquarters of the Royal Society, with Sir Joseph John Thomson, president of the society and an ally of Hale's, in the chair. In three days of plenary sessions Thomson apparently jeopardized the modified declaration against the Central Powers only once—when he yielded the chair to Picard. The Frenchman tried to provoke a revolt against the Anglo-American position, and the meeting erupted in a cacophony of indecorous debate. But of the thirty-three delegates, less than a quarter—nine—came from France and Belgium, and six more were scattered among Italy, Portugal, Serbia, Japan, and Brazil. The United States and Britain had all the rest, more than a majority, and with the two delegations commanding the meeting at least in voice, Picard's maneuver fizzled. In the end the conference unanimously adopted the revised British declaration as the political preamble to its organizational resolutions. Hale happily reported to the White House that the proceedings had been kept in harmony with the President's foreign policy.[37]

As for the organizational resolutions: the nations at war with the Central Powers were to withdraw from all existing international scientific associations and form new ones under a general scheme to be worked out by a Committee of Enquiry. The committee itself was to be the nucleus of a new international council, the council to be a federation of NRCs and the NRCs to be fathered by the academies represented at the conference. Particularly important, the new council was not only to advance pure and applied science in general; it was also to deal with matters related to national defense. In Hale's confident assessment, the new international council was "just the sort of vessel we wanted to build"—and it had been "safely launched."[38]

Some two weeks after the armistice the Committee of Enquiry met in Paris. The delegates christened Hale's vessel the "International Research Council," made it the flagship of the new scientific associations called for in London, and laid down the keels for two of them, an international union in astronomy, another in geophysics.[39]

[35] Copy of the Royal Society's proposed resolutions is in Hale MSS, Box 47.

[36] Hale to Wilson, Oct. 15, 1918, Wilson MSS, File VI, Case 206c; the resolutions as adopted are in Royal Society, "Preliminary Report of Inter-Allied Conference on International Scientific Organizations, Held at the Royal Society on October 9–11, 1918," Hale MSS, Box 48.

[37] Hale to Evelina Hale, Oct. 12, 1918, Hale MSS, Box 80; Royal Society, "Preliminary Report," Hale MSS, Box 48; Hale to Wilson, Oct. 15, 1918, Wilson MSS, File VI, Case 206c.

[38] Royal Society, "Preliminary Report," Hale MSS, Box 48; Hale to Evelina Hale, Oct. 12, 1918, Hale MSS, Box 80.

[39] "Inter-Allied Conference on International Organizations in Science," *Nature*, Dec. 26, 1918, pp. 325–327.

But in late November 1918 administrative matters were not the only items on the agenda. The coming of peace had raised the challenge which all along Hale and Schuster had known would confront the reorganization of international science—winning the allegiance of the neutrals.

At Paris the French and Belgians were not particularly helpful. Neither would tolerate inviting the neutrals immediately to membership in the International Research Council. While there had been an armistice, a state of war still existed, and the IRC might yet have to deal with military problems. The French and Belgians seem to have argued that incorporating any neutrals would be improper; and further, that bringing in Holland and Sweden, both friendly to Germany during the war, would be unsafe. Picard was sticky even about the new scientific unions, neither of which could conceivably deal with martial matters. When it was urged that the neutrals be admitted to them by a two-thirds majority of the member nations, Picard insisted upon three-quarters and won his point.[40]

By early spring of 1919 Hale was ruefully describing the action at Paris as "drastic." A good many neutral scientists had angrily taken the three-quarters rule as a clear affront. A Dutchman sneered at Hale for thinking that his distinguished academy would "accept an 'admission' to this company by allowance of a handful of votes from Portugese, Serbians, Brazilians, etc.!"[41] As the reaction mounted, both Schuster and Hale worried that the neutrals might boycott the International Research Council and its unions altogether. Both were certain that the insult of Paris had to be erased at the IRC's next meeting—its constitutive assembly—which was scheduled for July in Brussels. To that end they had the Royal Society and the National Academy go on record in favor of the admission to the IRC of the key northern European neutrals. Hale, unable to attend the assembly, was counting on W. W. Campbell, "a most able fighter," he assured Schuster, to keep the French in line.[42] If Picard and his allies balked, Hale was even ready to have the US delegation threaten American withdrawal from the inchoate council itself.

As the weeks passed, the French position softened. After all, by late spring 1919 it was clear that the military war would not resume. Moreover, the nucleus of the IRC's military possibilities, the Research Information Service, was losing its potency as an agency of defense; amid the swift corrosion of wartime military cooperation, armies and navies on both sides of the Atlantic were halting the open exchange of technical reports. Under the circumstances even the French and Belgians had to concede that restricting membership to the belligerent nations was no longer justified. So it seemed, at least, when in late May the IRC's Executive Committee met in Paris under Picard's chairmanship. Along with drafting the council's statutes and drawing up an agenda for the coming assembly, the committee recommended, by a unanimous vote, the admission of a number of additional countries, including Holland and Sweden.

But at the insistence of the French, no formal invitation was to be issued until the meeting at Brussels.[43] While the Versailles conference was not yet over, it seems that

[40] Hale to William W. Campbell, Jan. 21, 1919; to Jacobus C. Kapteyn, June 10, 1919, Hale MSS, Box 9, Campbell file; Box 24, Kapteyn file.

[41] Hale to Schuster, Apr. 8, 1919; Schuster to Hale, Feb. 12, 1919, Hale MSS, Box 37, Schuster file; F. M. Jaeger to Arthur A. Noyes, Jan. 13 1919, Hale MSS, Box 32, Noyes file.

[42] Hale to Schuster, Apr. 8, 1919, Hale MSS, Box 37, Schuster file.

[43] Charles E. Mendenhall to Hale, May 27, 1919, Hale MSS, Box 28, Mendenhall file.

Picard had more in mind than waiting until the legal end of the war. Once the neutrals were admitted, they might very well maneuver to bring in the Central Powers. Evidently to prevent the success of any such attempt, the Executive Committee had written into the statutes a weighted voting scheme and what amounted to a two-thirds majority rule for relaxing the bars against Germany.[44] Even if all the neutrals proposed for membership actually joined, just a few of the former belligerent nations would command a veto on all crucial administrative questions. By insisting upon delay until Brussels, the French were protecting themselves against the enlargement of the IRC until the adoption of the statutes was assured. The key to their reasoning was the agenda, which made the question of the neutrals virtually the last item of business on the last day of the assembly.

The assembly opened on July 18, its headquarters the splendid Palais des Academies, its inaugural session graced by the King of Belgium, its deliberations, in the proud remark of one of the 224 delegates, those of a "little peace conference."[45] Campbell, the able fighter, had arrived determined to press for early action on the neutrals; when he found the French and Belgians entirely unwilling to deviate from the agenda, he retreated. Not until the final day, after the adoption of the statutes as drafted, was the question of the neutrals called. Then, by a unanimous vote, the assembly invited to membership in the IRC and its unions thirteen additional nations. Not all of them, it seems, made the list on grounds of scientific worthiness. Only six—Denmark, Norway, Switzerland, and Spain, along with Holland and Sweden—had belonged to the International Association of Academies. Of the other seven, the new Czechoslovakia owed a good deal to the victors; China and Siam had both entered the war against Germany; together with Monaco, Argentina, Chile, and Mexico, they were likely, in the apparent thinking of the French, to help block any move to force the admission of the Central Powers.

Despite the invitation to the neutrals at Brussels and despite what Hale regarded as the moderation of the London conference, the National Academy won scarcely any support for its foreign policy from the American government. Hale was eager to maintain the scientific attachés in London, Paris, and Rome. So were the State Department and the offices of Military and Naval Intelligence, but no appropriation for the posts could be won amid the isolationist economizing on Capitol Hill. Congress aside, Hale never did manage to pry an official blessing for the International Research Council out of the architect of the League of Nations. At the end of 1919 the President was shut away by his stroke, and Colonel House happened to be out of grace at 1600 Pennsylvania Avenue.

Even had Hale managed to get to Wilson, he would undoubtedly have been rebuffed. The State Department's legal staff advised that the government owed the international dealings of the National Academy, which was strictly speaking a private organization, no more than "informal" cooperation.[46] More important than legal intricacies, Hale's foreign policy remained in conflict with the administration's. Whatever the outcome at Versailles, Wilson had gone there intending his League of Nations to be a concert of

[44] The statutes are in Arthur Schuster, ed., *International Research Council, Constitutive Assembly Held at Brussels, July 18th to July 28th, 1919, Reports of Proceedings* (London, 1920), pp. 222–226.

[45] *Ibid.*, p. 10.

[46] Memorandum, Office of the Solicitor, Sept. 17, 1920, State Dept. MSS, Entry 196, Doc. 592 D1/6.

the vanquished as well as the victors. If the President had acquiesced in Clemenceau's demand for the exclusion of the Central Powers, Wilsonians in the State Department were unwilling to sanction any new scientific organization that squeezed out Germany.[47]

Equally unwilling to do the same—and much more ominously for the International Research Council—were a number of prominent neutral scientists, particularly the distinguished Dutch astronomer Jacobus C. Kapteyn. A foreign associate of the Mt. Wilson Observatory and close friend of Hale's, Kapteyn was grateful for what the Americans had done to ease the way of the neutrals into the IRC. But since its founding, the Dutchman had objected to the council for a more bedrock reason—its exclusion of the Central Powers. At once the recipient, in 1914, of Germany's coveted cross Pour le Mérite, yet glad of the Allied victory, he was uncompromisingly loyal to the apolitical tradition of international science. For Kapteyn, the world's men of research ought to be helping to bind rather than aggravate the wounds of Europe. Shortly after the Brussels assembly he circulated a public letter of protest. By the late fall it had a formidable list of signatories, 278 in all, drawn from Denmark, Sweden, and Holland, Norway, Finland, Switzerland, and Spain.[48]

The document did more than call up the inherent universality of science—did not the founders of the IRC realize how "our grand governess, nature, mocks our petty hostilities"? With equal eloquence, it asked for simple compassion. Arguing how the inflamed atmosphere, not to mention the controlled press, of wartime must surely have affected the judgment of Germany's men of research, the letter urged that Hale and his colleagues not assign them to "another humanity" on grounds that their "morality" was "inferior to your own." Beyond ringing all the changes on the accustomed generosity and dispassion of the scientific community, in peroration the document importuned against what Kapteyn himself evidently feared as the ultimate disaster—the division of science, "for the first time and for an indefinite period, into hostile political camps."[49]

Kapteyn's plea stood no chance among the European members of the IRC, particularly the French and Belgians. If they remained implacably opposed to the admission of the Central Powers, it was not simply because they were in no mood to forgive Germany her sins during the war. It was also because, with Picard, they were sure she would scheme for "revenge" in the peace. Back in 1917 the Frenchman had cited a commonplace theme, that the war would continue with undiminished ruthlessness in the "economic and intellectual" arenas. In Picard's indictment, Germany regarded science as an instrument of "domination," and he had explicitly counseled the "danger" of separating science from "national" questions.[50] The French and the Belgians considered the IRC much the same way as did Clemenceau the League of Nations—as something like a mutual security pact. Kapteyn might plead for charity in the name of apolitical science, but his appeal could scarcely move the Allied victors,

[47] *Ibid.*

[48] Draft of letter, William W. Campbell to *Science* magazine, Dec. 1919, Hale MSS, Box 61, NRC Division of Foreign Relations file. For a brief appreciation of Kapteyn's life and work, see Arthur S. Eddington's obituary notice in *Proceedings of the Royal Society*, 1922, Ser. A, *102*: xxix–xxxv.

[49] "Aux membres des académies des nations alliées et des États-Unis d'amerique," attached to Kapteyn to Hale, Aug. 7, 1919, Hale MSS, Box 47.

[50] Picard to Hale, July 22, 1917, Hale MSS, Box 47.

determined, as they were, to safeguard themselves against the allegedly vengeful ambitions of the vanquished.

Kapteyn's appeal also failed to move the leadership of America's National Academy, particularly the father of the IRC. Whatever Hale's differences with Picard, he respected the Frenchman's political calculus. An emphatic advocate of American entry into the League of Nations, Hale regarded the IRC in the same way he viewed Woodrow Wilson's experiment in world order—as a peacetime coalition against the threat of revanche. Determined not to compromise the security of the Allies in science, he flatly considered the appeal of the neutrals impossible to heed. In mid-December 1919 Hale briefly discussed Kapteyn's letter before the National Research Council's Division of Foreign Relations. The members summarily agreed to the impossibility of conjoining with German scientists and quickly went on to other business.[51]

Still, whether the neutrals of northern Europe would join the IRC remained in doubt. The Swedish Academy defiantly awarded the year's Nobel prize for chemistry to Fritz Haber, who, it was widely known, had fathered the German army's gas warfare program. But the protests from the neutral camp happened to be more noisy than significant. Of the scientists who had signed Kapteyn's letter, 118, some 42 %, came from Finland, which had not been invited to membership; Norway, Switzerland, and Spain together had contributed a mere 32.[52] While the other 128 were from Holland and Sweden, the academies of both countries had their share of Hales, even of Picards. Apart from Finland, by the summer of 1920 every neutral country represented in the list of signatories had joined the IRC. Undoubtedly to appease their dissidents, most reserved the right to deal with the Central Powers independently. But whatever the caveats, Hale could be pleased that now 15 nations were aligned against Germany as members of his International Research Council.

The bars against the admission of the Central Powers were relaxed in 1926, but Germany refused to join either then or in 1931, when the IRC became the International Council of Scientific Unions. At the organizational level, science remained divided into hostile political camps until another postwar, when Hugo R. Kruyt, a Dutchman fittingly, once again sounded the warning of Kapteyn. If this time more scientists listened, it was not the least because they were determined to avoid another Versailles in science.

[51] "Report of the Meeting . . . , December 10, 1919," Hale MSS, Box 61, NRC Division of Foreign Relations file.

[52] Draft of letter, Campbell to *Science*, Dec. 1919, Hale MSS, Box 61, NRC Division of Foreign Relations file.

The Search for a Planet beyond Neptune

By Morton Grosser *

ON February 18, 1930, an assistant of the Lowell Observatory at Flagstaff, Arizona, found two tiny images 3.5 mm. apart on a pair of plates he was examining in a blink microscope. The dots were the photographic records of a moving object; their separation indicated that it was very distant, beyond the orbit of Neptune. The object was Pluto, the ninth planet of the solar system, and in finding it Clyde Tombaugh accomplished what his predecessors had been trying to do for more than eighty years. The planet he discovered is smaller than the earth and invisible to the naked eye; it moves around the sun at a mean distance of 3,666,000,000 miles, in unimaginable cold and darkness. Pluto is unique in the history of astronomy; before its discovery it was the subject of speculation and search lasting almost a century, and its reality was an article of faith for many astronomers who had no evidence that it existed at all.

The first recorded mention of a ninth planet was made twelve years before the eighth planet, Neptune, was discovered. On November 17, 1834, the Reverend Dr. Thomas Hussey, a British amateur astronomer, reported a conversation he had had with the French astronomer Alexis Bouvard to George Biddell Airy, the Astronomer Royal. Hussey wrote that when he suggested to Bouvard that the unusual motion of Uranus might be due to the perturbation of an unknown planet, Bouvard replied that "it had occurred to him, and some correspondence had taken place between Hansen and himself respecting it. Hansen's opinion was, that one disturbing body would not satisfy the phenomena; but that he conjectured there were two planets beyond Uranus."[1] When this was written, Peter Andreas Hansen was the director of the Seeberg Observatory in Gotha. Although his conjecture was correct, he was not pleased when it was later made public. Hussey's letter was published in 1846, following the discovery of Neptune. Several years later, when the American astronomer Benjamin Gould was commissioned to write a history of that discovery, Hansen sent Gould an emphatic denial that he had ever expressed a belief in the existence of a

* Boeing Scientific Laboratories, Seattle, Washington.

[1] Airy, George Biddell, "Account of some circumstances historically connected with the discovery of the planet exterior to Uranus," *Monthly Notices of the Royal Astronomical Society*, 1846, 7: 124.

ninth planet. He attributed Bouvard's remark to a misunderstanding, and his repudiation was published with Gould's report in 1850.[2]

The French physicist Jacques Babinet was less conservative than Hansen; in 1848 he published the first paper predicting the existence of a trans-Neptunian planet. Babinet based his work on the differences between the observed orbit of the planet Neptune and the orbits computed from the perturbation of Uranus by John Couch Adams and Urbain Jean Joseph Leverrier before Neptune was discovered in September, 1846. Although Leverrier's results had served to locate the planet, both his and Adams' figures for Neptune's orbital radius, eccentricity, inclination, and period differed markedly from the observed values. Babinet concluded from this that Uranus' perturbations were due to the combined effects of two planets, Neptune and an undiscovered body beyond it which he named Hyperion. His naïve calculations for Hyperion's elements (mostly simple addition and subtraction from Leverrier's figures for Neptune) gave the following results: [3]

Semimajor axis	47–48 astronomical units
Period	336 years
Mass	1/25900
Magnitude	10–11

These figures were sharply criticized by Leverrier, who found Babinet's reasoning drastically oversimplified. Leverrier had begun a similar project himself after Neptune was discovered, but had given it up as impracticable. In his opinion there was at that time " absolutely nothing from which one could determine the position of another planet, barring hypotheses in which imagination played too large a part." [4] He did, however, believe that a ninth planet would eventually be found. Shortly after Neptune's discovery he wrote that " This success allows us to hope that after thirty or forty years of observing the new planet, we will be able to use it in turn for the discovery of the one that follows it in order of distance from the Sun." [5]

The American mathematician Benjamin Peirce disagreed. In August, 1850, Peirce spoke at the American Academy of Arts and Sciences on the application of a " law of vegetable growth " to the periods of the planets. He showed that his formula described the arrangement of the planets more accurately than Bode's Law and concluded that " if this law is true, there can be . . . no planet beyond Neptune." [6] Peirce did not hesitate to present this empirical opinion to a meeting of scholars; as we have seen, Leverrier published his belief in a trans-Neptunian planet despite his own statement

[2] Gould, Benjamin Apthorp, Jr., *Report on the History of the Discovery of Neptune*, pp. 11–12 (Washington: The Smithsonian Institution, 1850).

[3] Babinet, Jacques, " Sur la position actuelle de la planète située au dela de Neptune, et provisoirement nommée Hypérion," *Comptes rendus*, 1848, *27*: 203.

[4] *Comptes rendus*, 1848, *27*: 209.

[5] Baldet, F., " Le corps céleste transneptunien," *Bulletin de la Société astronomique de France*, 1930, *44*: 228.

[6] Peirce, Benjamin, " On the law of vegetable growth and the periods of the planets," *Proceedings of the American Academy of Arts and Sciences*, 1852, *2*: 241.

that there was no evidence of its existence. Such subjective views about a ninth plant were to be commonplace during the next eighty years.

Peirce's opinions were respected by many astronomers, but his law of vegetable growth was an exception; within a year of its presentation, two searches for a planet beyond Neptune were begun. On October 21, 1850, James Ferguson, an assistant of the United States Naval Observatory in Washington, compared the asteroid Hygea with the 9th-magnitude star GR 1719k. He also observed the star on October 16, 19, and 22, but the observation of October 21 was published alone in January, 1851.[7] John Russell Hind, the director of a private observatory in London, was at that time mapping stars to 11th magnitude near the ecliptic in the 19th hour of right ascension. Despite a careful search, he could not find the star GR 1719k in the position reported by Ferguson, and in August, 1851, he reported the missing star to William Bond, the director of the Harvard Observatory, Cambridge, Massachusetts. Bond in turn notified Lieutenant Matthew Maury, the superintendent of the Naval Observatory.

On the night of August 29, 1851, Maury had Ferguson look for the star again. He could not find it, and on September 3, Maury wrote to William Graham, Secretary of the Navy: "The star of comparison with *Hygea* on the night of October 21, 1850, has disappeared. It is not now to be found where it then was. Hence I infer that it is probably an unknown planet."[8] On November 12, Hind made the same inference in a letter to Gould, and pointed out that if the object was a planet, its orbital radius would be 137 a. u., far greater than that of any other member of the solar system, and it would have a period of more than 1,600 years.[9] Hind had searched the suspected part of the sky minutely; he wrote Gould that "there is no planet hereabouts of the 9.10 magnitude," and asked that Maury publish the original observations. In March, 1852, all of Ferguson's positions for GR 1719k were published in the *Astronomical Journal,* together with the report that Ferguson had searched for the star from August to December, 1851, without success. Hind had no better luck and the matter was dropped at both the Washington and Regent's Park observatories. It was not revived until twenty-six years later.

In October, 1878, a description of the 1851 planet search appeared in the British periodical *Nature*; on December 2, C. H. F. Peters, the director of the Hamilton College Observatory at Clinton, New York, wrote to Rear-Admiral John Rodgers, the superintendent of the Naval Observatory, and explained the mystery once and for all. Peters was provoked by the anonymous writer who had brought up the missing planet of 1850 again,

> . . . using almost the identical, but very uncertain arguments of Mr. Hind of 18 [sic] years ago. In order, now, that nobody thereby might be induced to spend months and years upon a renewed search, I hasten to bring to your knowledge the errors I have detected in the "Washington Observations for

[7] Ferguson, James, "Observations of Hygea made with the filar-micrometer of the Washington equatorial," *The Astronomical Journal,* 1851, *1*: 165.

[8] *The Astronomical Journal,* 1851, *2*: 53.

[9] *Ibid.,* 1852, *2*: 78.

> 1850" (on pages 320 & 321), and which alone have given rise to the misconception.—This pseudo-planet indeed is nothing but a star observed already by Lalande, occurring besides three times in Argelander's zones and twice in Lamont's, — as I am now going to show.[10]

Peters showed that, by mislabeling one of three micrometer wires, Ferguson had incorrectly recorded the positions of the star GR 1719k taken on October 16 and on the first transit of October 19. He then changed the correct wire numbers used on October 19, 21, and 22 to conform with the mislabeled wire, thereby giving rise to the suspicion that the star had disappeared. Peters identified the star Ferguson had observed as Lalande's LL36613. Admiral Rogers published Peters' conclusions after having ordered a check which showed them to be entirely correct.

The alacrity with which Maury and Hind identified Ferguson's missing star as an undiscovered planet showed how acceptable the idea was. Many astronomers shared Leverrier's hope that perturbations of Neptune could be used to locate another planet beyond it. In 1866 the American astronomer Simon Newcomb published a definitive work on Neptune's motion, and because of his concern with the problem of a ninth planet, he made a thorough analysis of the tabular residuals of Neptune's position. He concluded that the nineteen years of observations were in such perfect accord with theory that they could not possibly be used to locate a trans-Neptunian planet. Newcomb's statement did not quench the enthusiasm for a ninth planet, but it did divert attention to another line of inquiry, the coincidence of cometary aphelia and planetary orbits.

It was known that the aphelion distances of many comets were clustered about the mean orbital radii of the outer planets, which were believed to have diverted the comets into elliptical orbits. In the 1879 edition of his *Astronomie populaire* the French astronomer Camille Flammarion pointed out that the comets 1862iii and 1889iii had aphelion distances of about 47 and 49 a. u., respectively, and suggested that they might indicate the orbital radius of an undiscovered planet.[11] Professor Hubert Newton, head of the mathematics department of Yale University, supplied a theoretical basis for this idea. On August 25, 1879, Newton read a paper at a meeting of the British Association for the Advancement of Science "On the Direct Motion of Periodic Comets of Short Period." During the discussion of his work, he remarked that it explained why the aphelion distance of a periodic comet is about the same as the distance of the planet that attracted it into an elliptic orbit. Thereupon Georges Forbes, a professor of astronomy at Glasgow University, announced that "there could be no longer a doubt that two planets exist beyond the orbit of Neptune, one about 100 times, the other about 300 times the distance of the earth from the sun, with periods of revolution of about 1000 and 5000 years respectively."[12]

10 *Astronomische Nachrichten*, 1879, *94*: 113.

11 Flammarion, Camille, *Astronomie populaire*, p. 661 (Paris, 1879).

12 Forbes, George, "On comets," *Proceedings of the Royal Society of Edinburgh*, 1880, *10*: 427.

Forbes was to be preoccupied with this idea for three decades. In his first paper, " On Comets," which was read to the Royal Society of Edinburgh in February, 1880, he showed that the aphelion positions of four comets grouped around 100 a. u. lay within 3° of a plane passing through the sun. He then calculated the orbit determined by the aphelion positions and the plane, and found that a planet with a mean distance of 98 a. u. and a period of 997 years would be within 9° or less of each aphelion position at the appropriate time.[13] A month later he published elements for a second, more distant trans-Neptunian planet, basing this prediction on six comets whose aphelion distances were about 300 a. u. Forbes was becoming obsessed with the idea of an undiscovered planet; during the next two months he published two more papers on the same subject. He reported that he had found " perturbations in the motion of Uranus, agreeing remarkably in character and period with those which would be produced by the new planet," and claimed that the orbit of the hypothetical planet computed from these perturbations agreed with the results of his comet calculations. He also published the positions of four stars stated to be missing by the Irish astronomer Edward Joshua Cooper; Forbes believed these might have been observations of the undiscovered planet.[14]

Soon after this article appeared Forbes received a letter from David Peck Todd, the chief assistant of the United States Nautical Almanac Office. The situation Todd brought to Forbes' attention was an embryonic duplicate of the Adams-Leverrier controversy of 1846. His letter was dated June 15, 1880:

> Dear Sir, — About a month ago I read with intense interest a copy of your memoirs " On Comets and Ultra-Neptune [sic] Planets," which came to this office. You cannot fail of understanding my enthusiasm about the matter, in so far at least as it relates to the prediction of position, when I tell you that some three years ago I was engaged on a provisional treatment of residuals of longitude of Uranus and Neptune, having in view the detection of possible exterior perturbations. On my first reading of your paper I took the notion in some way that your assigned longitude was a long way different from mine, and I thought nothing more about it for several days; but then, on referring to my papers, what was my astonishment and delight on finding that your position for the interior of the two planets differs *only four degrees* from the position which I had assigned from my own work, and marked upon a slip of paper on the morning of the 10th October 1877. Of course all my work was necessarily inconclusive, as there are not, even up to the present moment, any well-marked residuals in the case of either Uranus or Neptune; so I have never yet published the investigation. But, at the same time, I thought well enough of the work to attempt a practical search for a trans-Neptunian planet. It was conducted with the great refractor of the Naval Observatary [sic] during the latter part of 1877 and early in 1878 . . . I found the practical search much the most arduous task that I have ever set myself about, and the matter was the more aggravating because my regular day work

13 *Ibid.*, 428–429.

14 Forbes, George, " On an ultra-Neptunian planet," *Proceedings of the Royal Society of Edinburgh*, 1880, *10*: 636–637.

> could suffer no interruptions, and the great telescope was not at my service until after midnight. . . . The search was abruptly terminated, by circumstances beyond my control, at longitude 186°. But as I only took a narrow zone, at a given inclination so elliptic, I have never regarded the search definitive. I have not yet fully concluded what I shall do – if anything – with my investigation. I think, however, that if I have leisure during the next few months, I shall repeat the work quite independently of what I previously did, and publish at least the results, if there seems to be anything worth the while.[15]

Todd did not repeat the work during the summer of 1880, and in August he sent an account of his 1877 search to the *American Journal of Science*. His historical introduction and analytical procedure showed that he had been influenced by the story of Neptune's discovery. After comparing Adams' and Leverrier's predictions for Neptune's mass, orbital eccentricity, and longitude of perihelion with the values computed from observations, Todd decided not to predict these quantities. In October, 1877, he arrived at the following specifications for his hypothetical planet: [16]

Longitude, 1877.84	170° ± 10°
Mean orbital radius	52.0 a. u.
Period of revolution	375 years
Mean daily motion	9″.46
Angular diameter	2″.1
Stellar magnitude	13 +
Longitude ascending node	103°
Inclination	1° 24′

Todd made his telescopic search with the 26-inch Washington equatorial on "thirty clear, moonless nights, between the 3d of November, 1877, and the 5th of March, 1878." He swept a zone of the predicted orbit from longitude 146° to 186° with great care, but found nothing. Although even this negative result would have been worth reporting, Todd waited nearly three years before publishing his account. He had searched with the hope of finding the planet at the earliest possible moment; when he had no success, he lost confidence in his theoretical work. Now, in 1880, his hopes were rekindled by Forbes' independent confirmation, and he urged that other observers search for the planet which was, according to his calculations, approaching opposition.

The coincidence of Forbes' and Todd's results stimulated many observers. In January, 1881, Forbes reported that "a large number of excellent telescopes are now being employed in the search, and a considerable number

[15] Forbes, George, "Additional note on an ultra-Neptunian planet," *Proceedings of the Royal Society of Edinburgh*, 1882, *11*: 91–92.

[16] Todd, David P., "Preliminary account of a speculative and practical search for a trans-Neptunian planet," *The American Journal of Science*, 1880, *20*, 3rd series: 225–234. A modern astronomer has published similar conclusions about the simplification of unknowns in the problem of inverse perturbations. See R. A. Lyttleton's articles, "A short method for the discovery of Neptune," *Monthly Notices of the Royal Astronomical Society*, 1958, *118*: 551–559; and "The rediscovery of Neptune," *Vistas in Astronomy*, vol. 3, pp. 25–46 (London: Pergamon Press, 1960).

of charts of [the suspected] region have been forwarded to me." His own search was plagued by bad weather, but he did receive some theoretical encouragement during the course of it. In 1873 Pliny Earle Chase, professor of philosophy at Haverford College near Philadelphia, had evolved a simple formula for the mean orbital radii of some of the planets: $\pi^n/32$ = radius in astronomical units. Late in 1880 Chase notified Forbes that both of his hypothetical planets were represented by this series as follows: $\pi^7/32 = 92.79$ (Forbes 1, at 100 a. u.); $\pi^8/32 = 296.52$ (Forbes 2, at 300 a. u.). The agreement, however, remained a theoretical one, for none of the searchers discovered a planet at the points predicted by Forbes and Todd.

By the spring of 1881 a skeptical reaction had begun. The editors of the astronomical journal *Copernicus* expressed a common opinion when they wrote that Forbes' result could "hardly be entitled to much weight, as it depends entirely on the aphelion-distance of four comets. Now, that of the comet I. 1843 is more than doubtful, and Donati's orbit of comet II. 1855, does not represent the observations very satisfactorily, while the period of comet I. 1861 is very uncertain." [17] Todd's work was viewed more favorably, but the general attitude at this time was negative. One reason for the skepticism (besides the unsuccessful searches) was the trans-Neptunian planet's strong attraction for cranks and visionaries. While some astronomers computed, others spun fantasies; their writings run like bright, twisted threads through the sober fabric of scientific publication. An example was the Abbé François Constant Vassart of Catteniéres, France, who published in 1881 an essay on "The twelve constituent planets of our solar system, of which three are ultra-Neptunian." [18] Vassart's paper, the product of a fertile imagination, had no scientific basis whatever, and it was not unique. In 1882 the French astronomer Gabriel Dallet published several versions of an empirical formula predicting the orbital radii and periods of undiscovered outer planets; again, although Dallet had done some research on cometary aphelia, his "law" was entirely speculative.[19]

During the late nineteenth century the trans-Neptunian planet hypothesis went through several cycles of popularity. It was disparaged in 1881, but three years later a new wave of acceptance began. In 1884, although no new evidence had been discovered, Camille Flammarion wrote: "We can say today with certitude: *There is a planet beyond Neptune.*" [20] This opinion was shared by W. K. F. Zenger, professor of physics at the Böhmische Technische Hochschule in Prague, who published an article "On the possible existence of entirely unknown planetary bodies" in the same year.[21] Perhaps the most individual new contribution came from Count Oskar Reichenbach. In 1875 Reichenbach had privately published in London a pamphlet entitled *Two Planets beyond Neptune and the Motion of the Solar System: A*

17 "A record of the progress of astronomy during the year 1880," *Copernicus*, 1881, *1*: 68.

18 *Bulletin de la Société astronomique de France*, 1910, *24*: 174.

19 Dallet. Gabriel. "Note sur les planètes extremes de notre système solaire," *Revue scientifique*, 1882, *4*: 80–85.

20 Flammarion, Camille, *L'Astronomie*, p. 660 (Paris, 1884).

21 *Comptes rendus*, 1884, *99*: 290.

Speculation. This essay attracted little notice, but ten years later Reichenbach returned to the subject with an article on "The misapprehensions of Leverrier and Adams as evidence for the existence and position of two planets beyond Neptune." [22] In this paper he tried to prove that Adams and Leverrier had concealed the locations of two undiscovered planets by inventing fictions and combining the residuals of Uranus for maximum concentration of effect. Reichenbach's deductions were completely erroneous, but his paper illustrated the continuing subjective enthusiasm for additions to the solar system.

In 1886 David Todd, encouraged by the revived interest in undiscovered planets, published another account of his original sweep and proposed that a new search be made with a powerful telescope equipped for dry-plate photography. Todd argued that this project was well worth the expense and effort, for even if the search was unsuccessful, "there would still remain the series of photographic maps of the region explored, and these would be of incalculable service in the astronomy of the future." [23] Less than two years elapsed before this project became a reality.

In the early 1880's Isaac Roberts, a wealthy, middle-aged British amateur astronomer, had become interested in telescopic photography. After making tests with various lenses, Roberts commissioned the construction of a 20-inch aperture silver-on-glass photographic reflector. By 1886 he had taken more than two hundred photographs of stars and nebulae, and when these were exhibited at the Royal Astronomical Society in November of that year their quality and detail caused a sensation. During 1887 Roberts decided to make a photographic search for planets beyond Neptune and wrote to George Forbes for the positions of his hypothetical planets. Forbes promptly sent the necessary data.

Roberts retired from business early in 1888, planning to devote his full time to the search. He soon found that the climate near Liverpool, where his observatory was located, made the project all but impossible, and in 1890 he moved his home and observatory to Sussex, where he found clearer skies. The area he photographed lay between $11^h\,24^m$ and $12^h\,12^m$ right ascension, and between 0° and 6° north declination. This region was covered by eighteen pairs of plates; each pair was taken seven or more days apart and compared by superposition three times over. On May 13, 1892, Roberts reported the results of his search to the Royal Astronomical Society: "The whole of the plates covering the region were very carefully examined, and it now only remains for me to report that no planet of greater brightness than a star of the 15th magnitude exists on the sky area herein indicated, nor is there on the plates any abnormal appearance to which it is necessary here to draw special attention." [24]

The finality of this first photographic search must have disheartened

[22] *Journal of Science*, 1885, *22*: 1–13.

[23] Todd, David P., "Telescopic search for the trans-Neptunian planet," *Proceedings of the American Academy of Arts and Sciences*, 1886, *21*: 228–243; *Astronomische Nachrichten*, 1886, *113*: 153–166.

[24] Roberts, Isaac, "Photographic search for a planet beyond the orbit of Neptune," *Monthly Notices of the Royal Astronomical Society*, 1892, *52*: 501.

Forbes, who published nothing about trans-Neptunian planets for nearly ten years. When he returned to the subject in 1901, he dismissed Roberts' work with a single sentence: "Mr. Isaac Roberts made a search by photography but did not find the planet, possibly owing to my having indicated for his search an area that was too limited." [25] Forbes' attitude illustrated the powerful hold of the trans-Neptunian planet on men's imaginations. For the astronomers who became involved with it, it was a kind of celestial grail, and repeated failures to find it seemed to attract new searchers rather than to discourage those already seeking.

In 1899 the French astronomer A. Benoit proposed the construction of a new photographic telescope, mounted so as to be sensitive to extremely small motions in longitude in a narrow belt around the ecliptic. He claimed that by taking three exposures of two hours' duration, each month for a year, this instrument could find "every planet situated between the distance 30 and the distance 160 [a. u.] whose brightness was greater than 16th magnitude." [26] The telescope was to be set up in Algiers; it was not built, probably because of its cost, and Benoit's later trial of the method was defeated by poor weather.

Despite such disappointments publications on exterior planets continued to appear; the next contributor was Hans Emil Lau, a young Danish astronomer. Lau reviewed the observations of Uranus made between 1690 and 1895 and concluded that the hypothesis of a single trans-Neptunian planet was incompatible with the motion of Uranus. In 1900 he published the following data for the two planets he believed to be responsible for Uranus' perturbations: [27]

Interior planet:	Mean distance	46.6 a. u.
	Mean longitude	$274^\circ.6 \pm 180^\circ$
	Corrected mass	1/12900
Exterior planet:	Mean distance	70.7 a. u.
	Mean longitude	$343^\circ.9 \pm 180^\circ$
	Corrected mass	1/2800

Lau expected that the two planets would have respective apparent magnitudes of 10.0 and 10.5, and that they would be recognizable by their disks. Edward Charles Pickering, the director of the Harvard Observatory, made a photographic search for Lau's planets in the region near 275° longitude, but without success; at the same time, negative results were reported from another photographic search for Forbes' hypothetical planets by the amateur astronomer William Edward Wilson in Damarona, Ireland.

In May, 1901, Gabriel Dallet published a second paper on trans-Neptunian planets. This time he based his work on the perturbations of Uranus, but

25 Forbes, George, "Additional note on the ultra-Neptunian planet, whose existence is indicated by its action on comets," *Proceedings of the Royal Society of Edinburgh*, 1902, *23*: 370.

26 Benoit, A., "Planètes transneptuniennes," *Bulletin de la Société astronomique de France*, 1899, *13*: 494–497.

27 Lau, Hans E., "Planètes inconnues," *Bulletin de la Société astronomique de France*, 1900, *14*: 340–341.

he did not agree with Lau that two exterior planets were needed to account for the residual differences. Dallet's "Planet *X*" moved in an orbit of 47 a. u. mean radius and had a magnitude of 9.5–10.5; its probable longitude in 1900.0 was 358°.[28] Pierre Jules César Janssen, the director of the Meudon Observatory, promptly invited Dallet to make a photographic search for Planet *X*, and in America Pickering agreed to examine plates of the suspected region already taken. Again, despite much work, no new planet was found.

The theoretical bases for exterior planet searches alternated between the perturbations of known planets and the convergences of cometary aphelia. These two lines of inquiry sometimes triggered interdependent research. An example was the article published by Théodore Grigull in December, 1901. Grigull, who based his work on comets, said he had been stimulated by Dallet's paper of May, 1901, which was an analysis of Uranus' residuals. Grigull used the orbits of comets seen in 1532, 1661, and 1889 to calculate a longitude of 352°.6 ± 4°.2 in 1901.0 for his trans-Neptunian planet. His upper limit of position, 356°.8, was very close to the 358° Dallet had reached for 1900.0.[29] This agreement encouraged Grigull to make a more complete investigation; in October, 1902, he published a list of twenty comets seen between 1490 and 1898 that he used to determine precise elements for his hypothetical planet. His new results coincided even more closely with Dallet's figures: the longitude of the planet for the epoch 1902.0 was 357°.54 ± 1°.867 [!].

Some months before, the Vicomte du Ligondès had published a scheme in which the orbital radii of the planets varied as the surface of a conical solid of revolution with the vertex at the sun.[30] Ligondès only extended his model out to Neptune, but Grigull extrapolated it and found that Ligondès' predicted distance of 53.29 a. u. for a ninth planet agreed fairly well with the 50.61 a. u. he had calculated independently. In February, 1903, Ligondès allowed that the cones of revolution that determined the multiplicand of orbital radius in his system might be warped inward slightly, bringing his prediction into closer agreement with Grigull's.[31] Ligondès' scheme was one of a continuing line of exotic contributions to exterior planet research. In 1902 the French Academy of Sciences received a manuscript in which General Alexander Garnowsky of Nijni-Novgorod, Russia, claimed to prove the existence of *four* trans-Neptunian planets,[32] and in 1907 Le P. Choren M. Sinan sent from Trebizond his set of elements for a ninth planet.[33] Other

[28] Dallet, Gabriel, "Contribution à la recherche des planètes situées au dela de l'orbite de Neptune," *Bulletin de la Société astronomique de France*, 1901, *15*: 266–271.

[29] Grigull, Théodore, "Nouvelle contribution à la recherche d'une planète transneptunienne," *Bulletin de la Société astronomique de France*, 1902, *16*: 31–32.

[30] Ligondès, Raoul Marie, Vicomte du, "Sur les planètes téléscopiques," *Bulletin de la Société astronomique de France*, 1901, *15*: 358–361.

[31] Ligondès, Raoul Marie, Vicomte du, "Au sujet des planètes transneptuniennes," *Bulletin de la Société astronomique de France*, 1903, *17*: 122.

[32] Garnowsky, Alexander, "Sur l'existence de quatres planètes transneptuniennes," *Bulletin de la Société astronomique de France*, 1902, *16*: 484.

[33] Sinan, Le P. Choren M., "Recherches sur la planète transneptunienne," *Bulletin de la Société astronomique de France*, 1907, *21*: 122.

contributions came from even more remote sources, but the next major attack on the problem was made in the United States.

On November 30, 1908, the American astronomer William Henry Pickering announced in Harvard College Observatory Circular No. 144 that he had found evidence of the existence of an undiscovered planet beyond Neptune. He gave the planet's position for 1909.0 as right ascension 7^h47^m, declination $+21°$. Pickering had calculated these results in April, 1908, and reported them to the American Academy of Arts and Sciences on November 11; by the time the Harvard circular appeared, photographs of the suspected region had been taken at the Harvard Observatory station in Arequipa, Peru, and a second set of plates was being made by the Reverend Joel H. Metcalf at his private observatory in Taunton, Massachusetts. These were the first of many planet searches instigated by Pickering; he was eventually to publish more on the subject of trans-Neptunian planets than any other single investigator.

Pickering was born in Boston, Massachusetts, on February 15, 1858. He graduated from the Massachusetts Institute of Technology in 1879 and became an instructor there; later he was an assistant professor at the Harvard Observatory. Pickering's preoccupations were astronomy, mountain climbing, and travel in primitive and unsettled regions. He led solar eclipse expeditions to Colorado in 1878, to Grenada, West Indies, in 1886, to California in 1889, and to Chile in 1893. In 1891 he established the Arequipa station of the Harvard Observatory, and three years later he erected the Flagstaff, Arizona, observatory for Percival Lowell. In 1899 Pickering discovered Phoebe, the ninth moon of Saturn and the first satellite to be found by photography; in 1903 he published the first complete photographic lunar atlas.

The Arequipa photographs made for Pickering in 1908 were unsatisfactory, primarily because of the northern declination of the suspected region, and the search was made on twenty-eight plates taken by the Reverend Joel Metcalf. Each 8 × 10-inch plate covered an area 4°.5 in diameter and recorded the images of about 25,000 stars, to the 15th magnitude in the center of the plate and 14th magnitude at the edges. The region photographed lay between 7^h0^m and 8^h40^m along the ecliptic and extended 10° above and below it. Despite many hours of comparison and an individual star count of 300,000 images, no planet was found. Pickering attributed the failure to the possibility that the planet might have been more than 10° north or south of the ecliptic.[34]

The Harvard announcement of this search was followed by sensational newspaper reports that a new planet had been discovered. Pickering quickly contradicted these accounts and published his research. His objectives were "to determine if . . . a planet really existed beyond Neptune, and . . . to locate it by the simplest, shortest and most direct practical method." [35] He

34 Pickering, William H., "A photographic search for Planet O," *Annals of the Astronomical Observatory of Harvard College*, 1911, *61*, part 3: 369–373.

35 Pickering, William H., "A search for a planet beyond Neptune," *Annals of the Astronomical Observatory of Harvard College*, 1909, *61*, part 2: 113–162.

chose the method of graphical analysis described by John Herschel in his *Outlines of Astronomy* sixty years earlier. According to this system, if the longitude of a planet is plotted against its deviations from predicted positions, there will be short irregularities (or " kinks ") in the planet's motion at the points of conjunction with a disturbing body. If two such kinks can be found, the interval between them gives the period of the disturbing planet, which greatly simplifies the problem of finding it. A more difficult case is that of the planet observed only long enough for one kink to be recorded, so that the period of the perturbing body is unknown. Even in this case it is possible to locate the disturbing planet at conjunction, when maximum deviation is being recorded, but once conjunction has passed, the problem becomes steadily more difficult. Pickering used this technique to analyze the motion of Uranus and concluded that there existed an undiscovered " Planet *O*," whose mean distance was 51.9 a. u. and whose period was 373.5 years. This planet was the object of the unsuccessful photographic searches of 1908.

Pickering's investigation was hardly published before other theorists began to criticize his methods. The British astronomer P. H. Cowell announced his opposition to the whole hypothesis, and his opinion was echoed by an editorial in the *Observatory*. This article pointed out that Pickering had resolved the perturbative force on Uranus into two components, one tangential to the orbit and the other normal to the direction of motion, and plotted the intensities of these two forces against time independently. He assumed that the obvious kinks in the graphs near conjunction would occur in the resultant motion of the planet. The critic claimed that this assumption was erroneous because first, Pickering's force components were not actually interdependent; since the planet was not constrained to move in an idealized curve, extraneous forces acted on it independently. Second, the kinks in Pickering's force diagrams were produced by higher harmonics in the series for the longitudinal acceleration of Uranus; after integrating the series twice to obtain the planet's displacement in longitude, the coefficients of the higher terms were drastically reduced and the curves smoothed to the point where the kinks were imperceptible.[36] Pickering answered his critics in July, 1909. He wrote that despite the smoothing effect of integration on his graphs, " the observations of Uranus do actually show three equidistant maxima, i. e. forward displacements, corresponding to the maximum kinks in the curve of force, and also three well-marked minima where minima should occur; it does not seem likely that these results could be due to mere accident." [37]

In February, 1909, a new and unique hypothesis on trans-Neptunian planets was advanced by Thomas Jefferson Jackson See, an astronomer at the United States Naval Observatory on Mare Island, California. See believed that the solar system had been formed by the coalescence around the sun of bodies captured from outside because of the presence of a resisting

[36] *The Observatory*, 1909, *32*: 303–305.
[37] Pickering, William H., " The assumed Planet O beyond Neptune," *The Observatory*, 1909, *32*: 326–328.

medium. This hypothetical medium tended to decrease the eccentricity of planetary orbits, and since the density of the medium supposedly increased with proximity to the sun, the orbits of planets closer to the sun would be more circular than those distant from it. According to See, "To suppose the planetary system to terminate with an orbit so round as that of Neptune is as absurd as to suppose that Jupiter's system terminates with the orbit of the fourth satellite. . . . In all probability there are several more planets beyond the present boundary of the [solar] system, some of which may yet be discovered."[38] See gave orbital radii for three trans-Neptunian planets as 42.25, 56, and 72 a. u. He claimed to have calculated elements for his innermost planet in 1904 and gave its longitude at that epoch as 200° and its period as 272.2 years.

By 1909 so many different predictions for undiscovered planets had been made within a short time that a comparison was inevitable. In 1907 the French astronomer Jean Baptiste Aimable Gaillot had analyzed the residuals of Uranus and Neptune and found no evidence of an exterior planet. Two years later, provoked by the discordant predictions made by Forbes, Lau, and Pickering, Gaillot reviewed the problem and admitted the possibility that two planets might exist beyond Neptune, at distances of 44 and 66 a. u. Gaillot cautioned that both distances had wide margins of error, but noted that the position of his inner planet was remarkably close to Lau's: for the epoch 1900.0, Gaillot predicted a geocentric longitude of 271°; Lau, 269°. Gaillot's outer planet was completely different from Lau's, but coincided almost exactly with Pickering's. Gaillot offered neither correction nor confirmation of Forbes' work; he found that a planet as distant as 100 a. u. would have no measurable effect on Uranus. Further, after checking his earlier work on Neptune, Gaillot concluded "that the observations of Neptune can be of no real use in the investigation of a planet more distant from the Sun."[39]

In 1914, on the eve of World War I, Lau corroborated Gaillot's comparison by another method. Lau resolved his equations (as had Leverrier sixty-eight years before) to yield the mass of the hypothetical planet (s) for twenty-four positions, each 15° apart around the ecliptic, at a single epoch (1900.0). He found that the only positive values of mass occurred in the sectors 30°–120° and 210°–300°. According to this analysis the positions predicted by Todd, Forbes, Grigull, and See were inadmissable, but Pickering's and Gaillot's results were compatible with Lau's. Lau noted that the work done to that time showed "only that the hypothesis of two trans-Neptunian planets does not conflict with the observed facts, so that there *may exist two or more major unknown planets beyond the known limits of the solar system*."[40] One year after this was written, a monograph on a

[38] See, T. J. J., "On the cause of the remarkable circularity of the orbits of the planets and satellites and on the origin of the planetary system," *Astronomische Nachrichten*, 1909, *180*: 185–194.

[39] Gaillot, J. B. A., "Contribution à la recherche des planètes ultra-neptuniennes," *Comptes rendus*, 1909, *148*: 754–758.

[40] Lau, Hans E., "La planète transneptunienne," *Bulletin de la Société astronomique de France*, 1914, *28*: 276–283.

trans-Neptunian planet was published by the man whose efforts were eventually to lead to the planet's discovery.

Percival Lowell was born in Boston, Massachusetts, on March 13, 1855, and graduated from Harvard in 1876, distinguishing himself particularly in mathematics. He first became interested in astronomy in 1870; by 1877 he was spending many hours on the roof of his father's house observing the planets with a 2-inch telescope. His enthusiasm was intensified in that year by Giovanni Schiaparelli's discovery of " canali " on Mars; and in the early 1890's Lowell decided to give up his business career and devote his full time to astronomy. By that time Schiaparelli's eyesight had failed; Lowell determined to build an observatory to continue the work on Mars. Convinced that the first essential in planetary observation is good seeing, he spent some time investigating atmospheric conditions in various parts of the world. " A large instrument in poor air," he wrote, " will not begin to show what a smaller one in good air will. When this is recognized . . . it will become the fashion to put up observatories where they may see rather than be seen." [41] Following this principle, Lowell chose to locate his observatory on an isolated mesa above the town of Flagstaff, Arizona, at an altitude of 7,200 feet.

Observations were begun at Flagstaff in 1894 with an 18-inch refractor. This was replaced by a 24-inch refractor in 1896, and a 42-inch reflector was added later. Lowell's visual and photographic observations of Mars began to appear in the *Annals of the Lowell Observatory* in 1898; in 1902 he was appointed nonresident professor of astronomy at the Massachusetts Institute of Technology. Lowell did not confine his observations to Mars, and his studies of the solar system led him eventually to the problem of a ninth planet. He made his first calculations in 1906, but his " Memoir on a trans-Neptunian planet " did not appear until 1915.

Lowell attacked the problem from a new and rigorous standpoint. Unlike most of his predecessors, he made a careful review of earlier work and avoided repeating many errors. He also based his calculations on an unorthodox but sound philosophical position:

> The theory of a planet cannot in the nature of things be exact; and this for three reasons:
>
> (1) The observations on which it is founded are necessarily more or less in error;
> (2) The theory itself may be more or less imperfect;
> (3) An unknown body may be acting of which perforce no account has been taken.
>
> The latter is a particularly insidious pitfall inasmuch as the greater part of the disturbance it causes can be concealed by suitable changes of the disturbed's elements. . . . Consequently the residuals left by the theory are not at all the outstanding perturbations, but only such small part of them as cannot be got rid of by suitable shuffling of the cards. We have then no

[41] Lowell, Percival, *Mars*, p. v (Boston: Houghton, Mifflin and Co., 1895).

guarantee that our supposed elements are the real ones, but only the best attainable under the assumption *that no unknown exists.* Every theory of a planet is thus open to doubt, seeming more perfect than it is. It has been legitimately juggled to come out correct, its seeming correctness concealing its questionable character.

The test of the existence of an unknown body thus left out of account lies in the substantial reduction of the outstanding residuals of the theory by its subsequent admission. The reduction must be substantial to be indicative, because a slight betterment necessarily follows the introduction of another factor among the adjustable quantities.[42]

Lowell first adopted a method similar to that used by Leverrier in 1846, solving the equations of condition to find longitudes of perihelion for which the tabular residuals of Uranus were minimal. He used the residuals from Leverrier's 1873 theory of Uranus, brought up to 1906 with data from the Greenwich Observatory; the results indicated an unknown body in longitude 244° at 1912.0. In that year Lowell recalculated the problem using Gaillot's tables in place of Leverrier's, and found longitudes of 203° and 209° for 1912.0.

Lowell was unhappy with the disagreement between these predictions and resolved "to pursue the subject in a different way, longer and more laborious than earlier methods, but also more certain and exact: that by a true least square method throughout." This solution filled 115 pages of calculations, graphs, and explanatory text, and yielded a highly satisfactory result: admitting no errors of observation, the outstanding squares of the residuals of Uranus from 1750 to 1903 were reduced 71% by the admission of Lowell's hypothetical disturbing planet. With probable errors of observation allowed, the residuals were reduced nearly 95%. Like many previous investigators, Lowell arrived at two possibilities of position for his planet, 180° apart. The values for the two alternatives were as follows:

	Longitude near 0°	Longitude near 180°
Mean longitude of epoch	22°.1	205°.0
Semimajor axis (a. u.)	43.0	44.7
Mass (unit = 1/50000 × sun's mass)	1.00	1.14
Eccentricity	0.202	0.195
Longitude of perihelion	203°.8	19°.6
True longitude, July 0, 1914	84°.0	262°.8

The theoretical planet thus had a mass between Neptune's and the earth's, a disk more than 1″ in diameter, and a probable magnitude of 12–13, depending on albedo. From the approximate correlation between the eccentricity and inclination of other planets, Lowell inferred that his planet's orbit would be inclined about 10° to the ecliptic.

In 1906–1907 Lowell made his first search for an exterior planet. Under his direction Earl C. Slipher took about two hundred plates at 5° intervals

[42] *Memoirs of the Lowell Observatory*, 1915, *I*, no. 1: 4–5.

along the ecliptic with a 5-inch Brashear telescope; fifty additional plates were taken by Kenneth P. Williams. The plates were exposed for three hours, but although they recorded images down to the 16th magnitude in their centers, there was considerable loss of magnitude at the edges. Lowell began to examine the superimposed comparison plates with a hand magnifier; this method was slow and uncertain, and he substituted first a Hartmann comparator and then a Zeiss blink comparator, which gave the best results.

Despite excellent instrumentation, the first search was fruitless. Lowell was not discouraged; he had a new series of photographs taken by Slipher and Carl O. Lampland, using the Flagstaff 42-inch reflector. Although each of these plates covered only a small field, their ten-minute exposure time allowed them to be taken in rapid succession; they recorded stars to the 17th magnitude. Again no trans-Neptunian planet was found, and Lowell pressed on with a third search. This time a 9-inch Brashear telescope was borrowed from the Sproul Observatory, and during 1914–1916 many photographs were taken with it by T. B. Gill and E. A. Edwards. Lowell had only partially examined these plates when he died on November 12, 1916; his will specified that the search be continued at Flagstaff until a trans-Neptunian planet was found.

On October 30, 1916, a few days before Lowell's death, negative results were reported from a comprehensive planet search made by Alphonse Louis Nicolas Borrelly, an astronomer at the Royal Observatory in Paris. Borrelly was compiling an ecliptic atlas, and in the course of this work he had searched hundreds of plates taken over a span of almost fifty years. Although he had seen images of all the major planets among the stars near the ecliptic, he found no record of the bodies predicted by Pickering and Gaillot.[43] His sole reservation was that most of the plates had only recorded stars down to the 12th magnitude, and it was possible that the trans-Neptunian planet was fainter. After Borrelly's report, nothing was published on the problem for three years.

In September, 1919, Pickering notified the French Astronomical Society that Neptune had begun to deviate from its tabular longitude. According to measurements made by Professor Henry Russell at the Harvard Observatory, the difference was more than 2″ of arc in 1919.[44] Not unexpectedly Pickering attributed the discrepancy to the action of an undiscovered planet and made a new calculation, incorporating Russell's data and the deviations recorded by Professor Sydney Hough at the Cape of Good Hope. The result was to move the mean distance of his hypothetical Planet *O* out to 55.1 a. u., thus lengthening its period to 409 years. Its longitude for the epoch 1920.0 was 97°.8.[45]

[43] Borrelly, A. L. N., "Contribution à la recherche d'une planète transneptunienne," *Journal des Observateurs*, 1917, *1*: 126.

[44] Pickering, William H., "Perturbation de Neptune et planète transneptunienne," *Bulletin de la Société astronomique de France*, 1919, *33*: 393–394.

[45] Pickering, William H., "The transneptunian planet," *Annals of the Astronomical Observatory of Harvard College*, 1937, *82*: 49–59.

The deviations of Neptune persuaded some observers that Pickering's planet really existed; towards the end of 1919 Milton Lasell Humason, an astronomer at the Mount Wilson Observatory in California, began a search for it with a 10-inch Cooke astrographic telescope. Humason's search covered the latest position given by Pickering, and also those predicted by him in 1909 and by Lowell in 1915. The plates were exposed for two hours and showed stars as faint as 17th magnitude. On each pair of plates a zone 3° wide along the ecliptic was searched with a blink comparator, but though many planetoids and a few variable stars were found, no trans-Neptunian planet was discovered. The search was then extended, less exhaustively, to a zone about 10° wide, and during the following year a new series of plates was taken with the 60-inch reflector. No exterior planet was found on any of these photographs, but this time the failure was not one of prediction; the image of a trans-Neptunian planet was actually recorded on four of Humason's earliest plates.[46]

Humason's photographs were not to be re-examined for ten years, and in the meantime several new predictions for exterior planets appeared. In 1921 Théodore Grigull published the positions of two trans-Neptunian planets computed from cometary aphelia, one at 46.784 a. u. from the sun, the other at 90 a. u.[47] Grigull's methods differed little from earlier work on cometary orbits, but a more unusual hypothesis was advanced by Emile Belot, the vice-president of the French Astronomical Society, in the same year. In an article on "Repulsive forces in cosmogony" Belot presented an empirical law of mutual repulsion which placed an exterior planet at a distance of 62.44 a. u.; a month later he revised this distance to 56 a. u. "according to the same factor that . . . reduces the theoretical distance of Neptune from 33.25 to 30.1."[48] In 1927, six years after this prediction appeared, Belot restated it, adding the planet's mass (2.2 × the mass of the earth), its period (407 years), and its probable magnitude (12).[49]

Pickering had never abandoned his belief in an exterior planet. In April, 1928, after a nine-year lapse in publication on the subject, he issued a paper correlating the perturbations of Saturn, Uranus, and Neptune with the position of his theoretical Planet *O*.[50] He included the planet's geocentric coordinates for February 1, 1928, but these must have been previously sent to Professor Harlow Shapley at the Harvard Observatory, for Pickering reported in an appendix to his article that Shapley had taken photographs on the nights of January 20, 21, and 22, 1928, in an effort to find the planet. The region photographed was centered on right ascension 9^h00^m, declination $+16^\circ.5$, which Pickering had specified, but though the plates were examined carefully, no planet was found on them.

46 Nicholson, Seth B., and Nicholas U. Mayall, "Positions, orbit, and mass of Pluto," *Astrophysical Journal*, 1931, *73*: 1–12.

47 Grigull, Théodore, "Der transneptunische Planet," *Das Weltall*, 1921, *21*: 113–115.

48 *Bulletin de la Société astronomique de France*, 1921, *35*: 386–394, 437.

49 Belot, Emile, "Neptune et la planète transneptunienne," *Bulletin de la Société astronomique de France*, 1927, *41*: 19.

50 Pickering, William H., "The next planet beyond Neptune," *Popular Astronomy*, 1928, *36*: 143–165; 218–221.

Pickering quickly followed this paper with one in which he grouped the aphelion positions of many recently observed comets in order of radial distance. He assigned one group to Planet *O* and another group to Planet *S*, "whose mean distance is in the vicinity of 48 units." [51] Even more sensational, "there is a larger planet still further beyond ["*P*"], and after this and planet *O* have been found, and their perturbations allowed for, it will then be time for posterity to look for planet *S*." During the summer of 1928, Pickering tried to simplify this scheme by referring the perturbation effects of his two outermost planets to a single body, but he was not successful. He ended by reaffirming a hypothesis he had first proposed in 1911, that the solar system should be increased not by one, but by three undiscovered planets.[52]

Although Pickering continued to publish articles on hypothetical planets he became steadily more discouraged. By March, 1929, his old confidence had almost entirely disappeared: ". . . it has been suggested in some newspaper articles that all the large observatories would be hunting for planet *O* when it comes to opposition this February, I wish to state here that I myself have the gravest doubts on that point. Indeed if anyone is hunting for it I shall be much gratified." [53]

A month after this disheartened comment was published, the search for a trans-Neptunian planet was taken up again at the Lowell Observatory for the first time since Percival Lowell's death in 1916. The special photographic telescope built for this search had a 13-inch triplet objective and a focal length of 66.6 inches. Its field, though slightly curved, allowed the use of 14 × 17-inch plates covering 162 square degrees of sky. During a one-hour exposure the plates recorded stars of 17.5 magnitude in their centers, decreasing to 15.7 magnitude in the extreme corners. The number of images on each plate was, as one astronomer later commented, "appalling." In the least dense areas, near the galactic poles, there were 40,000 stars per plate; in the Gemini region of the milky way, each plate recorded about 400,000 stars, and in the Scorpio-Sagittarius region, plates were taken with 1,000,000 stars on them. The examination of these plates was assigned to Clyde W. Tombaugh, a young assistant at the Lowell Observatory. Tombaugh, who was born in 1906, had been a farmer, but he was also a skilled amateur astronomer. He built his first telescope, an 8-inch reflector, in 1926, and in 1928 he followed it with a 9-inch reflector. Early in 1929 he was hired by Dr. V. M. Slipher, the director of the Lowell Observatory.

The search was begun on April 1, 1929, when the Gemini region, in which Lowell's Planet *X* was suspected to lie, was past opposition. The first few plates of this region showed that it would be difficult to distinguish distant planets from planetoids near their stationary points. As a result, all subsequent plates were taken near the opposition point, where the angu-

[51] Pickering, William H., "The orbits of the comets of short period," *Popular Astronomy*, 1928, *36*: 274–281.

[52] Pickering, William H., "The orbit of Uranus," *Popular Astronomy*, 1928, *36*: 353–361.

[53] Pickering, William H., "Planet O," *Popular Astronomy*, 1929, *37*: 135–138.

lar motions of planetoids would distinguish them more clearly. The opposition point moves eastward about 30° each month, and good photographs could not be made in bright moonlight, so that the plates for each interval had to be taken in two weeks. Despite this handicap, the photographic work was maintained on schedule and caught up with the opposition point on the ecliptic in September, 1929.

That month Tombaugh began the blink examinations with the plates of the constellation Aquarius. He used a modified Zeiss blink-microscope comparator and blinked only negatives to minimize the spurious planet images produced by dust spots and minute silver deposits. It was a difficult job: "A pair of 14 × 17-inch plates in non-Milky Way regions could be blinked in two days of hard work at the rate of examining 30,000 stars daily."[54] As the search progressed eastward through the constellation Taurus, the number of star images on each plate increased to about 300,000, which multiplied the blink examination time almost fourfold. The plates of the Gemini region followed Taurus; Tombaugh found the photographs of the regions η and 36 Geminorum with their 400,000 star images so forbidding that he put them aside temporarily and blinked instead the plates of δ Geminorum, which showed about 160,000 stars. Two of these photographs had been taken on January 21 and 23, 1930, and a third check plate was taken on January 29. On February 18, when Tombaugh had examined about one-quarter of the δ Geminorum plates, he found the two faint images of an unknown planet.

The discovery was quickly confirmed on a pair of 5-inch Cogshall negatives taken on January 23 and 29, and on a new plate taken by Tombaugh on February 19. On February 20 the staff of the Lowell Observatory observed the planet with the 24-inch refractor, but despite good seeing they were unable to detect a disk. This was disappointing, because Lowell had expected Planet X to be similar to Neptune, which would have given it a disk 1″ in diameter and a magnitude of 13. Examination with the 42-inch reflector confirmed that the new planet was yellowish and not bluish like Neptune; its visual magnitude was slightly brighter than its photographic magnitude of 15.

Dr. Slipher waited until March 13, 1930, to announce the discovery. By that time the planet had been observed about seven weeks at Flagstaff, and Slipher was reasonably sure that Lowell's prediction had been fulfilled at last; he chose March 13 because it was both Lowell's birthday and the anniversary date of the discovery of Uranus. Professor Harlow Shapley, who was asked to communicate the discovery for the Lowell Observatory, sent the following telegram to the International Astronomical Union in Copenhagen:

> Systematic search begun years ago supplementing Lowell's investigations for Trans Neptunian planet has revealed object which since seven weeks has

[54] Tombaugh, Clyde W., "The Trans-Neptunian Planet Search," in Kuiper, Gerard P., and Barbara M. Middlehurst, eds., *Planets and Satellites*, p. 18 (Chicago: University of Chicago Press, 1961).

> in rate of motion and path consistently conformed to Trans Neptunian body at approximate distance he assigned. Fifteenth magnitude. Position March twelve days three hours GMT was seven seconds West from Delta Geminorum, agreeing with Lowell's predicted longitude.[55]

This telegram set off a worldwide rush to observe and photograph the new planet, and a flood of congratulatory messages began to arrive at Flagstaff. Miss Venetia Burney, an eleven-year-old girl of Eton, England, submitted the name Pluto for the planet. Professor H. H. Turner forwarded her suggestion to Dr. Slipher, who accepted it and noted later that he was guided by it in naming the planet.[56] William Pickering independently suggested the same name in a letter to Harlow Shapley. Many other nominations were received, but Pluto was adopted, and with it the symbol ♇, which combined an abbreviation of the planet's name with the initials of Percival Lowell.

The preliminary orbit calculated for Pluto indicated that in 1919 the planet had been within the field of four plates taken by Humason at the Mount Wilson Observatory. The plates were re-examined, and on June 7, 1930, four images of Pluto were found on them. Other observatories also found pre-discovery photographs of the new planet. It appeared on plates taken at the Königstuhl Observatory in 1914, at the Yerkes Observatory in 1921 and 1927, and at Uccle, Belgium, in 1927. These records allowed Pluto's orbital elements to be determined within narrow limits; it is interesting to compare them with some of the predicted elements for a trans-Neptunian planet.

As Table 1 shows, Lowell's 1915 elements were by far the most accurate. Pickering's 1919 prediction was also close, but not as consistent as Lowell's. Pickering's feelings were sorely tried by the discovery, and during 1930 and 1931 he published a series of politely rankled articles defending his own prediction. In April, 1930, two months after the discovery, both Lowell's and Pickering's predictions were rejected as worthless by another astronomer, E. W. Brown. Brown claimed that it was impossible for Pluto's existence to have been predicted from perturbations, and that the discovery of Pluto in Gemini was a complete coincidence.[57] Although both Lowell's and Pickering's work had defenders, Brown's opinion was widely accepted until 1951. In that year there appeared an English translation of an exhaustive investigation by V. Kourganoff, which had been published in France in the years 1940–1944 and received little publicity.[58] Kourganoff refuted Brown point by point; his detailed analysis left little doubt that Pluto's discovery was not an accident and that both Lowell and Pickering had detected genuine evidences of the planet's existence years before it was discovered.

[55] Whipple, Fred L., *Earth, Moon, and Planets*, p. 38 (Cambridge, Mass.: Harvard University Press, 1963).

[56] Crommelin, A. C. D., "The Discovery of Pluto," *Monthly Notices of the Royal Astronomical Society*, 1931, *91*: 385.

[57] Brown, E. W., "On the predictions of transneptunian planets from the perturbations of Uranus," *Proceedings of the National Academy of Sciences*, 1930, *16*: 364–371.

[58] Reaves, Gibson, "Kourganoff's contributions to the history of the discovery of Pluto," *Publications of the Astronomical Society of the Pacific*, 1951, *63*: 49–60.

TABLE 1.
Comparison of Predictions

	Pluto 1/1962	Todd 1877	Forbes 1887	Lau 1901	See 1904	Gaillot 1909	Lowell 1915	Pickering 1919
Mean distance, a. u.	39.29	52.0	104.4	46.5	42.2	44.0	43.0	55.1
Period, years	246.4	375	1066	317	272.2	292	282	409.1
Eccentricity	0.247	..	..	..	..	..	0.202	0.31
Inclination	17°.16	1°.4	..	..	..	..	10°	15°
Longitude, 1930.0	108°.5	221°.6	191°	308°.9	233°	308°.4	102°.7	102°.6
Longitude of ascending node	109°.8	103°	..	..	..	..	..	100°
Longitude of perihelion	224°.5	..	..	..	..	..	204°.9	280°.1
Date of perihelion	1989.8	..	..	..	..	..	1991.2	1720.0
Magnitude	15	13+	..	..	..	..	12–13	15
Mass (earth = 1)	1?	..	..	9	..	5	6.7	2

The Farm Chemurgic Council and the United States Department of Agriculture, 1935–1939

*By Carroll W. Pursell, Jr.**

ALTHOUGH THE FEDERAL GOVERNMENT has, from its earliest days, supported scientific research and development, the scale and direction of that support have seldom been important issues in the shaping of partisan political battles. Despite the fact that science and technology have an obvious contribution to make toward establishing justice, insuring domestic tranquility, providing for the common defense, and promoting the general welfare, criticism has more often been of frivolous waste of tax monies than of failure to harness these forces to national goals.

The New Deal was at least as vulnerable to attack on this score as have been most national administrations. Despite the imperial dreams of Harold Ickes, the scientific sophistication of Henry Wallace, and the larger hopes of some friendly scientists from the academic community, there was no new deal for the physical sciences until it was forced by the coming of World War II. This weakness was probed between 1935 and 1939 by the Farm Chemurgic Council, an organization which produced a bizarre though not unattractive alternative to the Agricultural Adjustment Act as a panacea for the nation's agricultural depression. Failing to carry its message to the people in the presidential campaign of 1936, the Council suffered serious setbacks in 1937 and by the end of the decade was overshadowed by well-publicized efforts of the Department of Agriculture in the same area.

The farm chemurgic movement was a legitimate child of the interwar years. World War I had been hailed as a war of chemistry, and chemists were treated with something of that exaggerated respect accorded after World War II to nuclear physicists. Germany's supposed ability to synthesize otherwise unobtainable strategic materials, the horror of gas warfare, and the forced growth of the American chemical industry all contributed to the heady faith that a new Chemical Age had dawned and that the miracles of the laboratory were only beginning. During the interwar years the campaign to "sell" industrial research, carried on by the National Research Council, Secretary of Commerce Herbert Hoover, and a host of individuals and agencies, all

* University of California, Santa Barbara.

served to convince the nation that better things for better living were possible through chemistry.[1]

Some of the postwar harvest reaped by the chemical industry was more substantial than merely the respect of a grateful nation. Claiming that theirs was an infant industry, chemical firms pressed for a protective tariff against the products of a reviving German chemical trust. A tariff directed against the Germans was urged "not only as punishment for their hideous crimes, but as a mark of respect and honor to the men who have given up their lives to the cause of freedom."[2] The tariff, finally passed in 1922, set up a special system of valuation of organic chemicals which provided a high degree of protection for the American industry.[3]

No small part of the credit for passing this tariff was given to the Chemical Foundation, Inc. During the war German chemical patents had been seized by the federal government and turned over to the Alien Property Custodian for administration and disposal. This official in turn sold the patents to the Chemical Foundation, a nonprofit educational corporation set up for the purpose of administering the patents and furthering the interests of academic and industrial chemistry in the nation.[4]

The origin and character of the Chemical Foundation was such that its legitimacy was soon questioned. It was controlled by the largest chemical firms in the country and could be considered a holding company for the American chemical industry. It acquired the four to five thousand German patents for a nominal sum, without the competition of public bidding. Finally, after disposing of the patents, the Alien Property Custodian resigned and became president of the Chemical Foundation, a position he held for the rest of his life. Members of Congress attacked the arrangement almost immediately.[5] Early in the 1920's the Harding administration brought action against the Foundation, but its president, Francis Patrick Garvan, charging that the White House was acting on information "furnished by German agents to an ex-German spy and interpreted by lawyers whose sole knowledge of the war and its lessons is derived from association with German clients," successfully defended himself before the Supreme Court.[6] Garvan, according to the Court, had neither broken any law nor exceeded his specific authority.

Among the many educational activities of the Chemical Foundation, one—its sponsorship of the Farm Chemurgic Council—brought it close to conflict with the New Deal during the 1930's. In April 1935 Garvan announced that a conference was planned for the following month in Dearborn, Michigan. At this meeting representatives of agriculture, industry, and science were "to survey the variety of farm products which through organic chemistry can be transformed into raw materials usable in industry, and to develop a plan for the joint cooperation of agriculture, industry and science for promoting in orderly fashion an increasing use of American farm products."

[1] For a brief discussion of the period of chemical ascendance see *Great American Scientists: America's Rise to the Forefront of World Science*, by the editors of *Fortune* (Englewood Cliffs: Prentice-Hall, 1961), pp. 35–59.

[2] Robert Hilton, "The Maintenance and Preservation of Our Dyestuff Industries," *Transactions of the American Institute of Chemical Engineers*, 1918, *11*:284.

[3] This tariff stood virtually unchallenged for over forty years, until the Kennedy round of tariff negotiations in 1964. *New York Times*, 13 Nov. 1964.

[4] Victor S. Clark, *History of Manufactures in the United States* (New York: Peter Smith, 1949), Vol. III, p. 318.

[5] *New York Times*, 12 July 1919.

[6] Francis P. Garvan, *The Chemical Question: An Open Letter to Warren G. Harding, President of the United States* (New York: Wynkoop Hallenbeck Crawford, 1922), p. 16.

The hope was expressed that "such cooperation will result in the gradual absorption of much of the domestic surplus by domestic industry."[7]

The Dearborn meeting opened on 7 May with Garvan as the elected chairman. Henry and Edsel Ford served as hosts to the meeting, which was officially sponsored by Edward A. O'Neal, president of the American Farm Bureau Federation, Louis J. Taber, master of the National Grange, Clifford V. Gregory, chairman of the National Agriculture Conference, and Garvan, as president of the Chemical Foundation.[8] After addresses of welcome and purpose, the morning session ended with the signing of a "Declaration of Dependence upon the Soil and of the Right of Self-Maintenance," carried out, appropriately enough, in the replica of Independence Hall which Ford had erected at his outdoor museum, Greenfield Village.

Using a desk that had belonged to Thomas Jefferson, a table and chairs once the property of Abraham Lincoln, and ink that exactly reproduced that used on the Declaration of Independence, the delegates signed a document which joyfully declared that "through the timely unfolding of nature's laws, modern science has placed new tools in the hands of man which enable a variety of surplus products of the soil to be transformed through organic chemistry into raw materials usable in industry."[9]

That afternoon and the following day approximately three hundred delegates heard many more speeches which sang the praises of modern chemistry, the virtues of the nation's farmers, and the willingness of everyone to cooperate in ending the Depression through the application of scientific research. On the last afternoon Harper Sibley, newly elected president of the United States Chamber of Commerce, moved that a committee be appointed to perpetuate the good that had been started. Fourteen delegates were selected for this assignment, including Dr. Roger Adams, president of the American Chemical Society, W. B. Bell, president of the American Cyanamid Company, Dr. William J. Hale, a chemist with the Dow Chemical Company, Col. Frank Knox, publisher of the Chicago *Daily News*, Dr. Charles M. A. Stine, a vice president of E. I. du Pont de Nemours & Company, Louis J. Taber, master of the National Grange, and eight others.[10] This group soon reorganized itself into the Governing Board of the new Farm Chemurgic Council, which became the sustaining body of the movement.

The program of the new Council was simplicity itself. Dislocations in agriculture, they believed, were at the heart of the Depression. The only real solution to the problem of overproduction of farm products was their utilization as industrial raw materials. Farms were, in fact, great outdoor chemical plants. The organic chemist could devise ways of taking cotton, for example, breaking it down into its constituent molecules, and then reassembling these into new and useful industrial materials and products. This would provide a market for the farm surplus, solving the basic problem of the Depression, and would also provide new jobs in industry, thus ending unemployment. To further this purpose the newly formed Council planned to provide an annual forum for the discussion of the problems of such utilization, to donate some modest

[7] *Science*, 1935, *81*:394–395.

[8] For a discussion of some of Ford's other activities see Reynold Millard Wik, "Henry Ford's Science and Technology for Rural America," *Technology and Culture*, 1962, *3*:247-258.

[9] "Declaration of Dependence upon the Soil and of the Right of Self-Maintenance," *Proceedings of the Dearborn Conference of Agriculture, Industry and Science, Dearborn, Michigan, May 7 and 8, 1935* (New York: Chemical Foundation, 1935), p. 32.

[10] *Ibid.*, p. 3.

financial support for the research and development of processes, and to attempt to educate the public to the potentialities of *chemurgy*, which was defined as "chemistry at work."[11]

The program seemed simple and straightforward in conception but in practice received eccentric emphasis from its leading protagonists. One of these was Dr. William J. Hale, who coined the word *chemurgy* in 1934. Hale had been born in 1876 and received his Ph.D. from Harvard in 1902. Like many chemists of the time, he made the grand tour of the Continent, visiting Berlin and Göttingen in 1902–1903 as a traveling fellow in chemistry. After teaching a year at the University of Chicago, he moved to the University of Michigan, where he rose to the rank of associate professor. In 1919 he was made director of organic chemical research for the Dow Chemical Company, a post which he still held in the mid-1930's.[12]

Hale contributed an authoritative scientific voice to the chemurgic movement, as well as that disdain for mere politicians which was then fashionable among scientists. In 1932 he wrote a small book entitled *Chemistry Triumphant*, in which he identified the key to the farm problem as overproduction. This was not, he said, the fault of the nation's farmers, who were only "following their natural bent. The fault lies," he maintained, "at the door of financiers and the agricultural departments of State and Nation." He praised the efforts of the Federal Farm Board, established in 1929 by the Hoover administration, but noted that "politics, of mean repute, evidently gained the upper hand." For one thing, "here was a magnificent chemical, physical, and biological problem and all chemists, physicists, and biologists [were] completely eliminated from decisions affecting its policies!"[13]

Declaring that "that exchange of goods which we call foreign trade is merely the international exchange of bundles of warm, moist air," Hale advocated something called "scientific nationalism." His argument was that wars are caused by quarrels over foreign trade, through chemurgy America can make everything she needs, hence there will be no need for foreign trade, and therefore we need never again become involved in war. One of his defenders protested that because he "insists that the nations of the world are becoming *less* interdependent and because this point of view is anathema to the so-called 'liberals' who still think in terms of the decadent Machine Age, Hale has been identified erroneously as a friend of Fascism."[14]

Hale's style was congenial to Francis Garvan, who, as president of the Chemical Foundation, was president of the Farm Chemurgic Council as well. Garvan had been born in 1875, taken an A.B. at Yale in 1897 and an LL.B. from the New York Law School two years later. He began practice in New York City in 1899 and the following year became an assistant district attorney. From 1917 to 1919 he served with the U.S. Bureau of Investigation (specializing in subversives and their activities) and as manager of the New York office of the Alien Property Custodian. In 1919 he was named to the post of Alien Property Custodian and was made an assistant Attorney General of

[11] Christy Borth, *Pioneers of Plenty: The Story of Chemurgy* (Indianapolis: Bobbs-Merrill, 1939), p. 23. The word *chemurgy* was apparently coined by combining the roots *chemi* and *ergon*.

[12] *Who's Who in America ... 1934–1935* (Chicago: A. N. Marquis, 1934), Vol. XVIII, p. 1047.

[13] William J. Hale, *Chemistry Triumphant: The Rise and Reign of Chemistry in a Chemical World* (Baltimore: Williams & Wilkins, 1932), pp. 86, 90.

[14] Borth, *Pioneers of Plenty*, pp. 81–82.

the United States. With the creation of the Chemical Foundation, he resigned to become president of that organization.[15]

Garvan shared Hale's enthusiasm for scientific nationalism and added to it a particular anglophobia which may have been the result of his Irish background. He believed that the United States was so deeply enmeshed in a web of foreign intrigue (woven in part by unpatriotic international bankers) that only science—and particularly the application of the scientific method to politics—could extricate her. During the presidential election of 1936 he declared that "no matter who is elected President this fall . . . that President and that Congress will be compelled to dance to the tune of 'God Save the King,' and J. P. Morgan & Co., head and control of our New York banks and head of the agents of foreign banking systems, will wield the baton."[16]

Other themes wove themselves into Garvan's thinking and the bias of the two organizations of which he was president. In a long and rambling address to the second Dearborn Conference in 1936 he chastised the international bankers, castigated Great Britain, touted alcohol as a fuel for automobiles, and cast serious doubts upon the democratic nature of a bicameral legislature. A lengthy condemnation of the bookkeeping procedures of the federal government led him to praise the study of arithmetic and indeed to declare that with educational emphasis on "arithmetic, athletics, and loyalty to one God, to one country and one family, . . . we can stop worrying about our children or about our beloved country."[17]

Although the public statements of Garvan were sometimes eccentric and even extreme, they presented a fairly coherent world view which contained elements by no means uncommon or unpopular during the 1930's. He believed that America had been betrayed by unpatriotic bankers and unscientific politicians into entering World War I to save England. All this country had gotten out of that war was a flourishing chemical industry which now provided the opportunity to prevent any more such disasters. If instead of destroying our agricultural surplus we converted it into products which we now imported, we could free ourselves of foreign entanglements and future wars, conserve our natural resources, make our farmers prosperous, stimulate industrial growth, and rid ourselves of those political and financial parasites which lived upon and reinforced the nation's woes.

All of this was not without implications for the current political situation in the United States. Early in 1936 the San Francisco *Chronicle* pointed out,

> . . . the program of the Farm Chemurgic Council . . . is diametrically opposed to the program of the Tugwellian theorists. Tugwell would help the farmers by making him plow under part of his crops and kill off the little pigs to create artificially higher prices. The Farm Chemurgic Council would help the farmer by finding new uses for his products so he can sell more of them. Which proposal makes the greatest appeal to common sense?[18]

When the Council published its résumé of the first Dearborn Conference it claimed

[15] *Who's Who in America . . . 1934–1935*, Vol. XVIII, p. 938.

[16] Francis P. Garvan, "Scientific Method of Thought in Our National Problems," *Proceedings of the Second Dearborn Conference of Agriculture, Industry and Science, Dearborn, Michigan, May 12, 13, 14, 1936* (New York: Chemical Foundation, 1936), p. 83.

[17] *Ibid.*, p. 78.

[18] San Francisco *Chronicle*, 28 March 1936, clipping in Records of the Office of the Secretary of Agriculture, Record Group 16, National Archives, Washington, D.C. Hereafter cited as NARG 16.

that that gathering was "not launched as a counter move against the prevailing situation nor to do battle with proponents of regimentation. It was a happy coincidence that it came into being at a time when the need for it seemed greatest."[19] The fact remained, however, that the Council's program was presented as a clear alternative to the "unscientific" farm program offered by the New Deal, and the prominence in the Council of such New Deal foes as Frank Knox, soon to be the Republican vice-presidential candidate, only sharpened the political threat.

Neither the Republicans nor the Democrats had mentioned the industrial utilization of farm products, the support of agricultural research, or indeed of any kind of science in their 1932 campaign platforms.[20] A resolution passed by the American Association for the Advancement of Science at its annual meeting in December of 1934, calling upon the government to support scientific research, caused only a flicker of interest in the Democratic party. An assistant to the chairman of the Democratic National Committee wrote to Secretary of the Interior Harold Ickes asking for "any information you can give me showing how the New Deal has aided science and assisted those engaged in scientific work." Ickes answered that such information was "difficult to summarize" and merely referred the committee to published reports of the Science Advisory Board and the National Resources Committee.[21]

This lack of interest on the part of the Democratic party to the political mileage which might be gotten from championing the cause of science was a chink in the armor of the New Deal. One Democratic congressman who realized this weakness was Jennings Randolph of West Virginia. As the sponsor of legislation which would have given the government a major responsibility for the subvention of research in the nation, Randolph was well aware of both the needs and the opportunities of science. In June 1936, shortly before the Democratic National Convention, he wrote the White House that "lest the Republicans seize upon this theme and since it is potentially a powerful weapon in our own hands may I urge that the Democratic Platform contain a provision *favoring the promotion of scientific research.*" He felt that the subject could have a "wide appeal" and that "convincing campaign arguments . . . can be made upon the basis of the promotion of scientific research for the purpose of developing new industries and creating employment."[22]

By this time, however, the matter was already being given some consideration in the Department of Agriculture. Chemurgic backers felt that it was "typical of our habit of overlooking the obvious that our politicians and press either overlooked this [Dearborn] conference entirely or dismissed it as merely another one of many wild schemes for 'saving the farmer.'"[23] In an attempt to stimulate some sort of response, Garvan had addressed a letter the previous November to the chief of the Bureau of Chemistry and Soils, inviting that agency to cooperate with the Farm Chemurgic Council in

[19] *Proceedings of the Second Dearborn Conference*, pp. 381–382.

[20] See Kirk H. Porter and Donald Bruce Johnson, *National Party Platforms, 1840–1964* (Urbana: Univ. Illinois Press, 1966), pp. 331, 333.

[21] For the text of the resolution see American Association for the Advancement of Science, *A Brief History of the Association from its Founding in 1848 to 1940* . . . (Washington: AAAS, 1940), pp. 109–110. For the reaction see Emil Hurja to E. K. Burlew, 29 Jan. 1935 and Ickes' answer of 7 Feb. 1935, in Records of the National Resources Planning Board, 801.1, Box 267, Record Group 187, National Archives.

[22] Jennings Randolph to Franklin D. Roosevelt, received 19 June 1936, Franklin D. Roosevelt papers (OF872), Franklin D. Roosevelt Library, Hyde Park, N.Y.

[23] Borth, *Pioneers of Plenty*, p. 27.

finding new uses for farm products. The letter drew no official reply until the following May, when Wallace himself sent Garvan his answer.

It was the second Dearborn Conference, coming as it did in an election year, which moved the Secretary to answer Garvan's challenge. Wallace had already been warned by his scientific adviser that "it seems doubtful that this movement can continue, but, inasmuch as they are well financed and the chief promoters are well paid, the Farm Chemurgic idea will continue to distract many from the main problems of the day."[24] The day before the conference was due to get under way the President's secretary warned Wallace that the White House had received word that "the Duponts, Ford, et al, were going to spring a big move," and that there were proposals to put the Farm Chemurgic program into the Republican platform.[25]

Wallace's problem was twofold: first, to decide whether the Council was in fact a political threat, and second, if it was, how best to deal with it. The Council itself took great pains to disavow any political pertinence or interests. At the midwestern regional meeting in March, preparatory to the national conference at Dearborn, spokesmen attempted to disassociate the Council from both the "personal" opinions of its members and the admittedly "Americanism" program of the parent Chemical Foundation.[26] The objective facts remained, however, that the Foundation was the sole source of funds for the Council, that they shared the same president and treasurer, and that the members of the Council's board of governors were avowedly conservative opponents of the New Deal. William B. Bell, for example, was not only the president of the American Cyanamid Company, but also the putative chief preconvention money-raiser for the Republicans and the boss of Lewis W. Douglas, Roosevelt's first but by then estranged Director of the Budget.

Such associations as these apparently led the Secretary to decide that the Council was a sufficient threat to justify some small action, although there was no evidence of panic or even of serious concern. Wallace cleverly adopted Garvan's own plan of attack in making his defense; that is, he regained the initiative by inviting the Farm Chemurgic Council to cooperate with the Department of Agriculture in the area of industrial utilization. *The New York Times* observed wryly that "with the advent of the political conventions and campaign, Secretary Wallace moved swiftly . . . to deprive Republican strategists of another 'cause' on which they might base an appeal for the farmer's vote." When reporters asked him why there had been a six-month delay in answering Garvan, Wallace said "smilingly that although the letter seemed at the time to require no reply, it was now thought propitious to answer and acknowledge the council's invitation to cooperate."[27]

Wallace's move was variously interpreted in the press. The Sioux City (Iowa) *Tribune* professed to see a change in departmental policy involving a partial repudiation of the Secretary's economic adviser, Mordecai Ezekiel. Others saw it as a purely political maneuver designed to outflank the Republicans. "For last-minute snatching," commented one journal, "Henry A. Wallace's effort to seduce the Farm Chemurgic Council, of which Francis P. Garvan is president, ranks as a major violation of the

[24] E. N. Bressman to Secretary of Agriculture, 15 March 1937, papers of E. N. Bressman, NARG 16.

[25] M. H. McIntyre to Secretary Wallace, 11 May 1936, FDR papers (OF2136), Roosevelt Library.

[26] Bressman to Secretary of Agriculture, 15 March 1937, Bressman papers, NARG 16.

[27] *New York Times*, 28 May 1936.

Federal Kidnaping Statute."[28] A columnist for the Washington *Daily News* accused the Secretary of being anti-science, a charge which he firmly denied:

> I am not suspicious of science, on the contrary, I am anxious that it shall not be discredited by heated propaganda or needless claims that tell only half the story. . . . [The] dominant thought in the Department of Agriculture today is to combine the advantages of research in biology, chemistry, engineering and similar fields with research in economics and to harness these advantages through farmers' cooperative programs which are based upon realistic measurement of market opportunities.[29]

Although research was certainly a long-term solution to the problem, it had little to offer bankrupt and starving farmers who desperately needed short-term relief.

Whatever may have been the intention of the Republicans in connection with a strong plea for scientific research as a farm relief measure, no such plank was fashioned at the convention in Cleveland. The party platform mentioned thirteen points concerning agriculture, the fifth of which stated simply that it was the object of the party to "promote the industrial use of farm products by applied science."[30] This was a pale reflection of what might have been a major concern. Even so, it was more pertinent than the response of the Democratic party. Apparently satisfied that Wallace had laid the ghost of research, the Democrats' 1936 platform made no mention of the subject.

The absence of a fight on the issue of agricultural research did not mean the end of interest in the problem. The Farm Chemurgic Council had been quite wrong in their implication that the Department of Agriculture had little interest in the problem of utilization. In fact, it had, by 1935, a long and commendable record of work in this field. Garvan's momentary initiative had been seized in the absence of departmental publicity rather than activity. After the presidential campaign of 1936 the problem was given greater prominence by the Department's scientists and publicists.

Some of the earliest research undertaken by the Department had concerned itself with the utilization of farm products in industry. Over the years work was concentrated in the leather and forest products industries, but the principle was capable of wide extension. When, in 1927, the old bureaus of chemistry and soils and the Fixed Nitrogen Laboratory were combined into the new Bureau of Chemistry and Soils, a separate Industrial Farm Products Division had been established. By 1933 Henry G. Knight, chief of the Bureau, could state that the first purpose of Chemistry and Soils was "to increase the financial return to farmers" by "solving certain technical problems which now stand in the way of greater or more profitable use of farm products, thereby expanding markets."[31]

Somehat similar efforts were also being carried on in the Bureau of Standards. In

[28] Sioux City *Tribune*, 15 June 1936, and *Today*, 20 June 1936, clippings in Bressman papers, NARG 16.

[29] H. A. Wallace to Raymond Clapper, 3 June 1936, subject file (Research Work), NARG 16. Wallace's record as a science administrator during these years is described in Carroll W. Pursell, Jr., "The Administration of Science in the Department of Agriculture, 1933–1940," *Agricultural History*, 1968, *42*:231–240.

[30] Porter and Johnson, *National Party Platforms*, p. 367.

[31] Gustavus A. Weber, *The Bureau of Chemistry and Soils: Its History, Activities and Organization* (Baltimore: Johns Hopkins Press, 1928), pp. 159–160; H. G. Knight, "Research in the Bureau of Chemistry and Soils," *Scientific Monthly*, 1933, *36*:308. The results of a survey made for the USDA are given in George M. Rommel, *Farm Products in Industry* (New York: Rae D. Henkle, 1928).

keeping with the popular crusade against waste and inefficiency, the Secretary of Commerce declared in 1929 that "waste is vast potential wealth. Research discloses its nature and measures its properties. To match these with some need of industry is most profitable." Despite worries over "unnecessary duplication of effort" by such Department of Agriculture defenders as the Association of Land-Grant Colleges and Universities, work carried on by Commerce clearly added to the government's ongoing commitment to research on the problem of industrial utilization.[32]

After the Farm Chemurgic challenge, the Department of Agriculture gave new emphasis and publicity to these traditional concerns. Within weeks of Wallace's answer to Garvan it was announced that the Department was spending $1,300,000 to test cotton fabric as "a reinforcing agent in low-type highway construction." The influential *Country Gentleman* wished the experiment godspeed and commented that "no one who understands the cotton growers' desperate need for new markets will cavil at the decision."[33] The most dramatic new program, however, was the building of four major regional laboratories authorized by Section 202 of the Agricultural Adjustment Act of 1938.

The regional laboratories grew out of efforts begun in 1937 by Senator Theodore Bilbo of Mississippi to procure a cotton research laboratory for his home state. With the tacit approval of the White House, the measure was transmuted into Section 202, which provided for four such laboratories, each to be placed in a different agricultural region and to concentrate on the particular surplus problems of that area.[34] The new laboratories were welcomed and were even seen by some as a victory for the idea of "chemurgy," but a departmental spokesman issued the familiar caution:

> A properly balanced program of fundamental and applied research, including chemical engineering developments, calls for years of effort before it can be operating at full effect. We ask your patience and your assistance in the development of this program, in which the role of the Department of Agriculture is to provide industry with new raw materials for new and valuable products and to furnish the farmer with a market for his special and surplus crops.[35]

If the events of 1936 spurred on the Department of Agriculture, they took some of the momentum out of the Farm Chemurgic Council. The Council had always declared itself to be nonpartisan, but its initial impetus was clearly due in large part to its potential for embarrassing the New Deal during an election year. When that move was countered by Wallace there was in fact nothing left but science, and that proved an insufficient basis. Furthermore, the failure of the Republican party to adopt a strong "chemurgic" position was only one of the blows which the Council suffered in the mid-1930's.

Perhaps the most serious setback was the death, on 7 November 1937, of Garvan. The removal of his very personal and strong leadership weakened the Council at a

[32] United States Department of Commerce, *Annual Report of the Secretary, 1929*, p. xxxiii; *Proceedings of the 45th Annual Convention of the Association of Land-Grant Colleges and Universities . . . 1931*, p. 465.

[33] *Country Gentleman*, 1936, *106*:20.

[34] "Farm Laboratories," *Business Week*, 5 March 1938, p. 37; Gladys Baker *et al.*, *Century of Service: The First 100 Years of the United States Department of Agriculture* (Washington: U.S. G.P.O., 1963), p. 228. The laboratories were built in Peoria, Ill., New Orleans, La., Wyndmoor, Pa., and Albany, Calif.

[35] H. T. Herrick, "Role of the Department of Agriculture," *Industrial and Engineering Chemistry*, 1939, *31*:144.

time when his qualities were much needed, for at this same time the Chemical Foundation, its chemical patents running out and its own income severely curtailed, cut the Council off financially.[36] Supporting itself from membership fees and small contributions, the Council struggled on, but without the support it had previously relied upon.

For a short time it appeared that the support withdrawn by the chemical industry and important Republican leaders might be made up for by the active participation of nationally known scientists. Karl T. Compton, perhaps the country's leading scientific statesman, served on the Council's board of governors and gave a number of speeches in which he drew upon the rhetoric of the movement. Writing in 1937 in the first volume of the Council's *Farm Chemurgic Journal*, he hailed "the phenomenal growth of this Farm Chemurgic movement . . . despite opposition from those who misunderstand it or who believe that their personal or political interests will be served by its failure." Describing the "Mission of Science," he condemned what he called the New Deal policy of "enforced scarcity" as being "neither an advantageous nor an ethical policy."[37]

This sort of talk rekindled resentment within the Department of Agriculture. It was charged that in an effort to wrap itself in the cloak of scientific legitimacy the Chemical Foundation had been buying support by giving critical financial aid to key chemistry departments in the leading universities—including Compton's own Massachusetts Institute of Technology. Another supporter of the Council was Roger Adams, the venerable chairman of the chemistry department at the University of Illinois. Adams was, like Compton, a member of the board of governors of the Council, and it was believed in the Department of Agriculture that his department had received a number of fellowships from the Foundation.[38]

One other major beneficiary of Chemical Foundation support was the American Institute of Physics, which had been financed at a critical point in its early life. Indeed, William W. Buffum, treasurer of both the Chemical Foundation and the Farm Chemurgic Council, was made the treasurer of the A.I.P., presumably to watch over the Foundation's investment.[39] This link between physics and chemurgy was strengthened when George R. Harrison, Director of Applied Physics at M.I.T., appeared before the second Dearborn Conference in 1936 and told the assembled delegates that not only was the physicist ready to join in abolishing farm poverty, but that even "war is going to be eliminated by science because war is economic in origin." This declaration of scientific nationalism was of course in line with the Council's basic orientation.[40] Henry A. Barton, head of the A.I.P., appeared at the third Dearborn Conference and urged that "we should train more physicists to deal with chemurgic problems."[41]

There was never any question of corruption involved; it was simply a fact that the close financial ties between the Chemical Foundation, the Farm Chemurgic Council, and centers of academic chemistry and physics made it more likely that prominent scientists would lend their names and prestige to a movement which otherwise would

[36] Wheeler McMillan, "Chemurgy's Challenge," *Farm Chemurgic Journal*, 1938, *1*:6.

[37] Karl T. Compton, "Mission of Science," *Farm Chemurg. J.*, 1937, *1*:21, 15.

[38] Memo by E.N.B., 7 June 1937, Bressman papers, NARG 16.

[39] "William W. Buffum, The Chemical Foundation, and Physics," *Review of Scientific Instruments*, 1938, *9*:211–212.

[40] George R. Harrison, "The Application of Physics to Agriculture," *Proceedings of the Second Dearborn Conference*, p. 322; republished in *Rev. Sci. Instrum.*, 1936, *7*:295–300.

[41] The text of the speech was published as Henry A. Barton, "Physics in the Production and Use of Bulk Crops," *Journal of Applied Physics*, 1937, *8*:644.

probably not have commanded their attention or interest. Indeed, after the demise of the Foundation and the Council's falling on evil times, the number of prominent scientists connected with the movement tended to shrink rapidly.

There were a number of small victories which the Council could count after 1937. In 1939 Ohio's Governor John W. Bricker established a chemurgic commission as a part of the state's government, and in the Congress Representatives Ben Jensen, Karl Mundt, and others began to introduce several bills to encourage the use of grain alcohol as a fuel for automobiles, a favorite chemurgic program. But these were only temporary victories; by the end of the decade the movement, as such, had all but disappeared.[42]

The Council might reasonably have expected to do better. Its program had a direct and powerful appeal to many Americans. The disillusionment over World War I and the consequent appeal of nationalism and isolationism, the growing concern over the exhaustion of natural resources, the popular antagonism toward politicians and financiers, an extravagant faith in the ability of science to produce the good life, and a still-strong agrarianism which put a special emphasis upon the importance of rural prosperity—these were powerful ingredients and, properly mixed, might have provided a bitter drink for the ruling Democratic party.

The advantages, however, were matched by real weaknesses. Foremost among them was the fact that farmers needed immediate relief just to survive, let alone prosper, and if given the choice, would probably prefer to have *both* subsidies *and* research programs, a combination which the Department of Agriculture was in fact already providing. A second handicap was that despite certain signal successes, such as the utilization of southern pine for wood pulp, years of research on industrial utilization had produced a disappointingly small record of accomplishment.

Like so many problems and enthusiasms of the 1930's, however, the farm chemurgic movement was never defeated in open contest; it merely became irrelevant. Leading foes of the New Deal found other and more attractive issues upon which to attack the Administration. Most importantly, the coming of World War II with one sweep eliminated the problems of farm surpluses and underemployed industry and provided more urgent activities for many of the scientists who might otherwise have pursued chemurgic tasks. Finally, the rise to popular acclaim of nuclear physicists in 1945 cut deeply into the glamour and influence of chemistry.[43] Both the Farm Chemurgic Council and the Department of Agriculture still work quietly on the problems of the industrial utilization of farm products, but robbed of the context of 1936, the subject has lost much of its potential for public controversy.

[42] Borth, *Pioneers of Plenty*, p. 46; Karl E. Mundt to author, 27 Sept. 1963. This legislation was still being sought by Rep. Jensen as late as 1963. See H.R. 8354, 88th Congress, 1 sess. The Council is still (1968) in existence, maintaining its national headquarters in New York City.

[43] A study by the National Academy of Sciences in 1965 termed chemistry a "little science," measured in terms of both public and private support. *Science*, 1965, *150*:1267.